The Importance of Aquatic-Terrestrial Ecotones for Freshwater Fish

Large rivers and floodplains

Management aspects of streams and rivers

Evolutionary aspects

Epilogue

Preface

Ecotones are interface zones between different ecosystems. Their ecological role and significance with regard to ecological management and conservation have become increasingly appreciated. For the management of freshwater resources, for example, an improved understanding of the role of land/inland water interfaces will be essential for reducing negative human impacts by engineering, nutrient loading, siltation etc. A management of ecotones, on the other hand, offers possibility to control aquatic system processes via stock control of fish populations. Fish, apparently, are both excellent indicators of ecotone quality as well as determiners of its structure and function.

The Scientific Advisory Committee of the UNESCO MAB programme "Role of Land/Inland Water Ecotones in Landscape Management and Restoration" has therefore initiated a subnetwork on fish (Fish and Land/Inland-Water Ecotones, FLIWE) covering various aspects of fish ecology within a broad context of aquatic habitat dynamics. A first meeting of the group, organized by M. Zalewski, was held in Krakow, Poland, between 5–8 March 1980. During this meeting, concepts, ideas and results of pilot investigations were presented and discussed (see Zalewski, Thorpe & Gaudin, 1991: "Fish and land/inland-water ecotones" UNESCO, MAB).

The present volume contains papers presented at the mid-term meeting of FLIWE organized by F. Schiemer & G. Bretschko at Lunz/Austria from 31.8.–2.9. 1992. The aims of the meeting were a discussion of the progress of the various projects, and a discussion on strategies for the final integration of results.

The relatively small number of participants – about 60 including guests – and the secluded location at Pension Stamminger at Lunz, near the Biological Institute of the Austrian Academy of Sciences – made it a meeting of strong scientific interaction and fruitful discussions. We would like to thank all those who helped in financing and organizing the meeting: Unesco, Paris (especially Fred Fournier for his help and enthusiasm); Ministry of Higher Education, Austria; Magistrate of Vienna; State Government of Lower Austria; Austrian Committee of the International Association on Danube Research a.o..

F. Schiemer
M. Zalewski
J. Thorpe

Hydrobiologia **303**, 1995.
F. Schiemer, M. Zalewski & J. E. Thorpe (eds), The Importance of Aquatic-Terrestrial Ecotones for Freshwater Fish.

Captions to four pages of photographs:

Page ix
The participants of the Unesco/MaB "Fish and Land/Inland-Water Ecotones" meeting in front of the siminar-hotel in LUNZ, where the meeting took place (1).
Fred Fournier, Gernot Bretschko, Fritz Schiemer, Maciej Zalewski and Henri Decamps pushing the "ecotone sledge" in front of the seminar-hotel (2).

Page x
A "directors" meeting: Maciej Zalewski, John Thorpe and Fred Fournier at a coffee break (3).
Peter Bíró (4).
Tadeusz Penczak, Rabeni & John Thorpe at the evening invitation at the Biological Institute, Lunz (5).
Geoff Power (6).

Page xi
Field demonstrations at the "Ritrodat" stream (experimental area of the Biological Institute, Lunz). A small bar is prepared besides the stream (7).
Watching the demonstrations: Henri Decamps (8).
Juray Kolasa and Armin Peter (9).
Ike Schlosser (10).
Eric Pattee (11).

Page xii
Field excursion to the restoration programme of the River Melk, Lower Austria: Matthias Jungwirth explaining results to Maciej Zalewski (12).
Josef Matena, Matthias Jungwirth, Susi Muhar (standby), Ike Schlosser, Yurij Dgebuadze, M. Nowak, John Thorpe, Luis Tito de Morais at Melk (from left, 13).
Some members of the workshop in front of the baroc monastary "Dürnstein" at the Danube from left. Armin Peter, Fred Fournier, Ike Schlosser, Mrs Fournier, Don McQueen, Tadeusz Penczak, John Thorpe, Mrs Mühlhauser (14).
John Thorpe, Ike Schlosser and Don McQueen enjoying the baroc architecture of the "Dürnstein" monastery (15).

1

2

3

4

5

6

7

8

9

10

11

12

13

14

15

Hydrobiologia **303**: 1–7, 1995.
F. Schiemer, M. Zalewski & J. E. Thorpe (eds). The Importance of Aquatic-Terrestrial Ecotones for Freshwater Fish

Notes on ecotone attributes and functions

Jurek Kolasa[1] & Maciej Zalewski[2]
[1]*Department of Biology, McMaster University, Hamilton, Ontario L8S 4K1, Canada*
[2]*Institute of Environmental Biology, University of Łódź, 12/16, 90237 Łódź, Poland*

Key words: ecotone, dynamics and composition, filter, heterogeneity, scale, patchiness, zonation

Abstract

We attempt to identify general properties of ecotones. Earlier attempts to do so encountered difficulties resulting from contradictory conceptions of ecotones. Thus, we begin with and center our discussion on a definition of ecotone. The definition is complex. It includes scaling, structural, and functional aspects. Based on this complex definition, we offer a brief review of what is an ecotone, what attributes it has, and how it influences other habitats of interest. We identify feedback as a possibly important but ignored function of ecotones. This discussion is presented in general terms which apply to a variety of ecological situations. We point out that results of an evaluation of ecotone attributes largely depends on the spatial and temporal scale at which ecotone is conceptualized and data are collected. We suggest that some of ecotone determinants scale naturally in a repeatable fashion among various aquatic systems. Finally, we point to the concentration of dynamic properties of ecotones as applied to land/water interface.

Introduction

Research on the role of ecotones in aquatic and terrestrial systems has acquired considerable momentum during the last five years. Conceptual developments have generally kept pace with the empirical studies. These developments have been refined to the point that new fundamental problems and opportunities have become apparent. We need to discuss these.

One problem, important both to research on lotic ecotones and to general understanding of ecology, is the relation between spatial (ecotones) and functional (ecosystems) attributes of ecological fabric. Shugart (1990) articulated an important conceptual inconsistency in one common definition of ecotone (if one interprets that definition in a particular way). He argues that if an ecotone is a boundary between ecosystems, and ecosystems are bounded in arbitrary ways, then ecotones would also be arbitrary. Indeed, studying arbitrary objects is ineffective because it relies on finding natural regularities by chance rather than in a systematic way. A definition by Holland (1988) removes this difficulty but creates another. According to Holland (1988) an ecotone is a zone of transition between adjacent ecological systems such as uniform macrophyte stand and open water. The problem arising here is that functional approaches to ecological systems originally developed around the ecosystem concept that do not require uniformity of appearance as postulated by Holland's definition. A watershed ecosystem, or a lake, is far from being uniform. These logical and terminological inconsistencies are important because they interfere with the evaluation of the role of ecotones in the maintenance of integrated, sustainable associations of organisms. Informal and inconsistent approaches are particularly confusing when cross-system comparisons are the goal (e.g., Downing, 1991). We feel however that there is a good possibility for reconciliation of the functional, dynamic perspective and the spatial aspect embodied in the ecotone concept.

We propose that such reconciliation will be best accomplished by a more systematic examination of the ecotone as concept. This examination, as it applies to fish ecology, is a major goal of this paper. Several recent publications are particularly helpful in this regard. We will intensively use and rely on the experience and ideas assembled in Naiman *et al.* (1992), Naiman & Décamp (1990), Hansen & di Castri (1992), and Zalewski *et al.* (1991). In our view status of a habitat as ecotone differs depending on the spatial scale considered. This, in turn, depends on the prior delineation of the system of interest. Such delineation is a necessary basis for asking valid research questions (Kolasa & Pickett, 1989; Kolasa & Rollo, 1981). If, as it may be the situation in some cases, the delineation coincides with multiple ecological discontinuities (cf. Shugart, 1990), then the ecotone is a 'boundary of the system' such as an ecosystem, population, or species assemblage. If, however, discontinuities appear on the inside of the system of interest, then ecotone becomes a 'part of that system'. Such a part may be linear or reticulate in shape and is defined by an abrupt change in parameter values when one measures them across it. This is simpler than it sounds. For example, let our system be a river drainage basin. Here individual branches of the river separate, most of the time, identical habitats. When viewed at this large scale, they are not boundaries. Such ecotones form the fabric of a larger system. When viewed at a scale of tens of meters, each branch or tributary appears associated with two parallel shoreline ecotones. These ecotones are boundaries (between the water course and adjacent patches of terrestrial habitats). The distinction in ecotone status with respect to the system is important. It is so if we wish to compare various ecotones and their role in ecology of freshwaters, and fish in particular, among various settings, geographical regions, and habitats. For example, one might need to apply different methods and interpretation to ecotones that are part of a system compared to ecotones that are boundaries. We intend to combine this scaled perspective with some ideas recently developed by others and add a few of our own.

We organize the discussion into various areas. Separately, we focus on the identification of ecotones, their function, and the aspects of ecotone study.

Ecotone morphology and anatomy

Ideally, an ecotone can be represented as a straight habitat band between two other adjacent habitats (Fig. 1A, upper left panel). The reality is always more complex and the spatial complexity of an ecotone (as a transitional zone) has at least two aspects: sinuosity and resolution.

The sinuosity aspect refers to the degree of the band contortion and fragmentation (Fig. 1A, from left to right). As the sinuosity increases, (i) the total length of the ecotone relative to either habitat increases, (ii) the width of the ecotone increases and, (iii) the nature of the ecotone changes. In the first panel it was a simple band or transition zone. In the third panel it becomes a new and wider transition zone consisting of a mosaic of patches, including the original transition zone habitat as well as patches of the two habitats that originally formed that transition zone. This widening of ecotone and increase in ecotone complexity are identified by brackets at the bottom of the panels (Fig. 1A). Indeed, for each ecotone identified at these different scales it is possible to identify, analyze and compare attributes represented in Fig. 1B and 1C (see below).

The resolution aspect identifies spatial scale at which ecotone appears. For example, a patch of the ecotone habitat (Fig. 1A; right panel) has itself an ecotone with the open water habitat at a smaller spatial scale. However, this new scaled down ecotone differs in many respects. If the large scale ecotone was a zone of macrophytes along a river or lake shore, the small scale ecotone may be the outer edge of this zone, with greater water exchange, higher light and nutrient availabilities, and greater hydraulic energy conditions. Both scales may interact in a variety of ways. The simplest would be a positive correlation between the total lengths of ecotones observed at each scale. For example, doubling the length of the ecotone at one scale may mean quadrupling the total length of ecotone at a lower scale. Naiman *et al.* (1988) analyze and illustrate various consequences of scale and habitat hierarchy in greater detail, including the predictability of states. Some scales and some ecotones have been long recognized and measured. Shore development index (e.g., Hutchinson, 1957) is a good example of standard procedure used to summarize spatial properties of the lake shores and can be used to quantify one scale of terrestrial aquatic ecotones. Interactions among various scales have not been given much attention but may be important to many vertebrate and invertebrate species that use land/water interface. For example, the edge diversity index applied to wetland vegetation (Harris *et al.* 1983) would be greatly affected by the choice of resolution even though in each case it would describe some properties of ecotones. Indeed, while which resolution is selected and quantified depends on a research question, existence of ecotones at different scales does not.

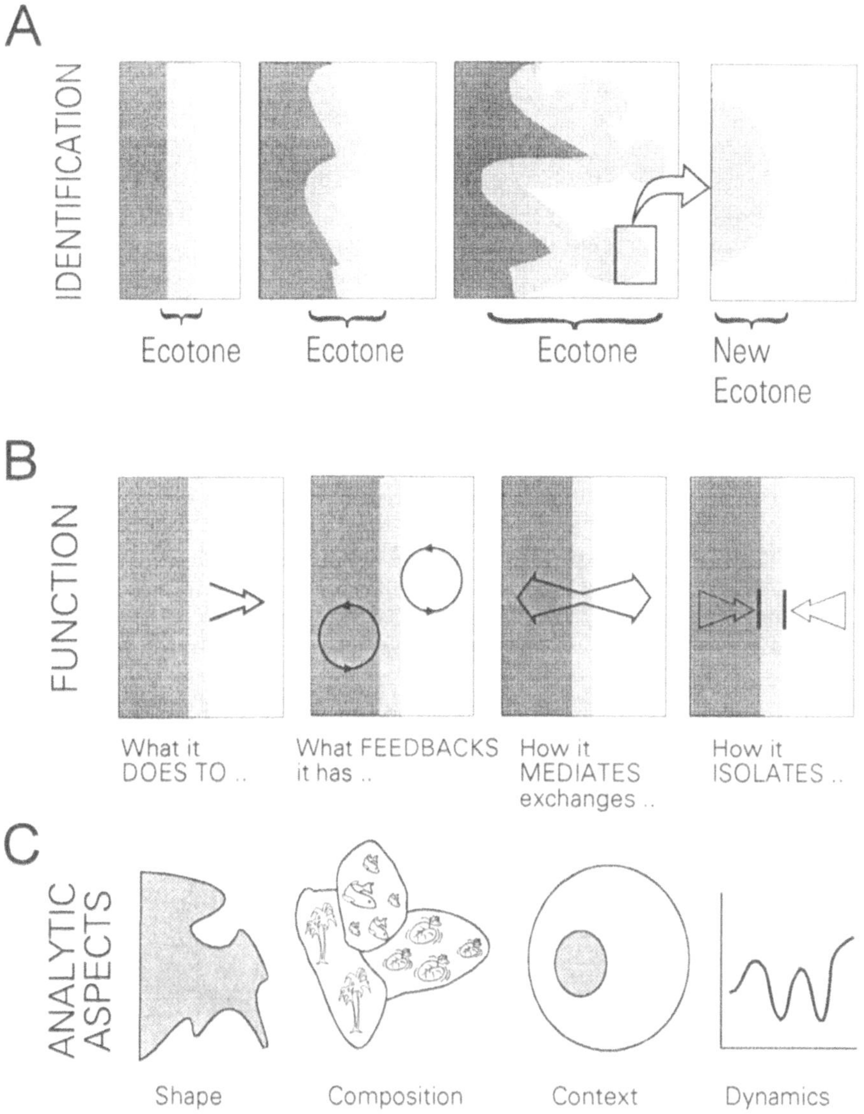

Fig. 1. A pictorial definition of ecotone. A – the scale dependence of the concept; the ecotone width changes with its shape and the change in shape may be associated with a change in composition, *i.e.*, the third diagram from the left includes 'dark', 'grey' and 'white' habitat types. B – the range of interactions (commonly referred to as functions) an ecotone may have with other habitats; the functions include, from left to right, directional impacts, feedbacks, modifications of flows or processes, and isolation or filtering out of processes. C – principal properties of ecotones that require systematic attention and specification include ecotone morphology, composition, embedding or context, and temporal behaviors. See text for the detailed discussion.

It is important to note that, while habitat identified as ecotone has in nature a third spatial dimension-height, that dimension is not a part of the ecotone delineation. This third dimension may become an important factor, however, when ecotone function is analyzed (see next section).

Finally, the temporal dimension should not be ignored any further. Identification of an ecotone depends on the time over which the interactions of two habitats are observed. A transition zone between land and water varies at the scale of days, seasons, years, and decades. A riverine ecotone may appear narrow during the summer, broader if an annual cycle is considered, and very large if a flood of a century and its effects are included. Many other, including spatial, characteristics will change accordingly.

Ecotone physiology

The primary interest in ecotones is motivated by the potential role they play in maintaining and shaping of various ecological process. Forman & Moore (1992) identify several ecotone functions which largely coincide with our diagram (Fig. 1B). They label these functions as conduit, filter or barrier, source, sink, and habitat. We propose to add feedback as a separate category. We summarize the role of ecotones graphically (Fig. 1B). The role of ecotone, or FUNCTION, appears under five headings: What it *does to* ..., what *feedbacks* it has ..., how it *mediates* exchanges ..., and how it *isolates* .... We differ here from Naiman *et al.* (1992) in breaking the function into more formal yet empirically analyzable components, instead of focussing on epiphenomenological descriptors such as their contribution to stability or to the maintenance of diversity.

What it does to (.. habitats). An ecotone may be a net source of material and energy in form of organisms, structural components such as logs, nutrients, plant carbohydrates, insect proteins, and many others or it may be a net sink. In addition to export and import inherently linked with the sink/source concept (Forman & Moore, 1992), an ecotone modifies habitat of interest. It may modify wind (wave energy, thermal regime), light conditions, or hydraulics (e.g., Bott, 1983).

Feedback. Ecotone may provide a feedback. The contrast between the feedback and sink/source concept is that a feedback amplifies or reduces the intensity of system processes without necessarily being a net sink or source itself. Feedback may occur whenever ecotone and one or both adjacent habitats are interacting synergistically in augmenting, reducing, or maintaining some ecological process. Most likely feedback operates in combination with sink or source process, at least in some dimensions. For example, carp foraging among cattails in a coastal wetland of Lake Ontario (Cootes Paradise) create local passages and open water patches among emergent plants. The created heterogeneity may, in turn, enhance survival and growth of carp fry and juveniles and thus close the feedback loop (see also the paper by Kolasa & Weber, 1994). An ecotone may have multiple feedbacks with both habitats it divides, and each of these sets of feedbacks may be entirely different in nature. The same wetland cattail zone may provide nesting and foraging sites for land animals (muskrats, redwinged blackbirds) whose activities contribute in various ways to the pattern of cattail distribution, reproduction, and growth.

How it mediates. Any transition zone is likely to let organisms, matter, or processes through differentially. It filters them. This aspect has been discussed by landscape ecologists in detail with the exception of, perhaps, one important facet. A filter is always associated with a change of frequency. Thus, any phenomenon crossing a boundary imposed by an ecotone changes its temporal scale of operation (Allen & Hoekstra, 1992). This is important because, in land/water ecotones, the filtering operation often means a smoothing of external impacts. For example, a mosaic of flood plain habitats of different duration, characteristics and connectivity may smooth variation in reproductive success of several fish species of the Danube River (cf. the hierarchy of generalized relevant factors is shown in Fig. 2). Finally, an ecotone is a constraint. The constraint can be thought of as a boundary condition, *i.e.*, a quantitative determinant of how much and which exchanges are allowed across an ecotone. By contrast to filtering function, the constraint refers to allowable magnitudes of phenomena.

How it isolates. There seems to be more to isolation than just being a simple barrier. An ecotone has two sides and the isolation on these two boundaries may be symmetrical but, more likely, it is not. The isolation may scale differently – an ecotone may be porous to particles/organisms of various sizes. For example, lush riparian vegetation of tropical streams and lakes may be penetrable by large animals only to act as an effective barrier to airborne materials. Indeed, the barrier properties of ecotone will differ depending on whether passive or active processes are involved and whether the barrier itself is homogeneous or heterogeneous (Wiens, 1992).

Habitat. Ecotones are indeed habitats. In order, however, to treat them as equivalent habitats to those they divide, or unite, a conceptual shift to another spatial scale must take place (see Fig. 1A, rightmost panel). By definition, a transition zone between two habitats is not at the same hierarchical level as those habitats. Still, a mechanistic understanding of the functions ecotones

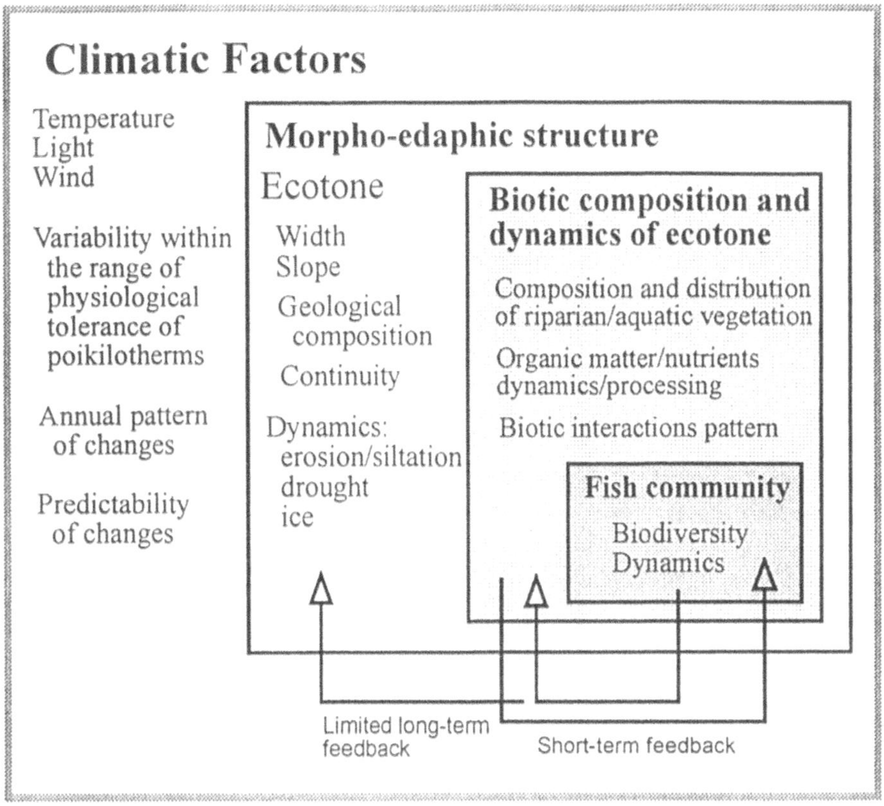

Fig. 2. Hierarchy of factors influencing fish communities in streams – an example of the context and role of ecotones (more explanation in text). Note that suites of factors are associated with different spatial and temporal scales, with nested boxes representing a progressively smaller scale.

provide may still require treating them as independent, self-maintaining entities.

After having reviewed various aspects of ecotone function, we illustrate some of the factors affecting stream fish communities (Fig. 2). We break these factors hierarchically into those that are clearly external and constraining, those that characterize ecotones from the morphological and geological perspective, and those that constitute biotic component, superimposed on and interacting with the former. Figure 2 also illustrates that some of the aspects of the ecotone definition (Fig. 1) fit into different levels of a specific analysis of stream ecotones. For example, 'Aspects to study-composition' belongs to the second innermost box in Fig. 2, while 'Context' appears at the two highest hierarchical levels, *i.e.* Climatic Factors and Morpho-edaphic structure.

Finally, we note that ecotones, being transitional zones, affect various aspects of local ecology. By definition, they change values of many variables such as humidity, temperature, sedimentation rate, and many others. More interestingly, ecotones often accelerate or intensify many processes. Land/lotic ecotones, for example, speed up mineralization of organic matter or accelerate sediment build-up (e.g., Ranwell, 1974). Concentration of energy and nutrients builds often up at ecotones. Lush riparian vegetation and nutrient retention are examples of such processes amply documented for lotic environments.

'Ecotonology' or what is there to study

The study of the role of ecotones ('What it does' component) can be enhanced and deepened by the analysis of the ecotone itself. Such an analysis might include (Fig. 1C) several standard angles or aspects. We group them in four obvious categories of shape, composition, context, and dynamics.

'Shape' may include specific information on the extent, width, sinuosity, continuity, or patchiness of the whole ecotone or one of its edges. By contrast, composition has more than just a spatial dimension. Whenever appropriate, the arrangement of components in space, their frequency and relative sizes, including height, become important attributes to consider. In addition, the biotic (taxonomic or functional) qualities described in terms of guilds, resource base, or impacts on habitat (Fig. 1C) may become subjects of analysis. For example, a stream community shaded by *Eucalyptus* trees may respond differently to CPOM than if trees were willows, maples, or conifers (e.g. Boulton & Boon, 1991). Composition may affect ecotone porosity or permeability to matter and energy fluxes as well as organism movements along the ecotone (conduit; e.g., Forman & Moore, 1992). Another example of the specific impact of composition on the ecotone dynamics is provided by carp (Kolasa & Weber, 1994). Several consequences of composition, spatial arrangements and temporal availability of components within a headwater stream are reviewed in this volume (Schlosser, 1995) and may serve as an example of the complexities involved. Context is yet another important aspect in the study of ecotones. For instance, the same ecotone such a strip of riparian vegetation may play a very different role in a naturally forested landscape from that in an intensely agricultural landscape. The size of the river and the landscape type (contexts) influence the importance of riparian ecotones for fish biomass and communities (Schiemer & Zalewski, 1992). Landscape heterogeneity has a direct impact on fish biomass, productivity and diversity when the landscape becomes a floodplain (e.g. Lowe-McConnell, 1987) and an indirect impact in other situations. Finally, ecotone dynamics is an important but largely neglected aspect. Ecotones do change and the analysis of the rate and conditions under which such changes occur are likely to become an integral part of the understanding of ecotones and their contribution to the maintenance of healthy aquatic ecosystems. In this volume (Kolasa & Weber, 1995) we provide an example of ecotone change and examine some of the possible consequences of this change for community level characteristics.

Conclusions

Understanding ecotones has progressed sufficiently far to justify attempts at a synthetic and general review of ecotone related problems. These problems include clarification of the ecotone concept, enumeration of ecotone roles or functions, and identification of major perspectives within which ecotones require investigation.

The most important aspects identified in this paper are the scale-dependence of ecotones and its significance for making comparisons among ecotones, feedback functions of ecotones, and concentration of ecological processes at ecotones. It appears that the actual expression of ecotone attributes is highly scale dependent. It is thus desirable to give careful consideration to identification of scale, ecotone, and to determination whether ecotone is a boundary *vs* a part of the system before proceeding with field data collection.

Acknowledgments

This study was financially supported by NSERC grant to JK. The UNESCO Man and Biosphere program provided additional support for the ecotone workshop held in Lunz where many of the ideas for this paper were conceived. We thank the organizers and participants for their stimulating talks and insights.

References

Allen, T. F. H. & T. W. Hoekstra, 1992. Toward a unified ecology. Columbia University Press, New York, 384 pp.

Bott, T. L., 1983. Primary productivity in streams. In Barnes, J. R. & G. W. Minshall (eds), Stream ecology, Plenum, New York: 29–53.

Boulton, A. J. & P. I. Boon, 1991. A review of methodology used to measure leaf litter decomposition in lotic environments: time to turn over an old leaf? Aust. J. mar. Freshwat. Res. 42: 1–43.

Downing, J. A., 1991. Comparing apples with oranges: Methods of interecosystem comparison. In Cole, J., G. Lovett & S. Findlay (eds), Comparative analyses of ecosystems. Springer, New York: 24–45.

Forman, R. T. T. & P. N. Moore, 1992. Theoretical foundations for understanding boundaries in landscape mosaics. In Hansen, A. J. & F. di Castri (eds), Landscape boundaries (Ecological Studies 92). Springer, New York: 236–258.

Hansen, A. J. & F. di Castri, 1992. Landscape boundaries. Springer, New York, 452 pp.

Harris, H. J., M. S. Milligan & G. A. Fewless, 1983. Diversity: Quantification and ecological evaluation in freshwater marshes. Biol. Conserv. 27: 99–110.

Holland, M. M. (compiler), 1988. SCOPE/MAB technical consultations on landscape boundaries: report of a SCOPE/MAB workshop on ecotones. Biology International, Special Issue 17: 47–106.

Hutchinson, G. E., 1957. A treatise on limnology – volume I Geography, physics, chemistry. John Wiley and Sons, New York, 1015 pp.

Kolasa, J. & S. T. A. Pickett, 1989. Ecological systems and the concept of biological organization. Proc. natl. Acad. Sci. USA 86: 8837–8841.

Kolasa, J. & C. D. Rollo, 1991. Heterogeneity of heterogeneity: a glossary. In Kolasa, J. & S. T. A. Pickett (eds), Ecological heterogeneity, Springer-Verlag, New York: 1–23.

Kolasa, J. & L. Weber, 1995. Relationship between the spatial scale and biotic variability in a wetland ecotone. Hydrobiologia 303 (Dev. Hydrobiol. 105): 61–67.

Lowe-McConnell, R. H., 1987. Ecological studies in tropical fish communities. Cambridge Tropical Biology Series. Cambridge Univ. Press, Cambridge, 382 pp.

Naiman, R. J. & H. Décamp, 1990. The ecology and management of aquatic-terrestrial ecotones. Man and Biosphere Series. UNESCO, Paris, 316 pp.

Naiman, R. J., H. Décamp, J. Pastor & C. A. Johnson, 1992. The potential importance of boundaries to fluvial ecosystem. J. North Am. Benthol. Soc. 7: 289–306.

Ranwell, D. S., 1974. The salt marsh to tidal woodland transition. Hydrobiol. Bull. 8: 139–154.

Schiemer, F. & M. Zalewski, 1992. The importance of riparian ecotones for diversity and productivity of riverine fish communities. Neth. J. Zool. 42: 323–335.

Schlosser, I. 1995. Critical landscape attributes that influence fish population dynamics in headwater streams. Hydrobiologia 303 (Dev. Hydrobiol. 105): 71–78.

Shugart, H. H., 1990. Ecological models and the ecotone. In Naiman, R. J. & H. Décamp (eds), The ecology and management of aquatic-terrestrial ecotones. Man and Biosphere Series. UNESCO, Paris: 23–36.

Wiens, J. A., 1992. Ecological flows across landscape boundaries: a conceptual overview. In Hansen, A. J. & F. di Castri (eds), Landscape boundaries (Ecological Studies 92). Springer, New York: 217–235.

Zalewski, M., J. E. Thorpe & P. Gaudin, 1991. Fish and land/inland water ecotones. Man and Biosphere Series, UNESCO, University of Łódź, Łódź, 102 pp.

Ecotones in lakes and reservoirs

Hydrobiologia **303**: 11–30, 1995.
F. Schiemer, M. Zalewski & J. E. Thorpe (eds), The Importance of Aquatic-Terrestrial Ecotones for Freshwater Fish.

Land/water ecotone effects in reservoirs on the fish fauna

A. Duncan[1] & J. Kubečka[2]
[1]*Department of Biology, Royal Holloway University of London, Egham, Surrey TW20 0EX, UK*
[2]*Hydrobiological Institute, Czech Academy of Sciences, Na Sadkach 7, 370 05 Ceske Budejovice, The Czech Republic*

Key words: fish, reservoirs, ecotones, littoral area, temperate, tropical

Abstract

The morphology and function of one tropical and 25 temperate reservoirs are examined in relation to their effect upon the nature of the land/water interface and, further, to what extent the features of these ecotones satisfy the ecological requirements of the reservoir fish species throughout their life cycle during spawning, larval, juvenile and adult stages. The two main conclusions are that (1) reservoir fish species are especially dependent upon land/water ecotones during their early life history and (2) there exists a strong relationship between the extent of the littoral area and the nature of the fish stocks. Several examples are given to show that manipulation of the land/water ecotone is a major tool for the management of reservoirs advantageously for their major functions.

Introduction

In this paper, reservoirs will be taken to be man-engineered water bodies with a volume greater than 10^6 m^3, a maximal depth greater than 5 m and a residence time greater than 10 days and whose water level and water volume can be manipulated according to human needs.

The 26 reservoirs included in this review conform to this definition and have been selected because of the authors' familiarity with their ecology and nature of fish stocks. No attempt has been made to review the literature comprehensively but rather to use our knowledge of the selected reservoirs to illustrate some ideas about land/water ecotones. Thus, the reservoirs chosen are largely European and only one tropical reservoir from Sri Lanka is considered, because we know it, and none of the African ones (Kapetsky & Petr, 1984) or other ASEAN (Association of Southeast Asian Nations) (Baluyut, 1983; De Silva, 1988) or Amazonian (Tundisi *et al.*, 1992) have been reviewed and information is available in the cited references.

Reservoirs have various kinds of ecotones, for example the water/water interface between inputs and the reservoir and between the reservoir and outputs. The land/water interface extending around its perimeter is no less significant to the reservoir's ecology because it incorporates the whole shoreline. Borrowing from classical limnology, we can apply Hutchinson's (1967) zonations of a lacustrine littoral to the reservoir ecotone. In reservoirs, the classical vertical zonation can be simplified into two major depth strata: (1) the all-important eulittoral defined as the area between the annual minimum and maximum water levels, and (2) the rest extending below the minimum water level, consisting of the infralittoral plus littoriprofundal which are indistinguishable in reservoirs with poor macrophyte development. Normally the lower boundary of the infralittoral (C/D in Fig. 1) is defined by the lower limit of light for plant growth but, in reservoirs, this becomes blurred by fluctuating water levels and changing transparency. Of the horizontal areas generated by the two depth zones, the area 'A' is characterised by any terrestrial plant cover present and is usually negligible in most reservoirs; areas 'B' and 'C' also usually do not occupy any great extent of the total reservoir surface area, or their volumes any large fraction of the reservoir volume, although there are some exceptions.

According to Fernando & Holcik (1991), the species composition of a particular reservoir is initially dependent upon the riverine fish populations of its catchment area. Which of these survive to establish stable populations depends on whether their ecological requirements are satisfied within the new reservoir and its subsequent development. The authors suggest that some of the original species may be sufficiently preadapted to lacustrine conditions but otherwise more riverine species are likely to stay close to the shore, shallow water or the inlets of tributaries.

As in any newly flooded territory, fish faunas in new reservoirs undergo successional changes in species composition and biomass. Conditions in newly built reservoirs are variable and unstable to which the fish respond until a stable fish stock is attained. During the initial flooding, the land/water ecotone changes continuously and dramatically and its effect overlaps with that of the now flooded forest or meadow which forms the bottom of the new reservoir. Because of this instability of new reservoirs, we have chosen older reservoirs well beyond the initial stage of succession.

In this paper, we aim to review the characteristics of ecotonal areas 'B' and 'C' (Fig. 1) in reservoirs of different morphology and function to determine how they satisfy the ecological requirements of different stages in the fish life

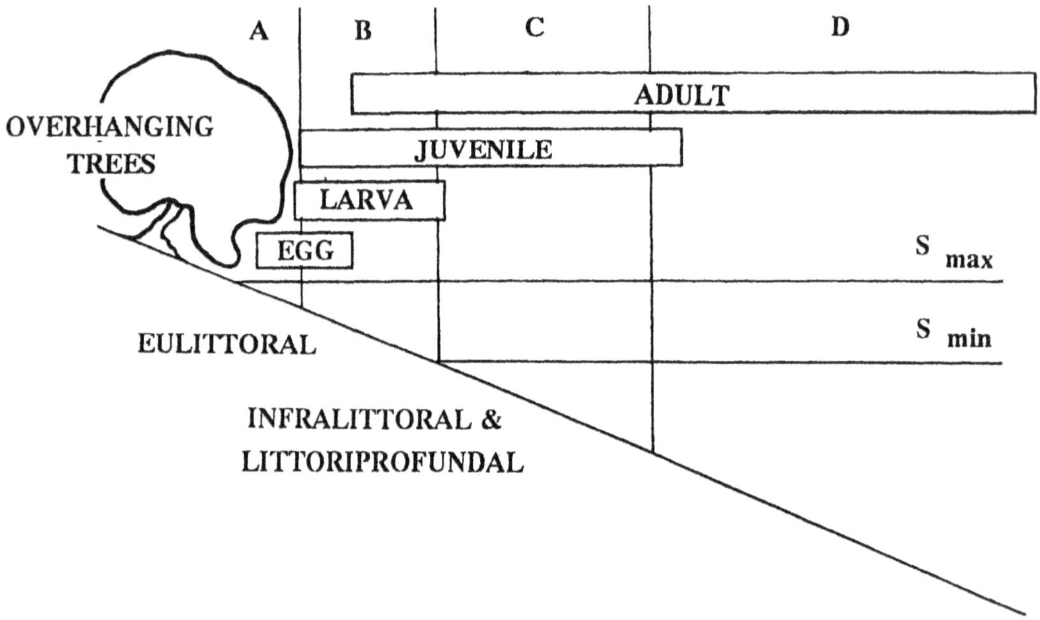

Fig. 1. A reservoir ecotone borrowed from the classical zonations of a lacustrine littoral (Hutchinson, 1967) together with a schematic diagram illustrating the relative dependence of different fish life cycle stages upon the reservoir ecotonal zones. A. overhanging tree; B. between maximum and minimum water surface and is the eulittoral zone; C. limit of significant light for primary production; D. the rest of the reservoir.

cycle. The riverine origin of most reservoir fish (Fernando & Holcik, 1991) pre-adapts them to an environment in which the land/water interface is physically more extensive than in most reservoirs, if ratio of shoreline to surface area is considered. We agree with Fernando & Holcik's (1991) view that few reservoir fish are entirely independent of shallow ecotonal areas. Here, we try to show that the success of fish in colonising reservoirs is related to the nature of this land/water ecotone and the degree to which the species are dependent upon it.

Features of reservoir ecotones

Fish faunas in relation to reservoir morphology have been compared for seventeen Russian, five Czech and three London reservoirs (Table 1) as well as one tropical reservoir which is a typical present-day component of the ancient Sri Lanka irrigation network designed by the early hydrologist kings. The Russian reservoirs were the largest in area, length and width and were like the Czech reservoirs in being longer than wide due to their origins as drowned river valleys. By contrast, the London and Sri Lankan reservoirs were embanked impoundments with more rounded shapes. The walls of the London reservoirs were built on impermeable clay strata on the banks of their supply river whereas the Sri Lankan one was constructed by embanking sloping terrain some distance from the riverine source. A comparison of the mean reservoir surface areas (Table 2) shows that the Russian reservoirs are much larger than the others, by two or three orders of magnitude.

Table 1. Reservoirs names and references.

Russian reservoirs from Isayev & Karpova (1980)
Bukhtarminskoye
Volgogradkoye
Gorkovskoye
Zaporozhskoye
Ivankovskoye
Kamskoye
Kanevskoye
Kievskoye
Kremenchugskoye
Kuybyshevskoye
Veselovskoye
Proletarskoye
Pavelovskoye
Rybinskoye
Saratovskoye
Uglichkoye
Tsymlyanskoye
Czech reservoirs from Kubečka (1992)
Rimov Reservoir
Klicava Reservoir
Lipno Reservoir
Jordan Reservoir
Slapy Reservoir
London reservoirs from Kubečka & Duncan (in press)
Queen Mary Reservoir
Queen Elizabeth II Reservoir
Wraysbury Reservoir
Sri Lankan reservoir from Schiemer (1983)
Parakrama Samudra

Reservoir morphology and the nature of their land/water ecotone

European reservoirs

The basin sides of the V- or U-shaped Russian and Czech reservoirs had very different degrees of steepness or slope, which is a feature of great significance for the nature of the land/water ecotone. In most V-shaped Russian reservoirs, with maximum depths of 10 m to 100 m, the sides were gently sloping. As most were built on the eastern European plain of fertile agricultural land by damming big rivers like the Volga, Don, Dniepr, a rich growth of littoral macrophytes developed everywhere in the extensive shallow littoral areas, except where the shoreline suffered erosion duc to wind and wave action. Many large North American reservoirs seem to be similar water bodies, with extensive shallow vegetated littorals (Ploskey, 1986; Dolman, 1990).

Most of the Czech Republic is geologically old with altitudes between 300–600 m above sea level.

Table 2. Reservoir ecotone features.

Examples	'Russian'	'Czech'	'London'
Numbers	17	5	3
Shape	V-shape	V- or U-shape	U-shape
Sides	Shallow	Steep 'Canyon'	Steep 'inverted trapezoids'
Type	dammed-river reservoirs	dammed-river reservoir	embanked impoundment
Previous land-use	agriculture human uses	little agriculture some human use	no agriculture no human uses
Littoral	vegetated or eroded	stones, rock gravel, sand	concrete slabs
% littoral of area	23.0 ± 11.0	11.6 ± 3.6	2.3 ± 0.7
Fish stocks			
% pike	2.74	0.39	0.00
% cyprin.	81.60	65.60	22.70
% percif.	7.90	31.0	75.60
Reservoirs			
Numbers	17	5	3
Area (1000 ha)	198.4 ± 193	1.23 ± 1.90	0.21 ± 0.07
Length (km)	264 ± 177	18.9 ± 20.5	3.3 ± 1.5
Width (km)	7.7 ± 9.1	0.47 ± 0.58	0.63 ± 0.05
Depths (m)	10–100		
Yearly water Fluctuations (m)	4.73 ± 2.6	2.86 ± 1.7	3–5
Age (years)	33.1 ± 12	50.4 ± 61	32 ± 29
Summer Total phosphorus (mg/m^3)	45 ± 23	38 ± 12	> 1000

The relatively flat countryside is interrupted by 50 m–100 m deep river valleys with forested or rocky slopes (Plate 1) which, when flooded, form 'canyon' reservoirs with V- or U-shaped basins whose steep sides supported little or no aquatic or terrestrial vegetation when the littoral was either flooded or exposed to air.

In contrast, the London reservoirs are embanked impoundments of simple morphology, built close to the source of the river from which abstracted water was pumped (Windle-Taylor, 1961–62; Steel, 1976). The basins were inverted trapezoids whose steep sides were lined with concrete slabs or cemented bricks down to the bottom depths of 12–22 m (Plate 2). Top water levels of these supply reservoirs are kept high above the surrounding land both to increase storage volume and to enhance full depth mixing by wind. These three littoral features – a hard substratum, steep sloping sides and wind-generated wave action – prevent any possibility of colonisation by aquatic macrophytes. Above the normal water level or further down when the water levels were lowered for repair work, a complex terrestrial plant community of about 25 species was allowed to develop by rooting in the interstices

Plate 1. Rimov Reservoir, Czech Republic, showing a shallowly sloping clear-cut bank and, in the distance, steep rocky margins.

between the bricks or concrete slabs (Plate 3). In recent times,removal of these plants, manually or by application of herbicides, no longer takes place.

Proportion of the littoral area in temperate reservoirs
In European reservoirs, the littoral area can be taken as the band of water extending down to 2 m depth around the perimeter and, as a proportion of the total reservoir area, was rather greater in the Russian (23%) than in the Czech (12%) or London (2%) reservoirs (Table 2). For temperate fish, a depth of 2 m has some significance as the lower limit for inshore spawning amongst aquatic macrophytes as well as for inshore fry assemblages. Hence, the proportion of the littoral area down to 2 m depth is a particularly suitable comparative measure of a reservoir ecotonal characteristic relevant to fish.

Sri Lankan lowland reservoir
The ancient Sri Lankan irrigation reservoirs were usually built by embanking sloping terrain of Dry Zone jungle and by transporting water, by gravity alone, from the large Mahaweli River along artificial channels. In the reservoir basin, the deepest water lay close to the embankment and depths became shallower away from it. Thus, in Parakrama Samudra, the reservoir area between maximum and minimum water levels (52.5–56.3 m above sea level) which contributes to the littoral ecotone can be as much as 40–50%. One might expect extensive development of aquatic macrophytes but only small patches appear because of the strength and frequency of wind-generated wave action and low transparency. Although true aquatics are not very abundant, a terrestrial vegetation develops during low water periods which is grazed daily by cattle and buffalo and which provides perching sites for the small bird fish-predators. When flooded, the vegetation provides small fish with a more diverse habitat and a periphytic food resource as an alternative to the meiofauna of the silty sand enriched organically by faeces from cattle and buffalo 'wallowing' in the flooded littoral. In this reservoir, the land/water ecotone is exploited equally and similarly by nature and by man and his domestic animals.

Other features of the European reservoirs
In other features, the three sets of European reservoirs show similarities (Table 2): the water level

Plate 2a

Plate 2b

Plate 2c

Plate 2. London reservoirs: (a) Queen Mary Reservoir with cemented bricks; (b) Queen Elizabeth II Reservoir with concrete slabs; (c) Wraysbury Reservoir with concrete slabs.

fluctuations are similar in magnitude; all are older than ten years so that their fish stocks will have stabilised (Kubečka, 1992). Although the London reservoirs differ in their hypertrophic levels of phosphate-phosphorus, biomanipulative control of the light regime by full depth mixing in two of

Plate 3. Queen Mary Reservoir with overhanging terrestrial vegetation where bream eggs have been colected.

them reduces excessive algal biomass by imposing high respiratory losses (Steel, 1972; Duncan, 1990). Consequently, phytoplankton biomasses attain similar levels in the Czech and London reservoirs.

The influence of reservoir function on the land/water ecotone

All reservoirs are subject to periodic fluctuations in water levels, due to rain, temperature changes and/or how they are operated for their functional role. Temperate lake and river fish species have become adapted to the seasonal pattern of high water levels that occur in winter and spring and lowered ones in summer and autumn and may become confounded by unexpected kinds of water level changes that can occur in reservoirs, such as greater amplitude which may also be out of phase seasonally. Large oscillations in year to year fish recruitment may be a consequence of such unpredictable water level changes.

Tyurin 's reservoir types
For Russian reservoirs, Tyurin (1961) established three regimes of water level fluctuations (Isayev & Karpova, 1980):

Type 1: High water levels in spring, decreasing in summer, autumn and winter. This pattern applies to Czech drinking water reservoirs, retention reservoirs and temperate irrigation reservoirs. The spawning substrata available to phytophilous fish species consist of terrestrial vegetation flooded briefly in spring and amphibious marsh and peat vegetation (Il'ina & Gordeyev, 1972; Spanovskaya *et al.*, 1980; Kubečka, 1990). Similarly, in tropical reservoirs, increased water levels during monsoon rains enable spawning on flooded terrestrial vegetation.

Type 2: Stable high water levels throughout the ice-free period which are lower in winter. This is common in Russian (Isayev & Karpova, 1980), American (Ploskey, 1986) and Czech reservoirs. Spawning conditions for phytophilous fish vary but are poorer than in type 1 reservoirs.

Type 3. Stable water levels throughout the year. This regime is characteristic of reservoirs built solely for navigation or leisure activities. A mature littoral hydrosere of macrophytes, similar to that in lakes, may develop within ten years after filling (Isayev & Karpova, 1980).

London reservoirs
Up to the 1980s, the London supply reservoirs would have been classified as of the Tyurin Type 3 but recent changes to a drier climate and increased human population numbers have altered them to Type 1 reservoirs.

Balancing and pump-storage reservoirs
Tyurin's classification ignored short-term fluctuations in water level which are important in smaller systems such as the balancing reservoirs below large hydroelectric or pump-storage reservoirs. Very little is known about the development of either littoral vegetation of the fish stocks under these extreme conditions. In Czech hydroelectric reservoirs, a daily level fluctuation of 1–2 m together with sparse soil ensures that the littoral area 'B' is virtually plant-free.

Sri Lankan reservoir
The seasonal patterns of wind, monsoonal rains and solar radiation in tropical Sri Lanka (Fig. 3) differ from the temperate patterns to which the European reservoirs are subject. In irrigation reservoirs like Parakrama Samudra, management of water outputs for cultivation of two crops of rice per year in the dry zone of the country are superimposed upon the climatic patterns. During the NE monsoon, plentiful rain enables inputs and outputs to be balanced whilst providing irrigation water for the Maha rice crop, so that stored water volume and levels remains high and the land/water ecotone has its greatest aquatic phase. If the SW monsoon rains fail when the Yala rice crop needs water, this need is met from the reservoir's store; then the land/water ecotone has its terrestrial phase and develops its flora. This seasonal rain and development cycle of the rice crops imposes a regular pattern of variation in reservoir volume, depth and flow rates as well as predict-

able flooding or exposure of 50% of the reservoir area (Fig. 3); in the northern basin of Parakrama Samudra, this combined pattern of events has occurred regularly since the 4th century.

Reservoirs and the ecological requirements of riverine fish species throughout their life history

As most reservoir fish are phytophil and phytolithophil riverine species (Fernando & Holcik, 1991), they are dependent upon the land/water interface for spawning and for their early fry stages. The dependence upon ecotones decreases during ontogeny as adults are fully capable of surviving without any interaction with the littoral (Fig. 2).

Spawning

Scarcity of suitable spawning conditions limits the successful establishment of reservoir fish species (Il'ina & Gordeyev, 1972; Kuznetsov, 1975; Summerfelt, 1986; Kubečka, 1992). In 78 randomly chosen Central and East European reservoirs, 69.2% of fish species were phytophils and 86.9% were phytophils plus phytolithophils, needing plants in some degree for spawning (Balon, 1975) (Table 3).

Using diverse finely structured plants for spawning substrate is an evolutionary adaptation to the regular spring flooding of terrestrial and semi-terrestrial marginal plants. Many fish spawn their eggs in as shallow water as possible to avoid invertebrate and fish predation (Zuromska, 1967). Perch (*Perca fluviatilis*), carp (*Cyprinus carpio*), ruffe (*Gymnocephalus cernua*), eel (*Anguilla anguilla*), tench (*Tinca tinca*) and even the usually planktivorous hybrid *Coregonus lavaretus x C. peled* migrate shorewards to the spawning grounds of roach (*Rutilus rutilus*) and bream (*Abramis brama*) in order to eat their eggs (Kubečka, 1990).

In Tyurin's type 1 reservoirs (Tyurin, 1961) the regime that is theoretically more favourable for spawning than in natural waters because of its high water levels in spring, and explosions of phytophil and phytolithophil fish populations may

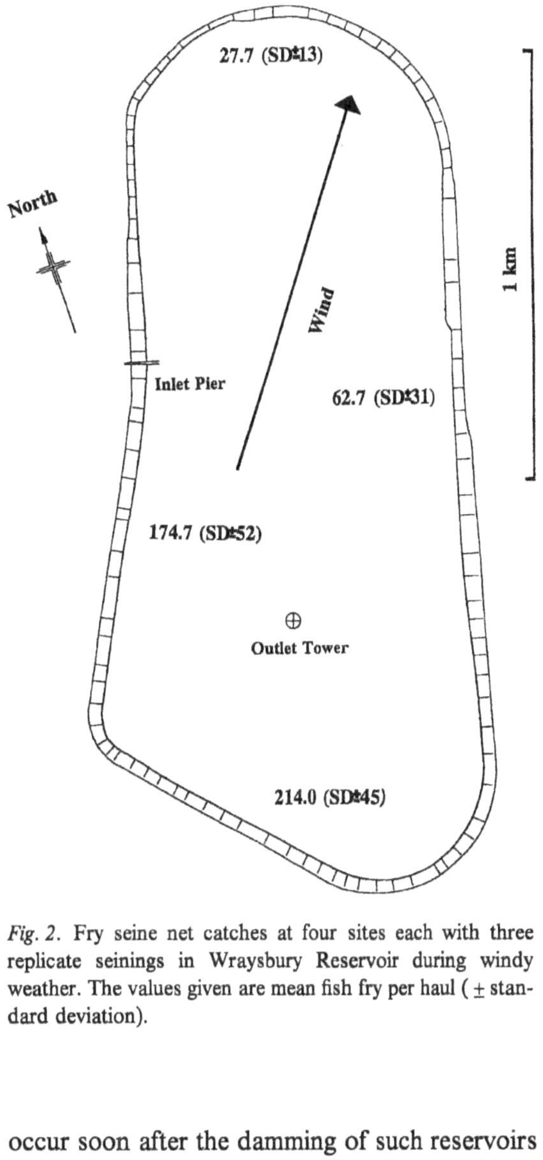

Fig. 2. Fry seine net catches at four sites each with three replicate seinings in Wraysbury Reservoir during windy weather. The values given are mean fish fry per haul (± standard deviation).

occur soon after the damming of such reservoirs (Kimmel & Groeger, 1986; Kubečka, 1992). In some Russian reservoirs, the survival rate of fish larvae is negatively density dependent because the carrying capacity of the shallows is exceeded in years of good spawning (Kuznetsov, 1975).

The numbers of phytophilous planktivorous fish can be manipulated, by lowering water levels in spring to dry their eggs (Kubečka, 1989). Any short-term fluctuation of about 50 cm during spawning time can almost eliminate whole year classes (Il'ina & Gordeev, 1972; Kutuzov, 1976; Hrbacek, 1984; Prosser, 1986; Zalewski *et al.*,

Table 3. Mean percentage frequencies of main fish species and reproductive guilds according to Balon (1975) in 78 randomly chosen reservoirs from Kubečka (1992).

Guild	Species	% freq. in group	% freq. in sum
A. Nonguarders			
A.1 Open substratum spawners			
A.1.1 Pelogophils	*Clupeonella* sp.	1.02	
	Pelecus cultratus	0.72	
	Ctenpharyngodon idella	2.97	
	Pelagophils total		4.71
A.1.3 Lithophils	*Coregonus* sp.	0.69	
	Aspius aspius	0.47	
	Leuciscus cephalus	0.33	
	Lithophils total		1.49
A.1.4 Phytolithophils	*Abramis brama*	21.28	
	Alburnus alburnus	3.47	
	Leuciscus leuciscus	0.61	
	Rutilus rutilus	26.33	
	Perca fluviatilis	15.02	
	Gymnocephalus cernua	2.31	
	Phytolithophils total		69.02
A.1.5 Phytophils	*Esox lucius*	4.14	
	Cyprinus carpio	2.46	
	Carassius sp.	1.37	
	Scardinius erythrophthalmus	1.64	
	Tinca tinca	0.40	
	Blicca bjoerkna	4.81	
	Phytophils total		14.83
A.2 Brood hiders			
A.2.1 Lithophils	*Salmo trutta*	0.33	
	Oncorhynchus mykiss	0.22	
	Brood hiders total		0.55
B. Guarders			
B.1 Substratum choosers			
B.1.2 Phytophils	*Silurus glanis*	0.27	
B.2 Nest spawners			
B.2.2 Phytophils	*Stizostedion lucioperca*	2.60	
	Guarders total		2.87
Overall sum of these species		93.48	93.47

1990). In the absence of plant structures, most phytophils resorb their eggs (Il'ina & Gordeev, 1972; Balon, 1975). Other fish use whatever is available in the littoral: pike (*Esox lucius*) in Klicava Reservoir spawn on submerged wood or behave like phyto-lithophils, like the rudd (*Scardinius erythrophthalmus*) and white bream (*Blicca bjoerkna*) in the same reservoir. In extreme scarcity of plant material, many fish spawn on the fine gravel on the littoral slopes of Klicava (Holcik &

Hruska, 1966). Away from the land/water interface, submerged trees, bushes or branches down to depths of 20 m have been used for spawning (Beliy, 1970) or fish spawn in tributaries (Il'ina *et al.*, 1987; Kubečka, 1990). An alternative strategy is intermittent spawning during the season that ensures reproductive success under reservoir conditions (Kubečka & Pivnicka, 1993).

The proportion of fish families in a water body with different spawning requirements reflects the existing spawning conditions. In Russian reservoirs with extensive littoral shallows on fertile soils, a macrophyte cover of 5% or more of the reservoir area ensures the success of phytophil populations (Il'ina & Nebolshina, 1976; Boitsov, 1983) and populations of pike and cyprinids are larger than in the poorer conditions of Czech reservoirs (Table 2). In the London reservoirs, where the margins are flooded infrequently, pike are usually absent and cyprinids relatively scarce. Perciformes, mainly ruffe and perch, being less dependent upon any land/water ecotone for reproduction (Kuznetsov, 1975; Kubečka, 1992a), dominate the fauna. In Wraysbury reservoir, cyprinids were almost absent but in Queen Mary Reservoir are more abundant because occasional inundation of terrestrial marginal plants enhanced their spawning possibilities (Kubečka & Duncan, in press). In June 1993, it was found that an algal covering of the littoral bricks in Queen Mary Reservoir successfully stimulated bream spawning in the absence of macrophytic plants (Seda, pers. comm.).

Larval and juvenile life cycle stages

The larvae of many phytophilous and phytolithophilous fish are best adapted, morphologically, ecologically and ethologically, to survive in littoral shallows with aquatic vegetation (Kryzhanovsky, 1948). Perciforme larvae usually have an offshore pelagic phase before becoming demersal in the littoral when 10–20 mm long (Craig, 1987; Matena, 1995). In all reservoirs, inshore fry communities (0 + to 1) inhabit a 'safe' area in shallow water up to 1.2 m deep, while offshore fry communities exist simultaneously (Ginzburg, 1958; Galkin, 1965; Forney, 1971; Konobeyeva *et al.*, 1980; Coles, 1981; Gliwicz & Jachner, 1992). The same species contribute to both communities but the relationship between the communities is not understood fully. The inshore fry communities are better known because they are easier to study. The mean densities of fish fry caught inshore lie within the same order of magnitude in different reservoirs, despite the large differences in littoral area (Tables 2 & 4). What limits the littoral carrying capacity of different types of reservoirs is not known.

Littoral regions with flooded aquatic macrophytes usually harbour the greatest fish species diversity and fry densities (Il'ina, 1968; Kundiev, 1979; Koval, 1979, 1981; Kuznetsov, 1973; Bryan & Scarnecchia, 1992; Jude & Pappas, 1992) because these offer suitable food conditions (Matena, 1995; Simonian *et al.*, 1995) and shelter from predation, about which less is known.

In the Czech Rimov reservoir, where flooding of the sparsely vegetated littoral zones of *Phalaris arundinacea* determined the degree of resource partitioning or competition for food between the

Table 4. Densities of inshore fry communities caught in summer from various European reservoirs. Sources: Zalewski *et al.* (1990); Kubečka (1993); Kubečka & Duncan (in press). Groups of reservoirs: L = London, R = Russian, C = Czech, P = Polish.

Group	Reservoir	N/100 m^2
L	Queen Mary	139.4
L	Wraysbury	275.8
L	Queen Elizabeth	295.6
R	Kegumskoye	620–1916
R	Kakhovskoye	124–1180
R	Kremenchugskoye	444–504
R	Kremenchugskoye	448–1412
R	Ivankovskoye	346
R	Kanevskoye	104–484
R	Kievskoye	308–772
R	Seligeevskoye	305
R	Mingechaurskoye	758
R	Mozhayskoye	76–5552
C	Klicava	456
C	Rimov	496
P	Sulejow*	202–5333

* Of the three reservoir types described in Table II, Sulejow Reservoir fits the Russian category best

two main fry species, roach and perch (Matena, 1995). In a flooded site, roach fed largely on littoral chydorid cladocerans and periphytic chironomid and oligochaete species whereas perch selected planktonic cladocerans, cyclopoids and calanoids. When not flooded in the following year, the roach became more planktonic, like the perch, and like the roach at a more steeply sloping unvegetated littoral site with more open conditions.

Northcott (1979) found a richer crustacean fauna (22 species, with three-fold densities and more than double the biomass) in the marginal *Typha*, *Sparganium* and *Elodea* weed-beds than in the open water (9 species). The small *Bosmina* (0.3 mm) dominated numerically in the open water compared with the larger *Ceriodaphnia* (0.45 mm) in the weed beds, where the large cladocerans *Sida* (1.0 mm) and *Simocephalus* contributed to the greater weed-bed crustacean biomass. Over 79% of the 0 + roach diet consisted of *Ceriodaphnia*, *Bosmina* and *Sida* (but not *Simocephalus*) with a numerical and biomass predominance of weed-dwelling forms.

Vegetated land/water ecotones are very important for providing a supply of good quality animal food to rapidly growing fish fry. Growth rate of young roach increases by an order of magnitude when they switched from an algal to a cladoceran diet (Mark *et al.*, 1989). These energy and shelter requirements as well as flowing water may explain the distribution of fry along the River Thames, where densities were *ca* 1000 per 100 m^2 in vegetated sites on slopes of 8° compared with 10 per 100 m^2 in unvegetated sites such as deep banks with 1 m water depths and slopes of 25°, or 500 per 100 m^2 in shallow sandy bays with slopes of 8°. They were densest in stands of *Scirpus* (*Schoenoplectus*), in mixed emergent *Typha* and *Sparganium*, under fringing herbs and amongst floating yellow water lilies (*Nuphar lutea*) (with submergent leaves) and *Sparganium emersum*. More detailed studies are needed to determine why certain plant habitats are more preferable to young fish than others.

In unvegetated margins of Czech reservoirs, fry communities prefer shallower more gently-sloping bottoms to deeper more open margins with steep gradients. In Klicava Reservoir (Table 5): more species of fry and greater fry densities were caught in a shallow littoral site overlying the former flood plains of the reservoir's tributaries, where bottom gradients were gentler (between 1°–7°) than in the deeper more lacustrine site (bottom gradients 9°–20°). Likewise, there were greater densities and diversities of older juveniles in their first autumn and second spring, but not of younger fish in their first spring and summer. Why the older fry prefer the shallow areas is not known. Schiemer *et al.* (1991) and Matena (1995) reported similar patterns of increasing habitat selectivity in juvenile fish during their development. In Klicava Reservoir, all the rare species of fish fry (white bream, asp (*Aspius aspius*), dace (*Leuciscus leuciscus*), pike, gudgeon (*Gobio gobio*) and nase (*Chondrostoma nasus*) were recorded only in the shallow tributary areas. For fry, life is probably more dangerous in steeply sloping littorals with more open lacustrine conditions. Such sites are normally occupied by larger individuals than are found in the more protected shallow areas (Mikheev, 1985; Kubečka, 1989A).

Species preferences for the littoral differ

Table 5. Shannon-Weaver diversity indices, densities and species number of 0 + fry caught by a fry seine in shallow and deep littoral sites of Klicava reservoir during 1981–83.

Littoral stations		
Period	Shallow	Deep
Shannon-Weaver's diversity index		
1st spring	1.041	0.985
1st summer	1.946	1.98
1st autumn	0.912	0.218
2nd spring	1.516	0.574
Catch per 100 m^2		
1st summer	232.4	231.0
1st autumn	155.0	15.3
2nd spring	112.7	72.0
No. of species	14	8
slope (degrees)	1 to 7	9 to 20
Shoreline netted (m)	1890	1222
Total No. fish caught	23869	3643

Table 6. Species composition of the inshore fry communities of the Klicava reservoir from 1981–1983 compared with the species composition of the adult fish during 1983–1985.

Species	Fry (1981–83) %	Adults (1983–85) %
Roach	29.8	55.7
Perch	18.3	30.8
Chub	9.1	5.0
Rudd	35.6	5.4
Bream	5.9	2.5
Pike	0.04	0.035
Total fish sampled	10920	
Source	Kubečka (1984)	Pivnicka (1992)

(Table 6). Rudd and chub (*Leuciscus cephalus*) fry occur in the inshore communities at higher proportions than their percentage frequency in the reservoir's adult populations whereas the reverse is true for roach and perch. This suggests that the fry of rudd and chub in Klicava Reservoir have a greater affinity for the littoral zone than do the fry of roach and perch. There was a similar finding for rudd fry in Rimov reservoir (Kubečka, 1990; Matena, 1995). In the open featureless littorals of the London reservoirs from mid July, fewer fry are caught by day than by night (Kubečka & Duncan, in press), probably due to a relatively poor littoral food supply compared with the zooplankton-rich open waters and to absence of shelter from adult fish predation. Wind-generated wave action on these exposed littorals result in fry densities only one tenth of those in sheltered lee shores (Fig. 2).

Mortality rates for perch, roach, rudd and chub fry in Czech reservoirs (Cerny & Pivnicka, 1973; Kubečka, 1991; Kubečka & Pivnicka, 1993) were greater than in the more protective fry-supportive marginal regions of natural lakes and rivers.

Adult stages

Kubečka (1993) distinguished, behaviourally, two groups of adult fish in reservoir littorals:

(1) Species whose inshore densities are the same throughout 24 hours (pike, rudd, carp, dace, tench, bleak (*Alburnus alburnus*), minnow (*Phoximus phoxinus*), three-spined stickleback (*Gasterosteus aculeatus*) and *Leucaspius delineatus*;

(2) Species which spend the day offshore and migrate inshore at night (all the most important reservoir species: roach, bream, perch, chub, asp, zander (*Lucioperca lucioperca*) and ruffe; Schiemer & Hofer (1983) found similar inshore-offshore migrations by *Barbus dorsalis* in the Sri Lankan reservoir Parakrama Samudra).

Inshore areas are visited by adult fish mostly at night except during spawning. Feeding occurs intensively at dawn and dusk and the night may be spent resting in the littoral regions. Very few adults occupy the land/water ecotone of reservoirs during the day.

How fish exploit the land/water ecotone of a tropical irrigation reservoir

Reproduction

In the absence of natural lakes and indigenous lacustrine fish species (Fernando & Indrasena, 1969), the ancient reservoirs have been colonised by indigenous riverine species and species invading from coastal regions along the hydrological network. In Parakrama Samudra, the empty pelagic niche has been occupied by the clupeoid *Ehirava fluviatilis* and the lacustrine niche has been filled by introducing the African cichlid *Oreochromis mossambicus*. These are the only two species of fish which reproduced in this reservoir during studies in 1979–82. The mature adults (> 16 mm length) and young stages (2–16 mm length) of *Ehirava* were caught in the open water only (Newrkla & Duncan, 1984) and it was taken to be a pelagophil spawner (Table 3). New nests of the mouth-brooder *Oreochromis mossambicus* were found at depths 0.8–1.0 m in August/September 1979 but not in March/April 1980 and it is known to be a nest-spawner (Table 3). The land/water ecotone of the reservoir was not exploited for spawning by the indigenous fish species of riverine origin and these may migrate back to the rivers for spawning before the monsoonal rains (De Silva, 1983).

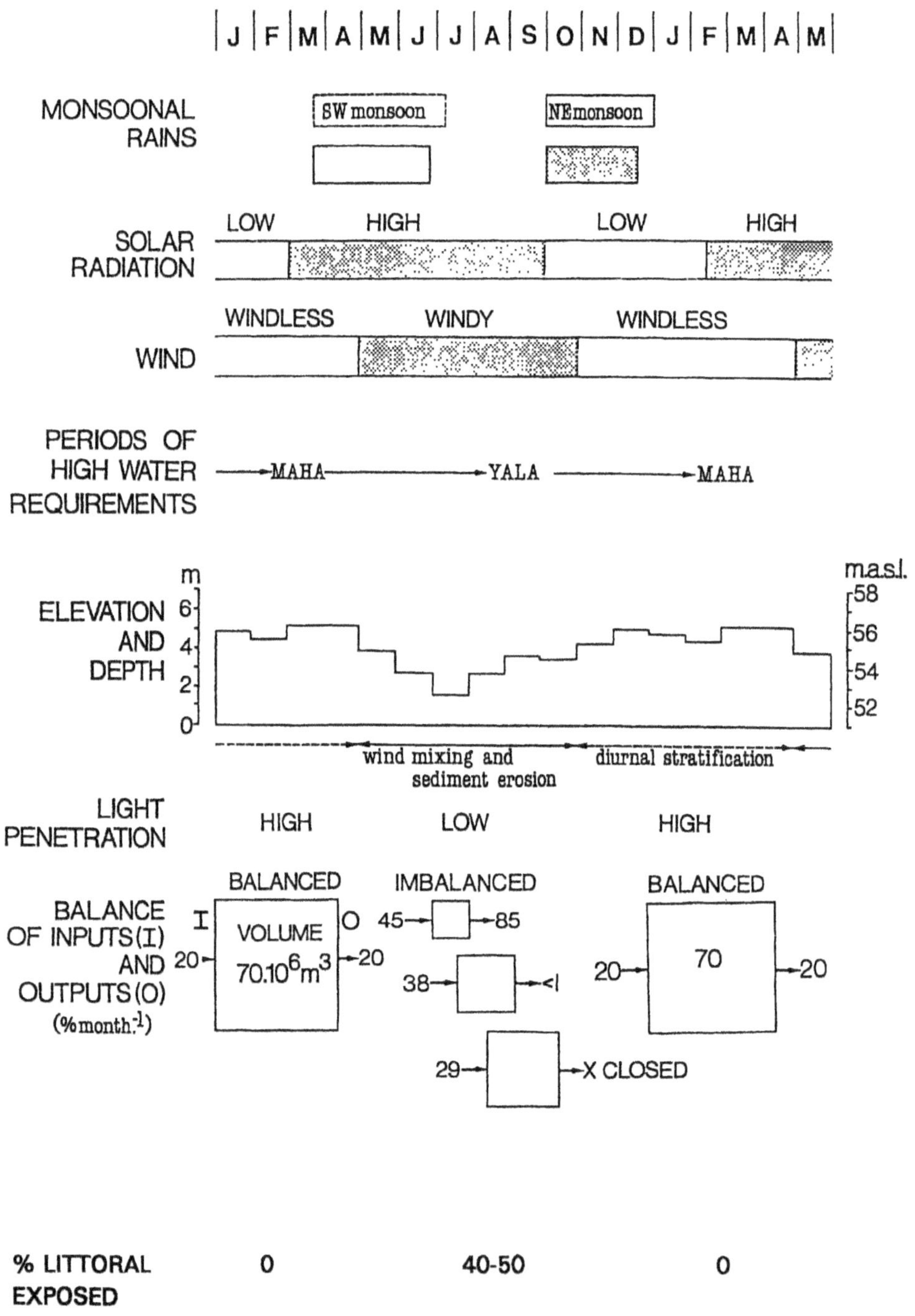

Fig. 3. The combined influences of climate and irrigation management upon the seasonal patterns of water volume, depth, water balance and extent of the exposed littoral in a lowland Dry Zone reservoir, Parakrama Samudra.

Small and young fish in the littoral area

Nevertheless, the shallow littoral area of this lowland reservoir was inhabited by small-sized fish species, only some of which were juveniles. Here in shallow water depths, the small fish were able to feed upon the abundant microbenthic food resource whilst sheltered by shallowness from predatory attacks of larger fish predators but also whilst vulnerable to predation by many of the 26 species of fish-eating birds (Winkler, 1980).

The marginal zone of the reservoir was the preferred habitat of several species of fish (5 out of the recorded 24). These were the fry of *Oreochromis* (8–33 mm), post-spawning adults of *Ehirava* (19–29 mm) and young cyprinid *Rasbora daniconius* in three size groups (10–17, 23–33 mm and 42–61 mm), together with a few small-sized *Barbus* species (14–35 mm) at the littoral end of their distribution (Duncan, unpublished; Schiemer & Hofer, 1983). When the reservoir was full, the *Oreochromis* fry patrolled along the shoreline in large shoals of one body size, the smallest (8–20 mm) in 10 cm of depth and the larger (26–33 mm) in 40–50 cm of water. Here they were heavily predated by several species of fish and birds. Dense concentrations of *Oreochromis* fry dried out in isolated 'wallows' when water levels were being rapidly dropped.

Figure 4 illustrates the characteristic community species and size structure of the littoral fish, with increasing body size associated with increasing depth and horizontal distance from the shore. When studied in 1980, the smallest tilapia, the smallest *Rasbora* and the post-spawning *Ehirava* living in the shallows were feeding on benthic microcrustaceans and early instar insects; in deeper littoral water, the larger *Rasbora* switched to feeding on *Ehirava* and the larger juvenile *Oreochromis* to a mixed diet of benthic diatoms and organic detritus (Duncan, unpublished).

In 1980, the overall abundance of fish in the non-vegetated littoral region was 5.83 fish per lift net haul (from 50 hauls of 1 m^2), with *Ehirava* contributing 84% and *Rasbora* 15%. This is roughly the same order of abundance as for inshore communities in European reservoirs (Table 4). These two species were more abundant within the *Ceratophyllum* patch growing in a sheltered bay (20 hauls; 7.2 fish per haul, 22% *Ehirava* and 66% *Rasbora*), as also occurs in the vegetated margins of European reservoirs. Almost all the *Rasbora* in the weed bed were too small to be piscivorous and both species were feeding on the rich crustacean fauna within the weed bed. In 1982, when the flooded littoral was greatly reduced in area, fish abundance was halved (3.04 per haul, mostly *Ehirava*), perhaps reflecting the impoverished food supply and/or greater vulnerability to predation by bigger fish.

Adult fish in the littoral area

In the deeper parts of the littoral, fish fauna included adults of smaller-sized species, some of which occurred here preferably and other adults of larger species which fed here, usually at night (Schiemer & Hofer, 1983).

Summary and management measures

From these examples, land/water ecotones may be larger and more significant ecologically for fish in reservoirs than in lakes. Their importance comes from greater water level fluctuations and the relatively short evolutionary time that has been available for riverine fish species to become adapted to reservoir conditions. It is difficult to distinguish the specific conditions in reservoir ecotones from those of lakes because of the widely ranging conditions existing in both kinds of water bodies. Amongst reservoirs, there are functional groupings with more specific ecotonal conditions.

Most of the indigenous fish species of Sri Lanka did not appear to spawn in the reservoir but appeared to migrate back to riverine spawning sites. The only fish species which exploits the littoral ecotone for nest-building and fry development is the introduced African cichlid. Probably the only other fish reproducing in the reservoir is a pelagophilic invading clupeid. One of these enhances local field yields and the other could be exploited commercially but, at present, both supports piscivorous fish and fish-eating birds maybe to the advantage of the indigenous minor cyprinids (Schiemer & Duncan, 1988; Sirisena & De Silva, 1988).

A clear relationship exists between the proportion of the total reservoir area that is occupied by the littoral land/water ecotone in twenty-five European non-salmonid reservoirs (Table 1) and

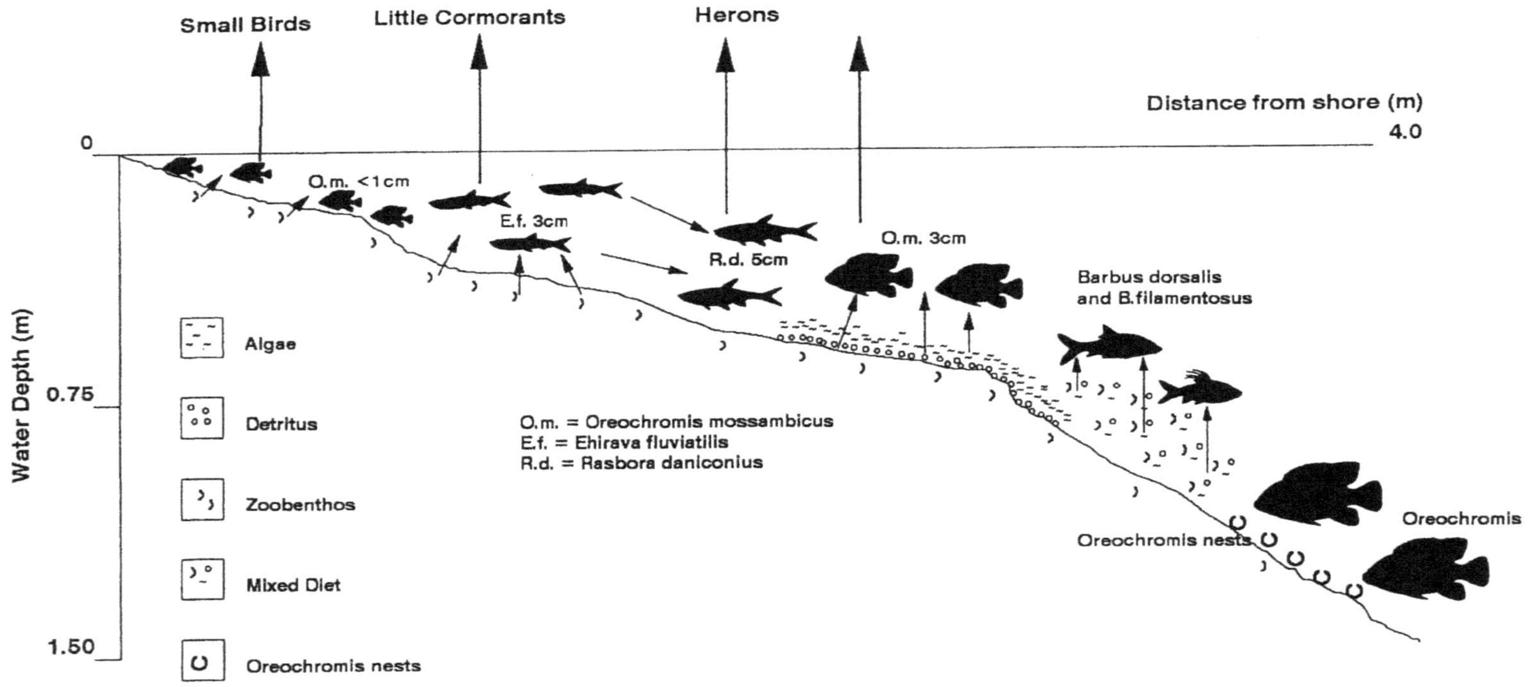

Fig. 4. The horizontal and depth distribution of small fish in the littoral of Parakrama Samudra. Om-*Oreochromis mossambicus*; Ef-*Ehirava fluviatilis*; Rd-*Rasbora daniconius*; Bf-*Barbus filamentosus*

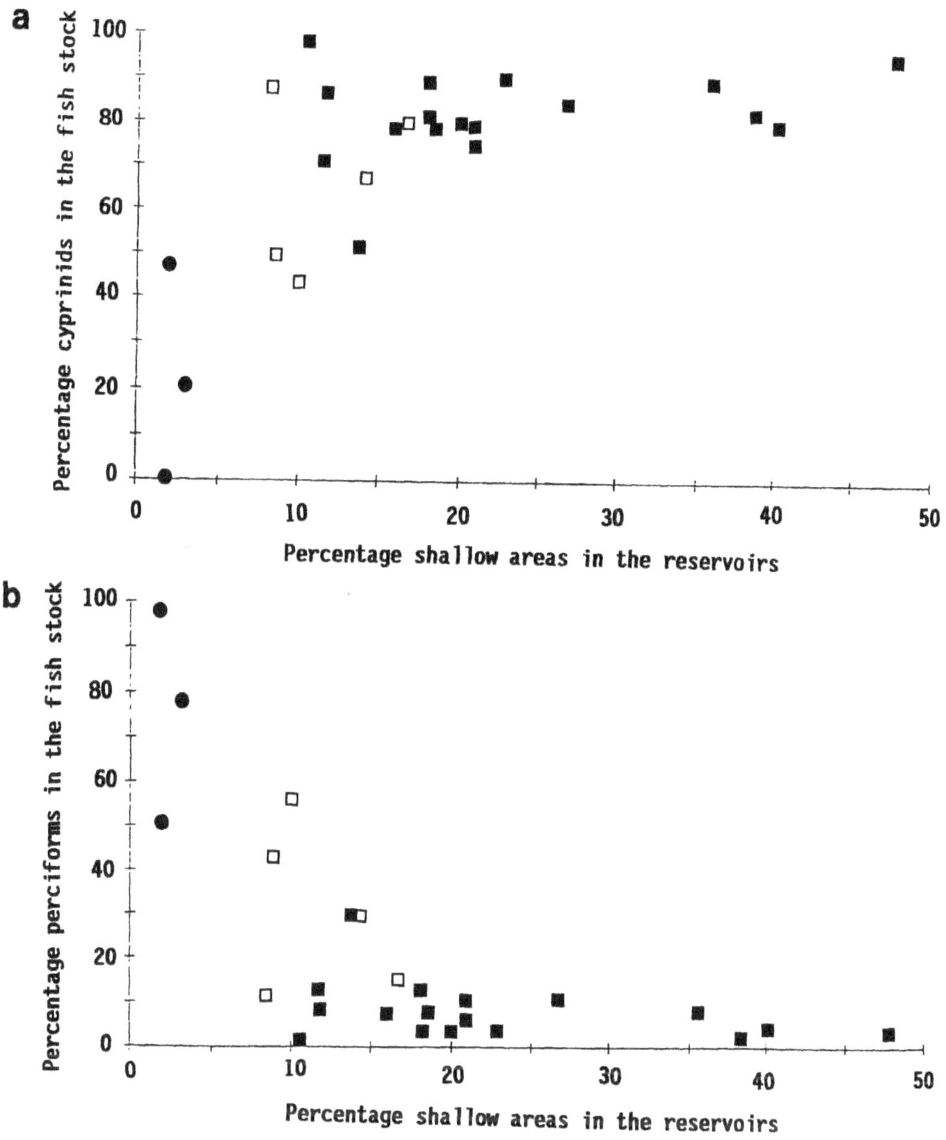

Fig. 5. Relationship between the proportion of (a) cyprinid and (b) perciform fish in the fish stock and the proportion of the reservoir area occupied by the littoral land/water ecotone in twenty-five European reservoirs. Black dots – London reservoirs; black squares – Russian reservoirs; empty squares – Czech reservoirs.

the proportion of their fish fauna which belongs to the Cyprinidae (Fig. 5a) or to the Percidae (Fig. 5b). A probable explanation for these relationships is as follows. The extensive littoral areas provide good spawning and rearing conditions for cyprinids enabling them to develop dense populations which are capable of affecting adversely co-existing perciform and other predatory fish species by competing more efficiently for zooplanktonic prey (Persson *et al.*, 1988) and thereby reducing water transparency for visual predation by ecosystem top-down effects: efficient cladoceran grazer populations are too reduced to be able to control phytoplankton levels. When availability of suitable substrata limits cyprinid spawning and juvenile survival, the competition for cladoceran zooplankton is weaker, grazing on phytoplankton is greater and water transparency more adequate for visual predation, then perciform survival is better and they attain a greater representation in the fish stock. The percentage of cyprinids, or of esocids (Table 2), can be used as a crude measure of adverse human management upon the land/water interface in older reservoirs.

The extremely poor quality for fish of the ecotonal littoral of the London reservoirs represents a trophic paradox. These reservoirs must be classified as hypertrophic, because of their continuously high levels of phosphate-phosphorus (Table 1). From their fish fauna, they should be considered the most mesotrophic of the three groups of reservoirs compared. In the generally adopted scheme of fish community succession from predominantly salmonid to predominantly cyprinid (Leach *et al.*, 1977; Colby *et al.*, 1987) the London reservoirs are considered as perciform or transient perciform-cyprinid (Kubečka & Duncan, in press). In terms of biomass, the fish stock of these reservoirs is low (6–49 kg ha^{-1}), corresponding rather to oligotrophic lakes (Kubečka & Duncan, in press).

Many guidelines exist for managing reservoir land/water ecotones for fisheries (Il'ina & Gordeyev, 1972; Prosser, 1986; Ploskey, 1986; Summerfelt, 1986; Willis, 1986). These demonstrate that human influences need not be adverse for reservoir fish populations although it is sometimes difficult to decide what kind of fish stock is most desirable in man-made reservoirs. Usually, reservoir managers want to create a fish stock most suitable for the purposes for which the reservoir was built. Ensuring an over-production of cyprinids by designing eulittoral ecotones which enhance their reproductive success is one strategy for reservoirs where this family of fish is to be exploited for human or animal food (Isayev & Karpova, 1980). Where no commercial fisheries exist, a large production of cyprinids is undesirable (Persson *et al.*, 1988) because of the ecosystem changes they generate. 'Anti-fish' eulittoral ecotones are highly desirable in drinking water supply reservoirs where the maintenance of low densities of planktivorous fish ensures low algal biomasses, high water transparencies and cheap water treatment costs.

References

Balon, E. K., 1975. Reproductive guilds of fishes: A proposal and definition. J Fish. Res. Bd Can. 32: 821–864.

Baluyut, E. A. 1983. Stockings and introductions of fish in lakes and reservoirs in the ASEAN countries. FAO Fish. Tech. Paper 236: 82 pp.

Beliy, N. D., 1970. The biology of the bream during early life stages in deep water conditions. Voprosy Ikhtiologii 10: 1047–1056 (in Russian).

Boitsov, M. P, 1983. (Abundance and growth of fish fry of the Uglichkoye reservoir). Sborn. Nauch. Trudov Gosniiorkh 202: 61–68 (in Russian).

Bryan, M. D. & D. L. Scarnecchia, 1992. Species richness, composition and abundance of fish larvae and juveniles inhabiting natural and developed shorelines of a glacial Iowa lake. Envir. Biol. Fish. 35: 329–341.

Colby, P. J., P. A. Ryan, D. H. Schupp & S. L. Serns, 1987. Interactions in north-temperate lake fish communities. Can. J. Fish. aquat. Sci. 44: 104–128.

Cerny, K. & K., Pivnicka, 1973. Abundance and mortality of the perch fry (*Perca fluviatilis*, Linnaeus 1758) in the Klicava reservoir. Vest. Cs. Spol. Zool. 37: 1–13.

Coles, T. F., 1981. Distribution of perch, *Perca fluviatilis*, L., through their first year of life in Llyn Tegid, North Wales. J. Fish Biol. 18: 15–30.

Craig, J., 1987. The biology of perch and related fish. Croom Helm, Timber Press, London, 333 pp.

De Silva, S. S., 1983. Reproductive strategies of some major fish species in Parakrama Samudra reservoir and their possible impact on the ecosystem – a theoretical consideration. In: F. Schiemer (ed.), Limnology of Parakrama Samudra, Sri Lanka. Dr W. Junk Publishers, The Hague: 185–191.

De Silva, S. S. (ed.) 1988. Reservoir fishery Management and development in Asia. IDRC, Ottawa: 246 pp.

Dolman, W. D., 1990. Classification of Texas reservoirs in relation to limnology and fish community associations. Trans. am. Fish. Soc. 119: 511–520.

Duncan, A., 1990. A review: limnological management and biomanipulation in the London reservoirs. Hydrobiologia 200/201 (Dev. Hydrobiol. 61): 541–548.

Duncan, A., A. Gunatilaka & F. Schiemer, in press. Limnological aspects of landscape management in Sri Lanka. In: W. Erdelen & Ch. Pren (eds), Ecology and landscape management in Sri Lanka. Verlag Josef Margraf Sci. Books.

Fernando, C. H. & J. Holcik, 1991. Fish in reservoirs. Int. Revue ges. Hydrobiol. 76: 149–167.

Fernando, C. H. & H. H. A. Indrasena, 1969. The freshwater fisheries of Ceylon. Bull. Fish. Res. Stat., Ceylon 20: 101–134.

Forney, J. L., 1971. Development of dominant year classes in a yellow perch population. Trans. am. Fish. Soc. 100: 739–749.

Galkin, G. G., 1965, (Species composition, distribution and growth of fish fry of the Gorkovskoye Reservoir in the first years of existence). Izvestiya Gosniiorkh, Leningrad 59: 98–112 (in Russian).

Gliwicz, Z. M. & A. Jachner, 1992. Diel migrations of juvenile fish: a ghost of predation past or present? Arch. Hydrobiol. 124: 385–410.

Ginzburg, J. I., 1958. (Biology and crop of fish fry in Tsymlyanskoye reservoir). Izvestiya Vniiorrkh 45: 111–141 (in Russian).

Holcik, J. & V. Hruska, 1966: On the spawning substrate of the roach – *Rutilus rutilus* (Linnaeus, 1758) and bream – *Abramis brama* (Linnaeus, 1758) and notes on the ecological characteristics of some European fishes. Vest. Cs. Spol. Zool. 30: 22–30.

Hrbacek, J., 1984. Ecosystems of european man-made lakes. In: F. B. Taub (ed.), Ecosystems of the world 23: Lakes and reservoirs. Elsevier, Amsterdam: 267–290.

Hutchinson, G. E., 1967. A treatise on Limnology. II Introduction to Lake Biology and the Limnoplankton. J. Wiley, New York: 1115 pp.

Il'ina, L. K. & N. A. Gordeyev, 1972. Water-level regime and the spawning of fish stocks in reservoirs. J. Ichthyol. 72: 373–381.

Il'ina, L. K., N. A. Gordeyev & L. N. Strizhnikova, 1987. (The role of the tributaries of Rybinskoye reservoir in the reproduction of phytophilous fish and the peculiarities of spawning areas in the years with low water level. Trudy Inst. Biol. Vnutr. Vod. 39: 124–135 (in Russian).

Il'ina, L. K. & T. K. Nebolshina, 1976. (Changes of the spawning conditions of phytophilous fish connected with the regulation of the Volga river). In Biological and production processes in the Volga catchment area). Leningrad: 113–124.

Isayev, A. I. & E. I. Karpova, 1980. (Fishery management of reservoirs). Legkaya i Pischvaya promyshlennost, Moscow: 235 pp (in Russian).

Jude, D. J. & J. Pappas, 1992. Fish utilisation of great lakes wetlands. J. Great Lakes Res. 18: 651–672.

Kapetsky, J. M. & T. Petr, 1984. Status of African reservoir fisheries. CIFA Technical Paper 10: 326 pp.

Kimmel, B. L. & A. W. Groeger, 1986. Limnological and ecological changes associated with reservoir aging. In: G. E. Hall & M. J. Van Den Avyle (eds), Reservoir Fisheries management: Strategies for the 80's. Am. Fish. Soc., Bethesda: 103–109.

Konobeyeva, V. K., A. G. Konobeycv & A. G. Poddubnyi, 1980. (About the mechanisms of the perch *Perca fluviatilis* L. fry accumulations in the open part of lake-type reservoir (on the example of Rybinskoye reservoir)). Voprosy Ikhtiologii 20: 258–271 (in Russian).

Koval, N. V., 1979. (Ecology of the fry of industrially important fry of the Kremenchugskoye reservoir). Gidrobiol. Zhourn. 15: 44–48 (in Russian).

Koval, N. V., 1981. (Conditions of fry existence in the Kanevskoye reservoir influenced by heated effluent from Tripolskaya plant) Gidrobiol. Zhourn. 17: 50–53 (in Russian).

Kryzhanovsky, S. G., 1948. (Ecological groups of fishes and the laws of their development). Izvest. Tikhookean. Nauchno-Issled. Inst. Rybn. Khoz. Okean. 27: 115 pp.

Kubečka, J., 1984. (Some aspects of the fry ecology of the Klicava reservoir). MSc. Thesis, Charles University, Prague: 207 pp. (in Czech).

Kubečka, J., 1989. Development of the ichthyofauna of the Rimov Reservoir and its management. Arch. Hydrobiol. Beih. Ergebn. Limnol. 33: 611–613.

Kubečka, J., 1989A. First year growth of the roach (*Rutilus rutilus* L.) fry and its comparison within geographical range. Acta Univ. Carolinae, Biologica 33: 94–105.

Kubečka, J. (ed.) 1990. Ichthyofauna of the Malse River and Rimov Reservoir. South Bohemian Museum, Ceske Budejovice, 151 pp. (in Czech with English summaries).

Kubečka, J., 1991. A case of an extremely low share of the 0-age group on the total biomass, production and ration of the perch (*Perca fluviatilis*) population. Acta Soc. Zool. Bohemoslov. 55: 28–33.

Kubečka, J., 1992. Succession of fish communities in reservoirs of Central and East Europe. In: M. Straskraba, J. G. Tundisi & A. Duncan (eds), Comparative Reservoir Limnology and Water Quality Management. Dev. Hydrobiol. 77. Kluwer Academic Publishers, Dodrecht: 153–168.

Kubečka, J., 1992a. Fluctuations in fyke-net catches during the spawning period of the eurasian perch (*Perca fluviatilis*) in the Rimov Reservoir, Czechoslovakia. Fish. Res. 15: 157–167.

Kubečka, J., 1993. Night inshore migration and capture of fish by shore seining. Aquacult. Fish. Manag: 5 pp.

Kubečka, J. & A. Duncan, in press. Low fish predation in the London reservoirs: Species composition, density and biomass. Int. Revue ges. Hydrobiol.

Kubečka, J. & K. Pivnicka, 1993. Numbers and production of juvenile cyprinids in the Klicava reservoir (Czech Republic). Acta Univ. Carolinae, Environmemtalica 5: 61–73.

Kutuzov, A. M., 1976. The effectiveness of reproduction of the white bream, *Blicca bjoerkna*, and the roach, *Rutilus rutilus*, in the upper section of Kuybyshevskoye Reservoir. J. Ichthyol. 15: 678–682.

Kuznetsov, V. A., 1973. Distribution of fish fry in the lower parts of Sviyaga Bay of the Kuybyshevskoye reservoir. Gidrobiol. Zhourn. 9: 76–80 (in Russian).

Kuznetsov, V. A., 1975. (Abundance and survival rate dynamics of freshwater fish fry). Kazan Univers. Press, Kazan: 71 pp. (in Russian).

Leach, J. H., M. G. Johnson, J. R. M. Kelso, J. Hartmann, W. Numann & B. Entz, 1977. Responses of percid fishes and their habitats to eutrophication. J. Fish. Res. Bd Can. 34: 1964–1971.

Mark, W., W. Wieser & C. Hohenauer, 1989. Interactions between development processes, growth and food selection in the larvae and juveniles of *Rutilus rutilus* (L.) (Cyprinidae). Oecologia 78: 330–337.

Matena, J., 1995. The role of ecotones as feeding grounds for fish fry in a Bohemian water supply reservoir. Hydrobiologia 303 (Dev. Hydrobiol. 105): 31–38.

Mikheev, V. N., 1985. (Connection between the feeding, distribution and downstream migration of fish fingerlings in the reservoir). In: D. S. Pavlov & A. G. Gusar (eds), Downstream migration of fish. Academy of Scienc. USSR, Moscow: 87–108.

Newrkla, P. & A. Duncan, 1984. The biology and density of *Ehirava fluviatilis* (clupeoid) in Parakrama Samudra, Sri Lanka. Verh. Int. Ver. Limnol. 22: 1572–1578.

Northcott, D., 1979 The importance of aquatic macrophytes in the provision of crustacean zooplankton food for young roach. Proc. 1st British Freshwater Fisheries Conf., Liverpool: 123–134.

Persson, L., G. Andersson, S. F. Hamrin & L. Johansson, 1988. Predator regulation and primary production along the productivity gradient of temperate lake ecosystems. In: S. L. Carpenter (ed.), Complex Interactions in Lake Communities. Springer, New York: 45–65.

Pivnicka, K., 1992. The Klicava Reservoir, Czechoslovakia: a 30 year study of the fish community. Fish. Res. 14: 1–20.

Ploskey, G. R., 1986. Effects of water-level changes on reservoir ecosystems, with implications for fisheries management. In: G. E. Hall & M. J. Van Den Avyle (eds), Reservoir Fisheries Management: Strategies for the 80's. Am. Fish. Soc. Bethesda: 86–97.

Prosser, N. S., 1986, An overview of reservoir fisheries problems and opportunities resulting from hydropower. In: G. E. Hall & M. J. Van Den Avyle (eds), Reservoir Fisheries Management: strategies for the 80's. Am. Fish. Soc., Bethesda: 238–246.

Schiemer, F. (ed.), 1983. Limnology of Parakrama Samudra, Sri Lanka. Dr W. Junk Publishers, The Hague: 236 pp.

Schiemer, F. & A. Duncan, 1988. The significance of the ecosystem approach for reservoir management. In: S. S. De Silva (ed.), Reservoir fishery management and development in Asia, IDRC, Ottawa: 183–194.

Schiemer, F. & R. Hofer, 1983. A contribution to the ecology of the fish fauna of Parakrama Samudra Reservoir. In: F. Schiemer (ed.), Limnology of Parakrama Samudra, Sri Lanka. Dev. Hydrobiol. 12. Dr W. Junk Publishers, The Hague: 135–154.

Schiemer, F., T. Spindler, H. Wintersberger, A. Schneider & A. Chovanec, 1991. Fish fry associations: Important indicators for the ecological status of large rivers. Verh. int. Ver. Limnol. 24: 2497–2500.

Simonian, A., I. Tátrai, P. Bíró, G. Paulovits, L.G.-Tóth & G. Lakatos, 1995. Biomass of planktonic crustaceans and the food of young cyprinids in the littoral zone of Lake Balaton. Hydrobiologia 303 (Dev. Hydrobiol. 105): 39–48.

Sirisensa, H. K. G. & S. S. De Silva, 1988. Non-conventional fish resources in Sri Lankan reservoirs. In: S. S. De Silva (ed.), Reservoir Fishery Management and Development in Asia IDRC, Otawa: 113–120.

Spanovskaya, V. D., V. A. Grigorash, T. V. Lebedeva & V. A. Rekurbratskii, 1980. Spawning efficiency of fish of the Mozhayskoye reservoir in different years (1969–1975). In: V. A. Bykov, I. M. Kisina & K. K. Edelstein (eds), Complex Investigations of reservoirs. Moscow, 5: 198–209 (in Russian).

Steel, J. A., 1972. The application of fundamental limnological research in water supply system design and management. In: R. W. Edwards & D. S. Garrod (eds), Conservation and productivity of natural waters. Academic Press, London: 41–67.

Steel, J. A., 1976. Eutrophication and the operational management of reservoirs of the Thames Water Authority. Inst. Publ. Hlth. Engs. symp. Eutrophication of Lakes and Reservoirs. Chameleon Press, UK, 1–9.

Summerfelt, R. S., 1986. Summarisation of the symposium. In: G. E. Hall & M. J. Van Den Avyle (eds), Reservoir Fisheries Management: Strategies for the 80's. Amer. Fish. Soc., Bethesda: 314–327.

Tundisi, J. G., T. Matsumura-Tundisi, M. C. Calijuri, 1992. Limnology and management of reservoirs in Brazil. In: M. Straskraba. J. G. Tundisi & A. Duncan (eds), Comparative Reservoir Limnology and Water Quality Management. Dev. Hydrobiol. 77. Kluwer Academic Publishers, Dordrecht: 25–56.

Tyurin, P. V., 1961. Influence of the water level regime upon the formation of fish stock. Izvestiya Gosniiorrkh 50 (in Russian).

Willis, D. W., 1986. Review of Water level management of Kansas reservoirs. In: G. E. Hall & M. J. Van Den Avyle (eds), Reservoir Fisheries Management: Strategies for the 80's: 110–114.

Windle-Taylor, E., 1961–62. Fortieth report of the Metropolitan Water Board, London: 48–49.

Winkler, H., 1980. Impact of fish eating birds. In: F. Schiemer (ed.), Parakrama Samudra Sri Lanka Limnology Project Interim Report. Vienna: 65–73.

Zalewski, M., B. Brewinska-Zaras & P. Frankiewicz, 1990. Fry communities as a biomanipulating tool in a temperate lowland reservoir. Arch. Hydrobiol. Beih. Ergebn. Limnol. 33: 763–774.

Zuromska, H., 1967. (Estimation of the mortality of eggs and larvae of roach (*Rutilus rutilus* L.) in lake spawning grounds). Roczn. Nauk Roln. Ser. H, 90: 539–556 (in Polish with English summary).

Hydrobiologia **303**: 31–38, 1995.
F. Schiemer, M. Zalewski & J. E. Thorpe (eds), The Importance of Aquatic-Terrestrial Ecotones for Freshwater Fish.

The role of ecotones as feeding grounds for fish fry in a Bohemian water supply reservoir

Josef Matěna
Hydrobiological Institute, Czech Academy of Sciences, Na sádkách 7, 370 05 České Budějovice, Czech Republic

Key words: littoral fry, perch, roach, diet overlap

Abstract

The food composition of young-of-the-year (YOY) roach, common bream and perch, the dominant fish species in Římov reservoir, was investigated during 1989 and 1990. The littoral fry assemblages were dominated by roach and perch, while bream larvae migrated into deeper water early in their development. The YOY roach fed opportunistically. Up to 15 mm length they remained close to the shoreline, preferring periphyton, *Chydoridae* and larval *Chironomidae*. Perch larvae between 10–15 mm long gradually became demersal but their diet continued to consist of the planktonic *Cyclops spp.* and *Diaptomus spp.* Resource partitioning was a characteristic feature of the littoral fry assemblages. The only significant overlap in diet occurred with *Polyphemus pediculus*, which was positively selected by both roach and perch fry.

Introduction

The ichthyofauna of reservoirs contains mainly riverine species that are able to adapt themselves to the lentic environment. True lacustrine species are very rare in reservoirs of the temperate zone (Fernando, Holčík, 1991). The land/water ecotones (Wetzel, 1990) serve as spawning areas for the dominant cyprinids in European reservoirs – roach (*Rutilus rutilus*) and common bream (*Abramis brama*) (Mark *et al.*, 1989). The land/water ecotones play further an important role as feeding grounds for fish fry in reservoirs. Cyprinids larvae form schools in the shallow littoral after hatching and begin to move offshore only after several weeks (Rheinberger *et al.*, 1986). Many authors investigated the food composition of YOY cyprinids (e.g. Townsend *et al.*, 1986; Weatherley, 1987; Mark *et al.*, 1987; Schiemer *et al.*, 1989). Perch larvae become limnetic after hatching and remain pelagic for about one month until full development of fins. The juveniles then move inshore to the ecotonal zone (Craig, 1987). The food of YOY perch was described for example by Spanovskaya & Grigorash (1977), Thorpe (1977), Guma'a (1978), and reviewed by Craig (1987).

The aspects of mutual relationships between planktivorous fish and zooplankton were reviewed by Lazzaro (1987).

The aim of this paper was to evaluate the role of the land/water ecotone in the Římov Reservoir as a feeding ground for the three dominant fish species – roach, common bream and perch.

Description of the reservoir

The Římov Reservoir (Fig. 1) on the Malše River in Southern Bohemia, was impounded in 1978.

Fig. 1. Map of Římov Reservoir showing the location of the littoral sampling stations A and B and the 10 m depth contour.

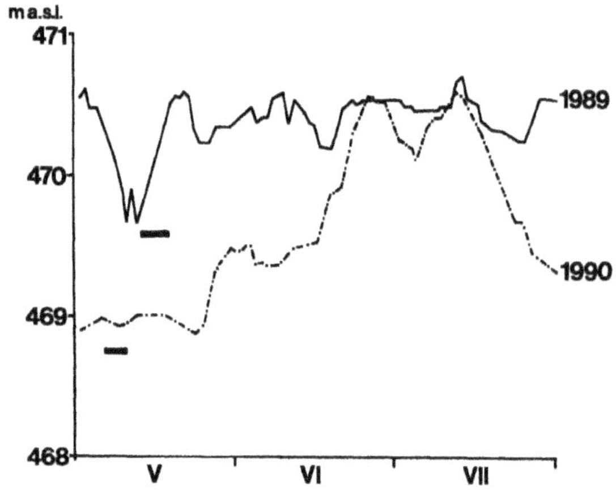

Fig. 2. Water level fluctuations in Římov Reservoir during 1989 and 1990. The horizontal bars indicate the mass spawning period of the cyprinids.

Its surface area is 210 ha, total volume 34×10^6 m^3, altitude 470 m above sea level, theoretical retention time 90 days, maximum depth 43 m, and average depth 16 m. Fluctuations of the water level during 1989–90 investigation are shown in Fig. 2. The slope of the banks is quite steep. A typical ecotonal zone is developed only at a few sites. Submerged vegetation is almost absent. A detailed description of the locality was given by Stach and Kubečka (1990). Development of the ichthyofauna was described by Kubečka (1989). The population of perch (*Perca fluviatilis*), the most abundant species during the first few years after impoundment, is still declining. The estimated total biomass was about 150 kg per ha in the investigated years.

Material and methods

The two years investigated, 1989 and 1990, differed with respect to the water level during the spawning period of cyprinids (Fig. 2). There was no flooded terrestrial vegetation in 1990. While in 1989 some areas had flooded vegetation. The water level of the reservoir was dropped by about 0.5 m after the mass spawning of cyprinids in 1989 and several preceding years. The aim was to destroy the eggs. Two sampling stations were chosen along the banks (Fig. 1). Station A represented a typical ecotonal zone with sandy bottom, moderate slope (<8%) and vegetation tolerant of flooding (mainly *Phalaroides arundinacea*). Station B had a steep slope (approx. 30%) and gravel bottom and thus served for comparison in 1990. Sampling was carried out from May to August. Hand nets (0.7 mm mesh size) were used during several days immediately after hatching. Later on, a seine net (10 × 1.8 m, mesh size 1 mm) was used. Samples were usually taken from 9 a.m. to 1 p.m.

In 1989 a 24-hour trial was carried out at sampling station A, with samples taken every 3 hours. The whole catch was immediately preserved in 10% formaldehyde. In the laboratory, the standard length of fry was measured and the gut contents of at least 10 specimens were analysed. Permanent mounts in polyvinylalcohol of the whole gut contents were prepared and counts were made of organisms present. The diet was designated using the numerical method (Hyslop, 1980).

Quantitative zooplankton samples were taken simultaneously with fish sampling by means of a 3 l Patalas sampler. The volume sampled on each occasion was 30–50 l. A net of 40 μm mesh size was used for retaining the zooplankton.

Ivlev's electivity index

$$E = \frac{r_i - p_i}{r_i + p_i}$$

was used for evaluation of selectivity of different fish species for particular food items, where r_i is the share of a given food item in the ration, and p_i is the share of the same food item in the environment.

The diet overlap between two species was calculated according to Zaret and Rand (1971)

$$C_{XY} = \frac{2 \sum_{i=1}^{S} X_i Y_i}{\sum_{i=1}^{S} X_i^2 + \sum_{i=1}^{S} Y_i^2},$$

where X_i, Y_i are the relative proportions (in %) of a given food item in the diets of the compared fish species and S is the number of distinguished food items.

The individual daily ration was estimated in the field on July 3–4, 1989, according to Elliott and Persson (1978). The instantaneous evacuation rate (R) (Persson, 1979, 1982) was used, adapted to the actual temperature during the experiment (19 °C).

Results

Composition of littoral fry assemblages

The ecotonal zone served as a spawning ground for roach and common bream. Both species spawned mainly in the flooded terrestrial vegetation in shallow water up to 0.5 m deep but bream eggs have been found even at a depth of 8 m. Mass spawning of roach coincided with the beginning of spawning of common bream (Fig. 2). In the absence of flooded vegetation in May 1990, due to low water levels, both species spawned successfully on rocks and gravelly bottom near the shoreline.

Table 1. Species composition of cyprinid larvae in the littoral zone on 18th May 1990 (%). Average percentages of 8 sampling stations.

Species	x	±	s.d.
Rutilus rutilus	88.2	±	5.9
Abramis brama	11.8	±	5.9

Immediately after hatching, the species composition of the littoral fry was homogeneous over the whole reservoir, from the dam to the tributary area (Table 1) with roach the more abundant. No perch larvae were found in the ecotonal zone at this time but they were present in the pelagial. Later, the perch larvae gradually became demersal on reaching a length of about 10 mm and a typical summer littoral fry assemblage developed which was dominated by perch and roach (Fig. 3). The percentage frequency of common bream larvae was generally low, with two peaks in 1989 corresponding with the hatching of new cohorts from successive spawnings. From June to August, at least three overlapping cohorts of bream could be distinguished.

By late summer, the abundance of fry gradually decreased in the ecotonal zone and, at the same time, the species composition of the 1989 littoral fry assemblage became dominated by rudd (*Scardinius erythrophthalmus*). No such corresponding change was seen in 1990 (Fig. 3).

The frequency of occurrence of particular fry species varied considerably between the subsequent samples during the 24-hour trial run on July 3rd–4th 1989. During daytime, perch (72.4%) predominated over roach (16.2%) but at night the reverse was seen, with 23.9% perch and 66.4% roach (Fig. 4).

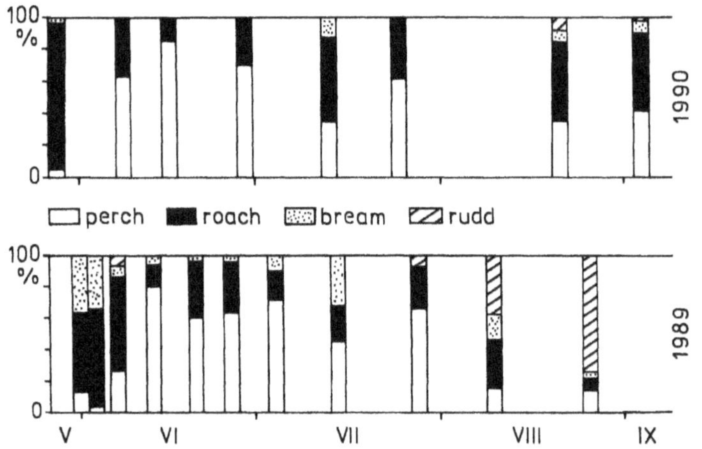

Fig. 3. The species composition of the littoral fry assemblages at station A in Římov Reservoir during 1989, when the vegetation was flooded, and in 1990, when the vegetation was not flooded.

Composition of the fry diets

Phytophilous and periphytic organisms were more evident in YOY roach diet in 1989 (Fig. 5a) than in 1990 (Fig. 6a, b). The ecotonal zone was well developed with flooded vegetation in 1989 and the phytophilous and periphytic fauna developing there was used effectively by roach larvae (Fig. 5) from an average length of 10 mm onwards. Chironomid larvae (*Cricotopus sp.*,

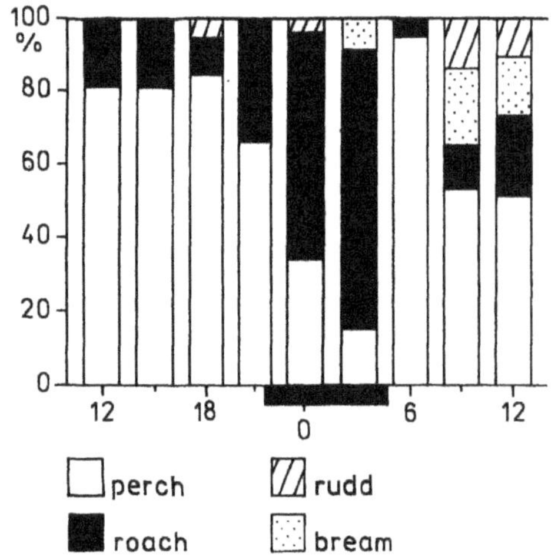

Fig. 4. Diurnal changes in the species composition of the littoral fry assemblage at station A in Římov Reservoir during July 3rd–4th 1989.

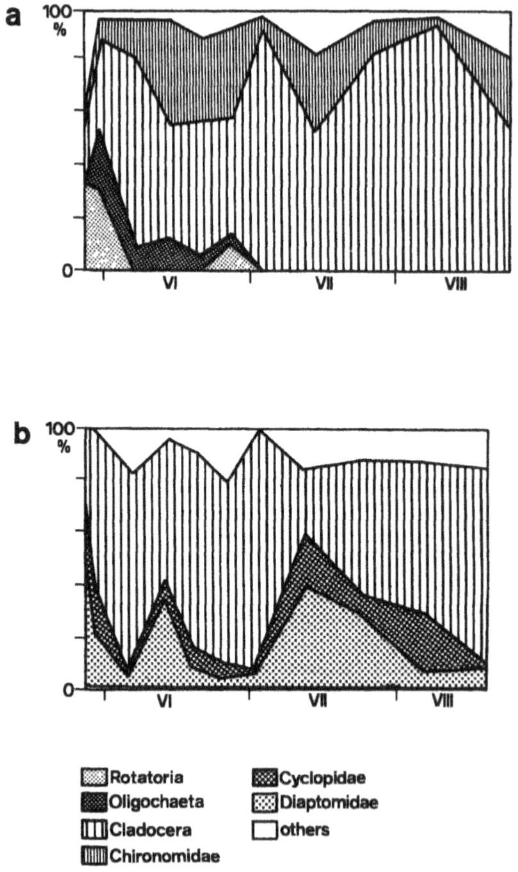

Fig. 5. The composition of the diet of YOY roach and perch at station A in Římov Reservoir during 1989: (a) roach; (b) perch.

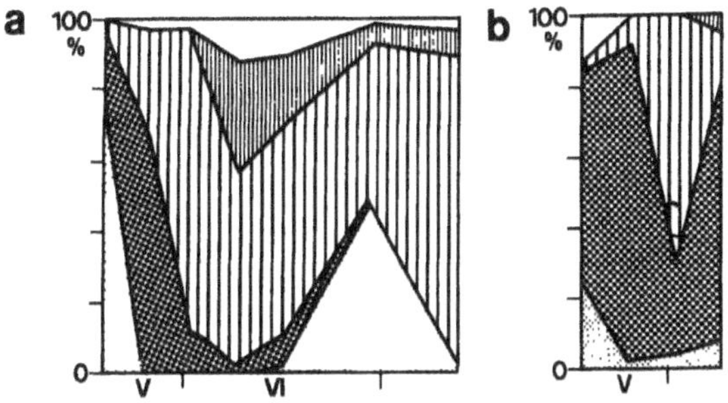

Fig. 6. The composition of the diet of YOY roach at stations A and B in Římov Reservoir during 1990: (a) station A; (b) station B. For key to diet organisms, see Fig. 5.

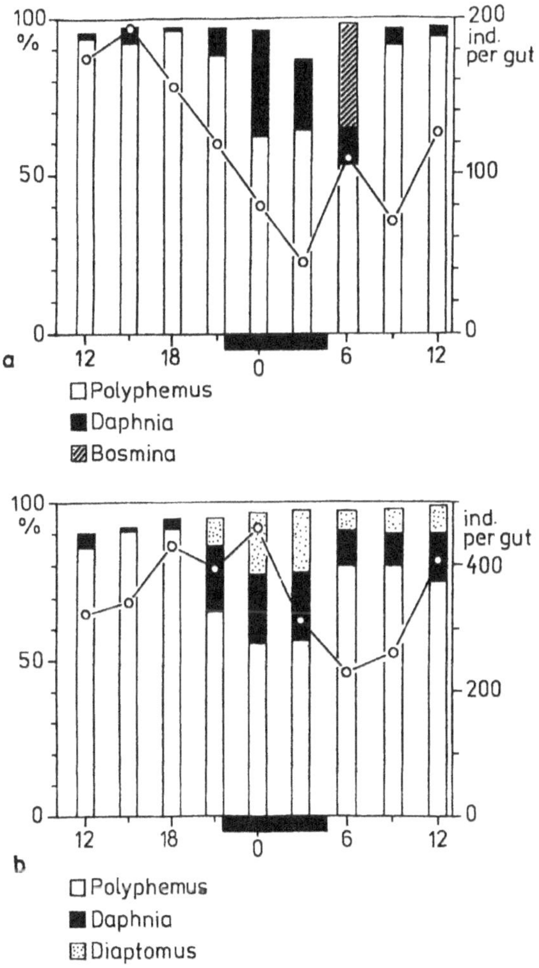

Fig. 7. Diurnal changes in the gut contents of YOY roach and perch at station A in Římov Reservoir during July 3rd–4th 1989. The histograms give percentage frequencies of the diet organisms given in the key. The plotted line gives the mean number of prey organisms per fish gut: (a) roach; (b) perch.

Corynoneura sp.) and *Oligochaeta* (*Stylaria sp.*) were the most frequently encountered food items of this group. Detailed analysis of Cladocera consumed by roach fry showed the prevalence of littoral Chydoridae till July. *Polyphemus pediculus* was preferred whenever this species occurred even in low numbers in the reservoir. *Daphnia galeata* appeared in significant numbers in the diet only after the fry reached an average length of 15 mm. The proportion of detritus and bluegreen algae in the diet increased during the summer.

The ecotonal zone was almost absent in 1990, due to the low water level (Fig. 2). The lack of flooded vegetation resulted in a higher frequency of pelagic organisms, mainly cyclopoid nauplii, in the diet of roach larvae at the beginning of exogenous nutrition (Fig. 6a). At sampling station B, without a developed ecotonal zone, the roach larvae were found only during the first six weeks after hatching. Their diet consisted mainly of cyclopoid nauplii and cladocerans (Fig. 6b).

The food composition of perch fry in the ecotonal zone consisted mainly of *Cladocera* and *Copepoda* (Fig. 5b). Compared with the roach diet, the lack of periphytic forms (*Chironomidae*, *Stylaria*, *Chydoridae*) is apparent.

The diurnal changes of gut contents (Fig. 7) and daily food ration of roach (mean length 20.5 mm) and perch fry (mean length 25.3 mm)

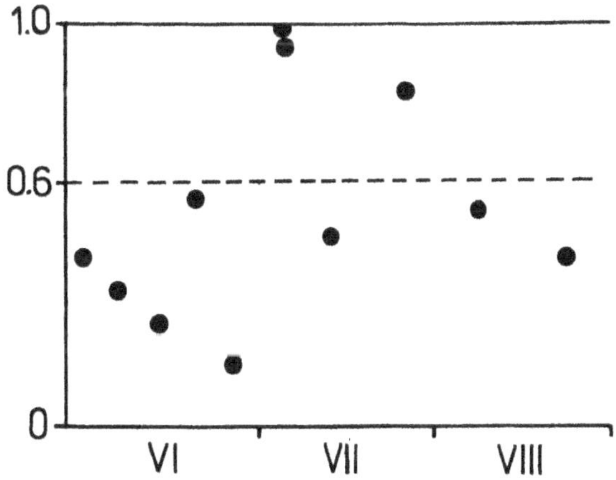

Fig. 8. Diet overlap index (C_{xy}) of YOY roach and perch in the Římov Reservoir, 1989. The value of 0.6 indicates significant overlap.

were estimated in the field in early July 1989. *Polyphemus pediculus* formed more than 90% of the diet in both species during daytime in spite of its low frequency in the nearshore zooplankton (16.6%) (Ivlev's electivity index: +0.69 for both species). Individual daily food ration was 615 *Polyphemus* in roach and 1680 in perch.

A significant diet overlap of both species, compared by Morishita's index (Zaret & Rand, 1971), occurred only when *Polyphemus pediculus* was eaten (Fig. 8). Despite forming mixed schools sometimes, a clear resource partitioning was present in the littoral fry assemblages. Perch favoured pelagic copepods, while roach avoided them, but used periphytic organisms such as chironomid larvae. Diet overlap of roach and common bream was estimated only in June and early July, 1989, due to irregular occurrence of bream in the ecotonal zone. The values of $C_{xy} = 0.6–0.8$ indicate significant diet overlap of these species at that time.

The role of the land/water ecotone can be dem-

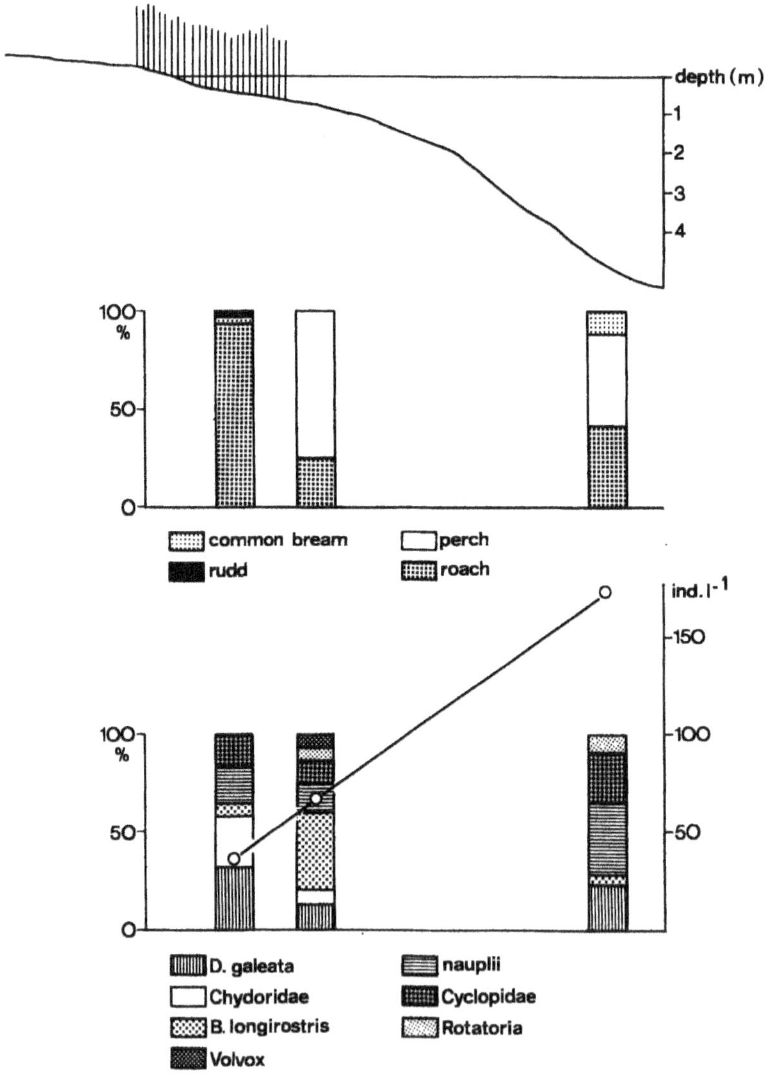

Fig. 9. Horizontal distribution of fish fry and zooplankton (abundance and composition) in the littoral zone, Římov Reservoir, 1.6.1990.

onstrated on the uneven horizontal distribution of zooplankton and the dominant fish species (Fig. 9). Distinct gradient in fry distribution was observed on most sampling dates. Perch fry were not found among flooded vegetation, corresponding with the lack of periphytic organisms in their diet. The decline of total zooplankton numbers in the shallow water nearshore was observed regularly on other sampling dates, too.

Discussion

Year-class strengths of cyprinids in reservoirs are highly variable (Kubečka & Böhm, 1991). The different water levels in 1989 and 1990 had no influence on the recruitment that was very high compared with several preceding years (Kubečka, unpubl. data). The drop of water level (Fig. 2), used as a biomanipulative too, was too low to reduce the recruitment of cyprinids in 1989 effectively.

Comparing the species composition of the ecotonal fry assemblages with that of older age classes, several differences are apparent. The numbers of common bream larvae in both years were very low compared with the share of this species found one year later during the fish population census in 1991. The roach: common bream ratio was 8:1 at the YOY age in 1990 and 2:1 in the 1+ age in 1991. It can be assumed that the first year mortality did not differ substantially between these two species. Fertilized eggs of common bream were found even at a depth of 8 m. It seems that spawning of this species is not confined to the very shallow parts of the ecotonal zone. Another possibility is the early migration of common bream larvae into deeper water. They were really caught in ichthyoplanktonic tows at depths of 6–8 m during the night (Matěna, unpubl. data).

The rudd fry showed a high affinity for the ecotonal zone in late summer. At that time rudd dominated the littoral fry assemblage in spite of forming only 0.2% of the fish stock in the reservoir.

The diurnal changes in the share of perch in the ecotonal zone (Fig. 4) might indicate the presence of diurnal inshore-offshore migrations, as described in yellow perch (*Perca flavescens*) by Post & Mc Queen (1988).

The diet of roach larvae in Římov Reservoir corresponds to the general scheme of succession in cyprinids: phytoplankton – rotifers – crustaceans – chironomid larvae. However, in 1989 Římov roach larvae shifted earlier toward larger phytophilous and/or periphytic organisms (*Oligochaeta*, *Chironomidae*) than those of studies by Hammer (1985) or Mark *et al.* (1987). This was probably caused by the low concentration of zooplankton (< 50 ind. l^{-1}) in the ecotonal zone of Římov Reservoir in June 1989. Similar results were obtained by Ponton & Stroffek (1987) in Lake Geneva. The diet of YOY perch in Římov Reservoir is in accordance with the results of Hartman (1983, 1986), who found a succession from nauplii to older copepods and to cladocerans in perch. Perch favoured pelagic *Copepoda*, while roach avoided them.

These findings differ from the results of Persson (1987), Persson & Greenberg (1990), and Diehl (1988). They found roach to be a more effective zooplankton feeder than perch. Perch, however, was more efficient than roach when feeding on macroinvertebrates. This caused juvenile perch to shift from feeding on zooplankton to macroinvertebrate prey. No such shift was observed in the Římov Reservoir, where the perch remained mainly planktivorous even in older age classes (Kubečka unpubl. data).

The land/water ecotones are of fundamental importance as feeding grounds for YOY roach in the Římov Reservoir. Perch, although abundant in this zone fed to a high degree on pelagic zooplankton. The ecotonal zone might serve as a shelter area in YOY perch to avoid predation during daytime. Common bream, the third abundant species in the reservoir, seemed to migrate offshore into deeper water early in its ontogeny.

Acknowledgements

The author thanks Z. Prachař for help in field samplings and J. Bartákova for technical assistance. Valuable comments on earlier drafts of this paper were given by Prof. C. H. Fernando and J. Kubečka. The author is further much obliged to Prof. F. Schiemer for his criticism and help with the figures. The work was partially supported by the Grant Agency of CSAS (Grant Nr. 61 701).

References

Craig, J. F., 1987. The biology of Perch and Related Fish. Timber Press, Portland, Oregon, 333 pp.

Diehl, S., 1988. Foraging efficiency of three freshwater fishes: effects of structural complexity and light. Oikos 53: 207–214.

Elliott, J. M. & L. Persson, 1978. The estimation of daily rates of food consumption for fish. J. anim. Ecol. 47: 977–991.

Fernando, C. H. & J. Holčík, 1991. Fish in reservoirs. Int. Revue ges. Hydrobiol. 76: 149–167.

Guma'a, S. A., 1978. The food and feeding habits of young perch, *Perca fluviatilis*. Freshwat. Biol. 8: 221–227.

Hammer, C., 1985. Feeding behaviour of roach (*Rutilus rutilus*) larvae and the fry of perch (*Perca fluviatilis*) in Lake Lankau. Arch. Hydrobiol. 103: 61–74.

Hartman, J., 1983. Two feeding strategies of young fishes. Arch. Hydrobiol. 96: 496–509.

Hartman, J., 1986. Interspecific predictors of selected prey of young fishes. Arch. Hydrobiol. Beih. 22: 373–386.

Hyslop, E. J., 1980. Stomach contents analysis – a review of methods and their application. J. Fish. Biol. 17: 411–429.

Kubečka, J., 1989. Development of the ichthyofauna of the Římov reservoir and its management. Arch. Hydrobiol. Beih. Ergebn. Limnol. 33: 611–613.

Kubečka, J. & M. Böhm, 1991. The fish fauna of the Jordan reservoir, one of the oldest man-made lakes in central Europe. J. Fish. Biol. 38: 935–950.

Lazzaro, X., 1987. A review of planktivorous fishes: Their evolution, feeding behaviours, selectivities, and impacts. Hydrobiologia 146: 97–167.

Mark, W., R. Hofer & W. Wieser, 1987. Diet spectra and resource partitioning in the larvae and juveniles of three species and six cohorts of cyprinids from a subalpine lake. Oecologia 71: 237–247.

Mark, W., W. Wieser & C. Hohenauer, 1989. Interactions between development processes, growth, and food selection in the larvae and juveniles of *Rutilus rutilus* (L.) (*Cyprinidae*). Oecologia 78: 330–337.

Persson, L., 1979. The effects of temperature and different food organisms on the rate of gastric evacuation in perch (*Perca fluviatilis*). Freshwat. Biol. 9: 142–147.

Persson, L., 1982. Rate of food evacuation in roach (*Rutilus rutilus*) in relation to temperature and the application of evacuation rate estimated for studies on the rate of food consumption. Freshwat. Biol. 12: 203–210.

Persson, L., 1987. Effect of habitat and season on competitive interactions between roach (*Rutilus rutilus*) and perch (*Perca fluviatilis*). Oecologia 73: 170–177.

Persson, L. & L. A. Greenberg, 1990. Juvenile competitive bottlenecks: the perch (*Perca fluviatilis*) – roach (*Rutilus rutilus*) interaction. Ecology 71: 44–56.

Ponton, D. & S. Stroffek, 1987. Régime alimentaire des alevins de Gardons (*Rutilus rutilus* (L.)) dans un port du lac Léman. Comparison avec nourriture disponsible. Schweiz. Z. Hydrol. 49: 329–342.

Post, J. R. & D. J. McQueen, 1988. Ontogenetic Changes in the Distribution of Larval and Juvenile Yellow-Perch (*Perca flavescens*). A Response to Prey or Predators. Can. J. Fish. aquat. Sci. 45: 1820–1826.

Rheinberger, V., R. Hofer & W. Wieser, 1986. Growth and habitat separation in eight cohorts of three species of cyprinidis in a subalpine lake. Envir. Biol. Fish. 18: 209–217.

Schiemer, F., H. Keckeis & J. Wanzenböck, 1989. Foraging in cyprinids during early development. Pol. Arch. Hydrobiol. 36: 467–474.

Spanovskaya, V. D. & V. A. Grygorash, 1977. Development and food of age-0 Eurasian perch (*Perca fluviatilis*) in reservoirs near Moscow, USSR. J. Fish. Res. Bd Can. 34: 1551–1558.

Stach, K. & J. Kubečka, 1990. Římov water supply reservoir – description of locality. In: Kubečka, J.(Editor). Ichthyofauna of the Malše River and Římov Reservoir. South-Bohemian Museum, České Budějovice, pp. 27–33.

Thorpe, J., 1977. Synopsis of biological data on the perch *Perca fluviatilis* Linnaeus, 1758 and *Perca flavescens* (Mitchill, 1814). FAO Fish. Synop. No. 113, 138 pp.

Townsend, C. R., I. J. Winfield, G. Peirson & M. Cryer, 1986. The response of young roach *Rutilus rutilus* to seasonal changes in abundance of microcrustacean prey: a field demonstration of switching. Oikos 46: 372–378.

Weatherley, N. S., 1987. The diet and growth of 0-group dace, *Leuciscus leuciscus (L.)*, and roach, *Rutilus rutilus (L.)*, in a lowland river. J. Fish. Biol. 30: 237–247.

Wetzel, R. G., 1990. Land water interfaces: Metabolic and limnological regulators. Verh. int. Ver. Limnol. 24: 6–24.

Zaret, T. M. & A. S. Rand, 1971. Competition in tropical stream fishes: support for the competitive exclusion principle. Ecology 52: 336–342.

Hydrobiologia **303**: 39–48, 1995.
F. Schiemer, M. Zalewski & J. E. Thorpe (eds), The Importance of Aquatic-Terrestrial Ecotones for Freshwater Fish.

Biomass of planktonic crustaceans and the food of young cyprinids in the littoral zone of Lake Balaton

Ashot Simonian[1], István Tátrai[2], Péter Bíró[2], Gábor Paulovits[2], László G.-Tóth[2] & Gyula Lakatos[3]
[1]*Institute of Hydroecology and Ichthyology, Armenian Academy of Sciences, 378610 Sevan, Armenia;*
[2]*Balaton Limnological Research Institute, Hungarian Academy of Sciences, H-8237 Tihany, Hungary;*
[3]*Institute of Ecology, Lajos Kossuth University, H-4010, Debrecen, Hungary*

Key words: Lake Balaton, littoral zone, crustaceans, feeding, cyprinids

Abstract

The littoral zone of Lake Balaton and its periphyton-zooplankton-fish communities have been investigated intensively in recent years. Total average number of crustacean plankton varied from 36 to 126 ind l^{-1}, their biomass from 0.49 to 1.86 mg ww l^{-1} month^{-1} at different areas of the littoral zone. In general, these values for the above parameters were higher in hypertrophic areas. 23 fish species occurred in the littoral zone with cyprinids dominating. The seasonal food spectra of Y–O–Y roach (*Rutilus rutilus*), white bream (*Blicca bjoerkna*) and bream (*Abramis brama*) were based mainly on planktonic crustaceans and benthic/periphytic invertebrates. According to the frequency of occurrence of crustaceans and other invertebrates, the food composition of young cyprinids differed significantly in the NE and SW-basins of the lake.

Introduction

Recently much attention has been paid to the role of the littoral zone of the highly eutrophic shallow Lake Balaton (593 km^2), having a longitudinal trophic gradient (Herodek, 1984). Earlier studies have proved the significance of the narrow littoral zone in the functioning of this lake ecosystem. Multivariate relationships between chlorophyll-*a* content of the water, the shoreline-length to water area ratio (S/A) and the annual total fish yield showed significant dependence of fish yield on morphometry of different lake areas. Accordingly, fish carrying capacity of the open water areas, calculated from chlorophyll-*a* content and S/A, ranged from 12 to 34%, but that of the littoral zone between 66 and 88% of the total fish production (Bíró & Vörös, 1988, 1990). In spite of the fact that only 0.5% of the littoral zone (1/6 of the total lake area) is covered by reed communities (Lakatos *et al.*, in press), the rich periphyton provides the bulk of fish food supply (Bíró & Vörös, 1990; Lakatos & Bíró, 1991). High primary production, however, is coupled with low efficiency of energy transfer to fish: 0.04–0.1%, and the produced energy dissipates within the ecosystem (Bíró, 1984; Bíró & Vörös, 1990).

There are sporadic data (Bíró, 1991; Bíró *et al.*, 1991) on the food webs of young cyprinids and their regulating effects on the zooplankton assemblages. The aim of this study was to assess the biomass of crustacean members in the littoral habitats. Preliminary data are presented for the density and species composition of fish assemblages, as well as their food composition.

Materials and methods

Zooplankton was sampled in four areas of the littoral regions monthly from May till October 1991 (Fig. 1). Three samples were taken from each area along the northern shoreline of Lake Balaton. Samples were collected from 1.5 m water depth at a distance of 3–5 m from the reed-belt using a mud-surface tube sampler of 56 mm in diameter (suitable for whole water column sampling), and filtered through a net of 60 μm mesh size. Samples were preserved in 75% ethanol.

The dominant zooplankton members were identified to species and enumerated in two or three subsamples, the volumes of which had been adjusted so that at least 50 individuals of each taxon were counted. The average length of each taxon was determined by measuring several hundred individuals per sample under a dissecting microscope. Biomass values for the dominant taxa were determined from the regression of wet weight on the length of each species (Salazchin, 1984).

Fish were collected by electric shocking in parallel with zooplankton sampling. Individuals were identified taxonomically and their lengths and weights measured. Altogether 127 Y–O–Y fish gut contents (55 roach, *Rutilus rutilus*; 42 white

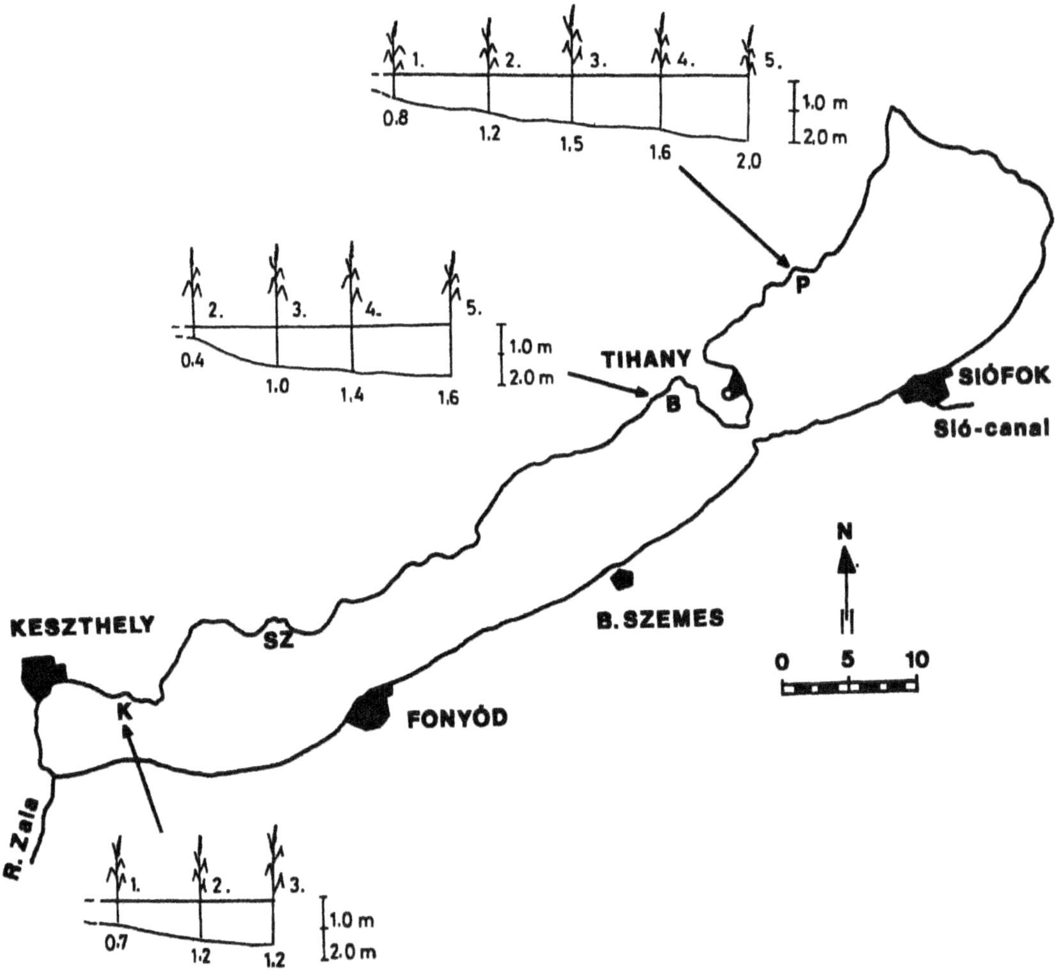

Fig. 1. Sampling areas in the littoral zone of Lake Balaton. P = Palóznak Bay; B = Bozsai Bay; Sz = Szigliget Bay; K = Keszthely Bay. Ecotone structure of the stations showing the transects of *Phragmites* stands are taken from Lakatos & Bíró (1991). For more details see text.

bream, *Blicca bjoerkna*; and 40 bream, *Abramis brama*) were analyzed. Individuals were dissected, their intestinal contents removed and fixed with 75% ethanol. Food composition was studied under a light microscope and their spectra were drawn according to the frequency of occurrence (%) of the main food items. Selectivity indices were obtained according to Dodson (1975).

Results

Conditions of fish habitats

The littoral area of Lake Balaton, according to the bed profile, is a relatively narrow zone. However, due to the long shoreline, this area (*ca* 100 km^2) is significant and about 25% has been covered by *Phragmites australis* (CAV.) Trin. ex Steudel stands. This relatively homogeneous vegetation (reaching 1–1.5 m depth) forms a more or less continuous belt along the northern and eastern bays (Fig. 1), where several streams flow into the lake. Different conditions of these reed stands have been shown by botanical and periphyton analyses.

Both the offshore and inshore *Phragmites* stands in the eutrophic Keszthely Bay, being heavily loaded by nutrients, are most exposed to human impacts. There is a significant inorganic sedimentation as well as reed decline in parallel with the intensive spreading of bulrush at the offshore area. In the eutrophic Szigliget Bay (NNE of Keszthely Bay) just the same conditions exist, however the sedimentation rate is higher and the marsh-formation greater. The reed stand at the offshore side has started to get thinner, with thick, high stems that bear rich periphytic communities below the water surface.

For the reed stands in the mesotrophic Bozsai Bay close to the western part of Tihany peninsula, more intact conditions are characteristic. At their inshore part, due to intensive sedimentation, a marshy vegetation (*Stratiotes*, *Sium*, *Carex*, *Gallium*, *Scutellaria* etc) appeared among the reed stands at the rear. In the middle of the reed belt, there is a narrow zone of *Fontinalis antipyretica*.

The oldest reed stands are located in Paloznak Bay. The water quality in the NE-basin has the lowest (mesotrophic) nutrient loadings. In the middle of the dense reed stands consisting of thinner stems, the *Fontinalis* zone is wide. In shallower inshore water, a dense macrophytic vegetation (*Utricularia*, *Ceratophyllum*, *Hydrocharis*, *Potamogeton*, *Najas*) can be found among the scarce reeds. No significant alterations took place during the last decade, and these stands can be considered as the most natural ones in the littoral zone.

Dynamics of planktonic crustaceans

The average monthly values of crustacean plankton density, biomass and production rates for the period May to October varied between different basins. However, these changes cannot be related always to differences in trophic conditions of each study area (Fig. 2). For example, while the biomass and the production rates in the littoral of the mesotrophic parts of Lake Balaton (e.g. Palóznak and Bozsai Bay) did not differ significantly between basins, the numbers of crustaceans were quite different. In the most eutrophic habitats, the crustacean plankton was represented almost equally by cladocerans and copepods by number. In the mesotrophic areas copepods dominated by number, but the biomass of cladocerans was higher in both meso- and eutrophic areas of the lake (Fig. 2).

Cladocerans in the littoral zone of Lake Balaton were dominated numerically by *Diaphanosoma brachyurum* and *Bosmina longirostris*, attaining maximal densities in July-August and May-July, respectively. *Bosmina* was dominant in the eutrophic, and *Diaphanosoma* in the mesotrophic parts of the lake. Daphnids were always less abundant, but their biomass proved to be as large as that of *Diaphanosoma*. *Leptodora kindtii* occurred in samples from the mesotrophic regions at a density of 100 ind m^{-3} and was five times more abundant than in the eutrophic areas.

Both the abundance and biomass of copepods were almost equally represented by cyclopoids

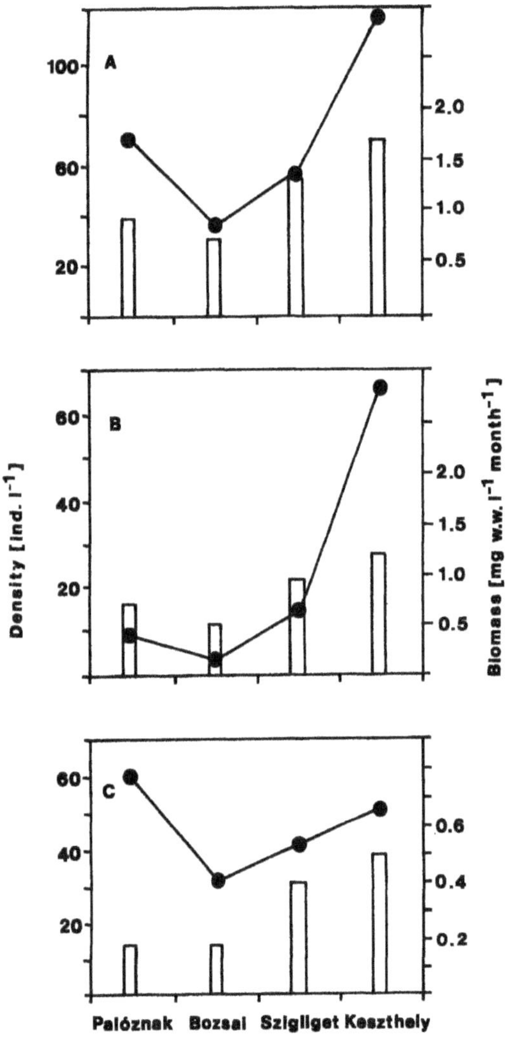

Fig. 2. Density (ind. l^{-1}) and biomass (mg wet weight l^{-1} $month^{-1}$) of crustacean plankton at different areas of the littoral zone of Lake Balaton. A = total Crustacea; B = Cladocera; C = Copepoda. -●- = density; □ = biomass.

and calanoids in the mesotrophic areas. However, cyclopoid copepods have dominated in the eutrophic parts of the lake. The average biomass of the copepods was about twice as great in the eutrophic littoral parts of Lake Balaton as in its mesotrophic regions. Cyclopoid copepod densities did not show any seasonal fluctuations during the late spring-summer period. However, a rapid numerical increase of *Cyclops* spp. was observed during autumn in the two extremely different sampling sites. On the other hand, the biomass of *Cyclops* spp. exhibited strong fluctuations especially in the eutrophic areas peaking in the spring and the fall. The density of *Eudiaptomus gracilis* was temporally more variable. The fluctuations in the biomass of calanoids were not as great as those of abundance and did not change in parallel with the trophic conditions of Lake Balaton.

Total average number of crustaceans varied from 36 to 126 ind l^{-1}, and their biomass from 0.49 to 1.86 mg ww l^{-1} $month^{-1}$ (3.08–11.7 kJ m^{-2}), in different littoral areas. The proportions of non-predatory forms were much higher than those of predators.

Fish species assemblages in the littoral zone

Twenty three fish species occurred in the littoral zone of Lake Balaton during May-October 1991. Altogether 5256 specimens were collected along the reed-belt at different areas of the lake. Cyprinids dominated by number (92.5%) all the time with the majority being bleak (*Alburnus alburnus*: 38.7%), bitterling (*Rhodeus sericeus amarus*: 15.6%), roach (15.1%), rudd (*Scardinius erythrophthalmus*: 12.6%), and bream (5.5%). The proportion of percids was 5.8% and other taxa comprised 1.8%.

The species composition varied slightly along the trophic gradient and differed from that of the open water (commercially exploited stocks). Total fish biomass in the mesotrophic areas (Palóznak and Bozsai bays) varied from 71.6 to 90.4 kg ha^{-1} (Table 1), while in the eutrophic areas (Szigliget and Keszthely bays) it ranged from 72.4 to 80.5 kg ha^{-1} (Table 2).

Food composition of roach, white bream and bream

Young-of-the-year (Y–O–Y) roach (TL = 8.5–10.2 cm, W = 3.8–9.1 g in the NE-basin, TL = 7.3–11.7 cm, W = 4.8–18.4 g in the SW-basin), white bream (TL = 9.6–12.1 cm, W = 9.2–13.7 g

Table 1. Fish species assemblages in the littoral zone of Lake Balaton.

Species	Biomas in kg ha^{-1}			
	Palóznak Bay	Bozsai Bay	Szigliget Bay	Keszthely Bay
Esocidae				
1. *Esox Lucius*	–	3.0	0.5	–
Cyprinidae				
2. *Rutilus rutilus*	1.4	9.0	23.3	15.8
3. *Scardinius erythropthalmus*	8.7	31.3	22.2	9.9
4. *Aspius aspius*	0.1	0.2	0.5	0.2
5. *Tinca tinca*	0.01	3.6	0.4	–
6. *Alburnus alburnus*	4.8	5.9	12.4	10.5
7. *Blicca bjoerkna*	19.5	3.0	0.5	5.9
8. *Abramis brama*	15.9	5.3	1.7	5.7
9. *A. ballerus*	–	–	–	0.01
10. *Pelecus cultratus*	–	–	0.2	–
11. *Rhodeus sericeus amarus*	0.04	2.0	3.7	2.0
12. *Carassius carassius*	0.2	–	–	–
13. *C. auratus gibelio*	8.3	3.9	0.1	0.3
14. *Pseudorasbora parva*	0.01	–	0.1	0.7
Cobitidae				
15. *Noemacheilus barbatulus*	–	–	–	0.1
Siluridae				
16. *Silurus glanis*	–	–	0.6	–
Anguillidae				
17. *Anguilla anguilla*	2.3	11.5	0.3	0.1
Centrarhidae				
18. *Lepomis gibbosus*	9.0	11.0	0.6	0.5
Percidae				
19. *Stizostedion lucioperca*	0.3	–	–	0.1
20. *S. volgensis*	0.1	–	–	–
21. *Perca fluviatilis*	–	–	0.1	0.003
22. *Gymnocephalus cernuus*	0.9	0.7	0.1	–
Gobiidae				
23. *Neogobius fluviatilis*	–	–	5.1	0.4
Total	71.6	90.4	72.4	80.5

in the NE-basin, TL = 8.8–11.2 cm, W = 7.7–14.7 g in the SW-basin) and bream (TL = 4.3–5.9 cm, W = 2.0–2.6 g in the NE-basin, TL = 9.4–14.2 cm, W = 9.0–25.9 g in the SW-basin) were the common phytolithophil cyprinids in the littoral of the lake. The average seasonal composition of gut contents (in percent by numbers) of these three species, inhabiting different ecotonal habitats in Lake Balaton, have been summarized in Tables 2–3. Planktonic crustaceans (mainly large sized cladocerans, e.g. *Daphnia, Diaphanosoma*) were the most frequent items in the fish guts (60–70% planktonic crustaceans and 30–40% benthic/periphytic invertebrates) from the mesotrophic Palóznak and Bozsai Bays independent of season (Table 2). In the eutrophic SW-basins (Szigliget and Keszthely Bays) fish preyed mostly on benthic/periphytic invertebrates (50–

Table 2. Seasonal composition of gut contents (% by number) of roach (*Rutilus rutilus*), white bream (*Blicca bjoerkna*) and bream (*Abramis brama*) inhabiting mesotrophic habitats in the NE-basin of Lake Balaton.

Food items	Cladocera	Copepoda	Amphipoda	Chironomidae	Insecta	Others*
Palóznak Bay						
Roach						
Spring	59.5	17.7	22.8	–	–	–
Summer	52.7	–	18.7	–	–	18.6
Autumn	40.1	20.2	25.6	–	6.4	7.7
White bream						
Spring	70.2	15.1	11.4	3.3	–	–
Summer	37.5	9.1	17.3	3.6	22.3	10.2
Autumn	58.1	17.8	12.1	1.8	–	–
Bream						
Spring	51.0	9.3	24.1	15.6	–	–
Summer	47.0	12.0	29.0	12.0	–	–
Autumn	66.1	16.9	9.7	7.3	–	–
Bozsai Bay						
Roach						
Spring	47.3	16.7	21.0	73.0	–	–
Summer	54.1	10.2	12.3	2.1	–	21.3
Autumn	49.6	19.4	16.7	–	–	–
White bream						
Spring	94.4	–	4.5	1.1	–	–
Summer	43.4	5.1	21.0	–	–	22.6
Autumn	24.6	–	44.1	27.3	4.0	–
Bream						
Spring	38.1	28.6	14.9	9.9	8.5	–
Summer	27.5	–	51.0	21.5	–	–
Autumn	17.3	7.8	28.6	42.1	–	4.2

* = Ostracoda, undetermined animals, detritus, algae.

70% amphipods and chironomids) with planktonic crustaceans making up only 30–50% (Table 3). Small-sized cladocerans (mainly *Bosmina*) were consumed in the western bays. Copepods were much less important in the diet of all three fish species in both mesotrophic and eutrophic areas (Table 4).

There was no clear seasonality in relation to the main diet components of roach. However, the plankton consumption of white bream and bream in the eutrophic habitats decreased, and feeding on amphipods and chironomid larvae increased, toward the autumn.

Discussion

In Lake Balaton the main food resource for fish stocks is concentrated in the macrophyte-dense littoral zone (Bíró & Vörös, 1988). This zone represents a rich food supply which benefits the inhabitants of the open water area also (Sebestyén, 1943; Bíró & Gulyás, 1974; Ponyi *et al.*, 1983). The lake's biota depends on the structure and functioning of this highly productive zone. The amount of periphyton is usually greater on the old than on the green reeds. The periphyton biomass expressed as dry weight/substratum ranged from

Table 3. Seasonal composition of gut contents (% by number) of roach (*Rutilus rutilus*), white bream (*Blicca bjoerkna*) and bream (*Abramis brama*) inhabiting eutrophic habitats in the SW-basin of Lake Balaton.

Food Items	Cladocera	Copepoda	Amphipoda	Chironomidae	Insecta	Others*
Sigliget Bay						
Roach						
Spring	27.2	22.5	39.4	–	–	
Summer	39.8	–	44.9	–	–	15.3
Autumn	32.3	9.3	48.2	–	–	10.2
White bream						
Spring	33.7	12.1	29.8	16.3	3.6	5.1
Summer	26.9	–	47.8	10.5	–	14.8
Autumn	6.6	5.9	42.7	44.8	–	–
Bream						
Spring	40.0	20.0	26.7	6.3	7.0	–
Summer	33.7	12.8	36.2	8.8	–	8.5
Autumn	6.6	5.9	42.7	44.8	–	–
Keszthely Bay						
Roach						
Spring	23.1	–	29.6	47.3	–	–
Summer	32.0	7.2	53.1	–	–	7.7
Autumn	36.5	10.7	50.3	–	–	2.5
White bream						
Spring	29.4	12.3	37.9	16.6	1.0	2.8
Summer	31.7	–	48.4	–	–	19.9
Autumn	17.5	–	50.7	29.1	2.7	–
Bream						
Spring	46.3	22.6	20.5	4.9	5.7	–
Summer	39.6	13.5	22.3	18.3	–	6.3
Autumn	10.2	8.3	37.1	41.5	–	2.9

* = Ostracoda, undetermined animals, detritus, algae.

4.1 to 36.4 g m^{-2} in 1991 (Lakatos *et al.*, in press). Earlier studies confirm that phytoplankton production in the open water has been utilized by fish with a very low degree of efficiency. An increase in the phytoplankton biomass in the eutrophic areas of Lake Balaton was not followed by a significant increase in zooplankton biomass as occurred in other lakes (Herzig, 1979).

Crustacean communities in the mesotrophic littoral zones of Lake Balaton were dominated numerically by copepods, while cladocerans were dominant in the most eutrophic area. Among cladocerans, small-sized *Bosmina* dominated in the eutrophic areas. Daphnids were virtually always less abundant being relatively more common in the eutrophic habitats. Density and biomass of the planktonic crustaceans in the littoral zone were half and two-thirds, respectively, of value in the open water (G.-Tóth, 1992). The high production of the littoral crustacean community came from the prolific cladoceran populations.

Predation by fish led to different outcomes depending on the fish faunal composition and the trophic conditions of the land/water ecotonal habitats. In the SW-basins fish preyed most heavily on amphipods. Also they preyed intensively on the most abundant smaller (*Bosmina*) and larger (*Daphnia*) planktonic crustaceans until

Table 4. Seasonal selectivity indices of some cyprinid fish for planktonic crustaceans in the littoral zone of Lake Balaton (Dodson, 1975; 0.00-0.5: no slection).

Food items	*Daphnia spp*	*Diaphanosoma brachyurum*	*Bosmina longirostris*	*Eudiaptomus gracilis*	*Cyclops spp.*
NE-basins					
Roach					
Spring	0.04	0.51	0.02	0.08	0.09
Summer	0.10	0.62	0.00	0.00	0.00
Autumn	0.05	0.52	0.03	0.07	0.06
White bream					
Spring	0.07	0.75	0.00	0.00	0.00
Summer	0.13	0.33	0.02	0.01	0.09
Autumn	0.21	0.56	0.18	0.09	0.14
Bream					
Spring	0.05	0.35	0.00	0.16	0.11
Summer	0.12	0.46	0.00	0.22	0.09
Autumn	0.08	0.69	0.06	0.09	0.06
SW-basins					
Roach					
Spring	0.08	0.06	0.07	0.00	0.00
Summer	0.18	0.11	0.12	0.03	0.02
Autumn	0.21	0.08	0.44	0.05	0.03
White bream					
Spring	0.39	0.17	0.26	0.12	0.06
Summer	0.51	0.11	0.43	0.00	0.00
Autumn	0.19	0.23	0.02	0.00	0.00
Bream					
Spring	0.28	0.21	0.36	0.12	0.08
Summer	0.53	0.10	0.47	0.39	0.13
Autumn	0.09	0.26	0.02	0.19	0.23

the summer, as long as they were abundant. After that, when the density of those cladocerans dropped, fish switched from larger and energetically more valuable food (amphipods and chironomids) to ingesting copepods in increased numbers.

In the mesotrophic littoral zones fish preyed mostly on planktonic crustaceans, chiefly on the dominant, large-sized *Diaphanosoma*, while that was abundant until the early autumn. But following a decrease in abundance of the large-sized cladocerans, fish extended their consumption to amphipods and chironomids as well.

According to Dévai (1984) biomass and production of the benthic chironomid larvae in the eutrophic lake areas exceeded the mesotrophic values by about 4–7 times. So fish might feed more intensively on benthic as well as on periphytic organisms in the eutrophic areas, because there the biomass of the benthic invertebrates is relatively higher than that of the crustaceans.

Crustacean plankton in the eutrophic area was almost equally represented by cladocerans and copepods by number, and cladocerans dominated in biomass. However, in the mesotrophic areas the majority of crustaceans was made up of copepods and cladocerans were represented only by a few individuals. According to Ponyi & Zánkai (1982) and Zánkai & Ponyi (1986), the open water zooplankton was generally dominated by copepods (e.g. *Eudiaptomus gracilis*, and cyclopids). Some defined trends of this structural difference might be related to the differences in the composition and quantity of phytoplankton (Vörös, 1982), and to the temporal limiting effect of the high abioseston on the feeding and development of daphnids in the open area (G.-Tóth, 1992).

The effects of fish predation on restructuring the crustacean communities might be similarly considerable, however these processes are complex. The strength of the positive and negative effects of fish predation on planktonic crustaceans will depend on the selectivity and intensity of predation, the proportion of total mortality resulting from fish predation and the ability of crustaceans to respond to alterations by reproducing (Vanni,1986; Tátrai & de Bernardi, 1992). For example zooplankton numbers under existing predation tended to be relatively high in the presence of underyearling bream of moderate biomass in the enclosure studies in the mesotrophic part of Lake Balaton (Tátrai *et al.*, 1991). From data on the composition of fish populations and their biomass in the littoral zone of Lake Balaton, the pattern of crustacean composition in different habitats of the lake might be attributed to the direct action of planktivorous fish of dense populations (and more probably that of the invertebrate predators such as *Leptodora*, whose density might be sufficient to account for some crustacean mortality) removing larger species especially in the west basins (e.g. absence or very low number of large-sized daphnids and *Diaphanosoma*). However, these changes may be attributed to a direct effect through the modification of the phytoplankton community (DeMott & Kerfoot, 1982). This effect will depend on the demographic response of the zooplankton to changes in food availability (G. Tóth, 1992). The changes of the littoral zooplankton in Lake Balaton have also been regulated by various age-groups of cyprinids (fry, Y–O–Y and older fish). In the mesotrophic areas young roach, white bream and bream fed mainly on planktonic and periphytic crustaceans, directly regulating their dynamics, biomass and production. Their food composition was fairly characteristic, but differed in the NE and SW-basins of the lake. Frankiewicz *et al.* (1991) described the food composition of various cyprinids from northern streams of Lake Balaton. In these streams an upstream-downstream diversification of fish community structure was observed. The increase in fish density in the downstream section might be explained as resulting from the ecotone effect (Przybylski *et al.*, 1991). The littoral area is the buffer zone of land/water interface. Because a number of streams flow into the area of *Phragmites* stands, their functioning (as natural filters, shelters, etc,) is influenced by the nutrient input from the drainage (food-webs, energy-flux). The divers planktonic, benthic, periphytic and fish communities here, as well as their responses are influenced by the ecotone effect more sensitively than in any other parts of the shallow Lake Balaton. However, all of these questions need further clarification.

Acknowledgements

This work was supported financially by the National Research Fund (OTKA), Project # 467 & # 1881). P. Bíró thanks UNESCO-ROSTE (1262 A DORSODURO, Venice, Italy) for their financial support for his participation at the UNESCO/MAB Fish Ecotone Symposium (31 August-5 September, Lunz-Wien, Austria)

References

Bíró, K. & P. Gulyás, 1974. Zoological investigations in the open water *Potamogeton perfoliatus* stands of Lake Balaton. Annal. Biol. Tihany 41: 181–203.

Bíró, P., 1984. Lake Balaton: a shallow Pannonian water in the Carpathian Basin. In F. B. Taub (ed.), Lakes and Reservoirs. Ecosystems of the World 23. Elsevier, Amsterdam: 231–245.

Bíró, P. 1991. Food resource partitioning between bream (*Abramis brama*) and razor fish (*Pelecus cultratus*) in Lake Balaton (Hungary). Verh. int. Ver. Limnol. 24: 2513–2516.

Bíró, P., Suhad E. Sadek & G. Paulovits, 1991. The food of bream (*Abramis brama* L.) in two basins of Lake Balaton of different trophic status. Hydrobiologia 209: 51–58.

Bíró, P. & L. Vörös, 1988. Relationship between the yield of bream, *Abramis brama* L., chlorophyll-*a* concentration and shore-length: water area ratio in Lake Balaton, Hungary. Aquacult. Fish. Mgmt 19: 53–61.

Bíró, P. & L. Vörös, 1990. Trophic relationships between primary producers and fish yields in Lake Balaton. Hydrobiologia 191: 213–221.

De Mott, W. R. & W. C. Kerfoot, 1982. Competition among cladocerans, nature of the interaction between Bosmina and Daphnia. Ecology 63: 1949–1966.

Dévai, Gy., 1984. Studies of the ecological effects of Lake Balaton and River Zala sediments on chironomids (Diptera: Chironomidae). Acta Biol. Debrecina Oecol. Hung. 1: 1–183.

Dodson, S. I., 1975. Predation rates of zooplankton in arctic ponds. Limnol. Oceanogr. 20: 426–433.

Frankiewicz, P., M. Zalewski, P. Bíró, I. Tátrai & M. Przybylski, 1991. The food of fish from streams of the northern part of the catchment area of Lake Balaton (Hungary). Acta Hydrobiol. (Kraków) 33: 149–160.

G.-Tóth, L., 1992. Limiting effect of abioseston of food ingestion, postembryonic development time and fecundity of daphnids in Lake Balaton (Hungary). J. Plankton Res. 14: 435–446.

Herodek, S., 1984. The eutrophication of Lake Balaton: measurements, modelling and management. Verh. int. Ver. Limnol. 22: 1087–1091.

Herzig, A., 1979. The zooplankton of the open lake. In H. Löffler (ed.), Neusiedlersee: the Limnology of a Shallow Lake in Central Europe. Dr W. Junk Publishers, The Hague: 281–335.

Lakatos, G. & P. Bíró, 1991. Study on chemical composition of reed-periphyton in Lake Balaton. BFB-Bericht (Illmitz) 77: 157–164.

Lakatos, G., P. Bíró, I. Mészáros, J. Padisák & L. G.-Tóth, (in press). Temporal changes in the structure of reed periphyton in the lakeshore of Lake Balaton. Lakes and Reservoirs: Research and Management. ILEC, Kusatsu, Shiga, Japan.

Ponyi, J. E., I. Tátrai & A. Frankó, 1983. Quantitative studies on Chironomidae and Oligochaeta in the benthos of Lake Balaton. Arch. Hydrobiol. 97: 196–207.

Ponyi, J. E. & N. P. Zánkai, 1982. Population dynamics, biomass and biomass production of *Eudiaptomus gracilis* (G.O. Sars) in two water areas of differing trophic state of Lake Balaton (Hungary). Acta hydrochim. hydrobiol. 10: 597–610.

Przybylski, M., P. Bíró, M. Zalewski, I. Tátrai & P. Frankiewicz, 1991. The structure of fish communities in streams of the northern part of the catchment area of Lake Balaton (Hungary). Acta Hydrobiol. (Kraków) 33: 135–148.

Salazchin, A. A., 1984. Metodicheskiye rekomendatsii s zooplankton i ego produktsiya. Izd. ZIN Akad. Nauk SSSR, Leningrad, 33 pp.

Sebestyén, O., 1943. Significance of the littoral in the life of lakes. A Magyar Biológiai Kutatóintézet Munkái 15: 301–308.

Tátrai, I., J. Padisák & A. Simonian, 1991. Impact of juvenile cyprinid fish predation on the food availability and zooplankton community. Verh. int. Ver. Limnol. 24: 2519–2522.

Tátrai, I. & R. de Bernardi, 1992. Indirect impact of cyprinid fish fry on the growth and fecundity of *Daphnia obtusa*. Arch. Hydrobiol. 125: 371–381.

Vanni, J. M., 1986. Fish predation and zooplankton demography: indirect effects. Limnol. Oceanogr. 67: 337–354.

Vörös, L., 1982. Quantitative and structural changes of phytoplankton in Lake Balaton between 1965–1978. Aquacult. Hung. 3: 137–144.

Zánkai, N. P. & J. E. Ponyi, 1986. Composition, density and feeding of crustacean zooplankton community in a shallow, temperate lake (Lake Balaton, Hungary). Hydrobiologia 135: 131–147.

Hydrobiologia **303**: 49–60, 1995.
F. Schiemer, M. Zalewski & J. E. Thorpe (eds), The Importance of Aquatic-Terrestrial Ecotones for Freshwater Fish.

Biomanipulation by ecotone management in a lowland reservoir

M. Zalewski, P. Frankiewicz & M. Nowak
Department of Applied Ecology, University of Łódź, Poland

Abstract

The comparative studies on fry of dominant fish species in the lowland Sulejow Reservoir have demonstrated the importance of an unstable ecotonal shoreline zone in whole system dynamics, which affects water quality. Ecotone zone utilization by fish determines their reproductive success and consequently pressure of fry on a zooplankton community. Density of perch, which dominates in the fry community of the Reservoir, influences the utilization of ecotonal resources by dominant cyprinid fry-roach. A reverse relationship between perch and roach fry densities has been demonstrated, which in turn, affects utilization of zooplankton by roach fry. At stations of highest perch density roach reduces feeding on zooplankton and eats more algae and insects. The knowledge of competitive relations between perch and cyprinid fry has been an important aspect of application of the biomanipulation concept in the reservoir ecosystem, to improve fish yield and water quality. Hypothetical models of energy flow through the reservoir's ecosystem with poor and well developed flooded riparian ecotonal zones are established.

Introduction

In many reservoirs one of the main limiting factors for the fish community is scarcity of spawning areas with vegetation in the littoral zone (e.g. Nelson & Walburg, 1977), which is usually a result of high fluctuations of water level. These fluctuations during the spawning season play also an important role in regulation of fish eggs and larval survival (Ploskey, 1985). When water level is stable during spring the reproductive success of many fish species is highly related to the quantity and quality of flooded spawning places along the reservoir shoreline. Proper management of this unstable ecotonal zone may be a main factor in controlling the strength of the juvenile fish population, thus their pressure on the zooplankton, and in consequence water quality and fish yield. The potential influence of fry density regulation by active shaping of the littoral zone on the top-down effect in the Reservoir, may be predicted on the basis of competitive relations between dominated fry species – perch and roach, in years of their low and high density.

Material and methods

The lowland Sulejow Reservoir, situated in Central Poland on the Pilica River, has an average area of 20 km^2 and a mean depth of 3.3 m (max. 11 m). Sampling stations were distributed in the littoral zone around the reservoir and represented the main types of habitats from the point of view of heterogeneity, character of the land/water ecotone zone, and the state of eutrophication (Fig. 1, Tab. 1).

At each site fish were collected with a 15 m long seine net with 1 mm mesh size. Samples have been taken each year since 1983 in mid July between 9 a.m. and 3 p.m. In the case of diurnal

Plate 1. Structure of ecotone zone at station 5. A. sand beach uncovered during low water levels; B. terrestrial vegetation flooded in springs during high water levels.

feeding pattern investigations fish were caught every 3 hours.

After preservation in 10% formaldehyde fish were transported to the laboratory where their length, weight, and gut content were analyzed. To compare perch (*Perca fluviatilis*) growth rate in different years the instantaneous coefficient of growth (Chapman, 1978) was calculated. The index of stomach fullness was expressed as a ratio between weight of food (w_f) and weight of fish body (w_b). The volumetric method (Hyslop, 1980) was used to analyse fish food. Additionally, if possible the number of prey in a given food category was counted. If the sample size was larger than 30 specimens of a given species, subsamples of 30 fish were taken randomly for food analyses. More detailed description of materials and methods (especially critical evaluation of sampling methods) were done elsewhere (Zalewski *et al.*, 1990a, b; Frankiewicz *et al.*, 1986).

Results and discussion

Trophic relations between perch and roach

In periods of low water levels the littoral zone in the Sulejow Reservoir is almost completely deprived of vegetation (Phot. 1). The importance of flooded areas with vegetation for fish reproductive success is presented by strong correlation between water level and fry density (Fig. 2). Excessive numbers of either perch and roach (*Rutilus rutilus*) – two species dominating the fry community of the Sulejow Reservoir, lead to rapid deterioration of water quality. As was shown in

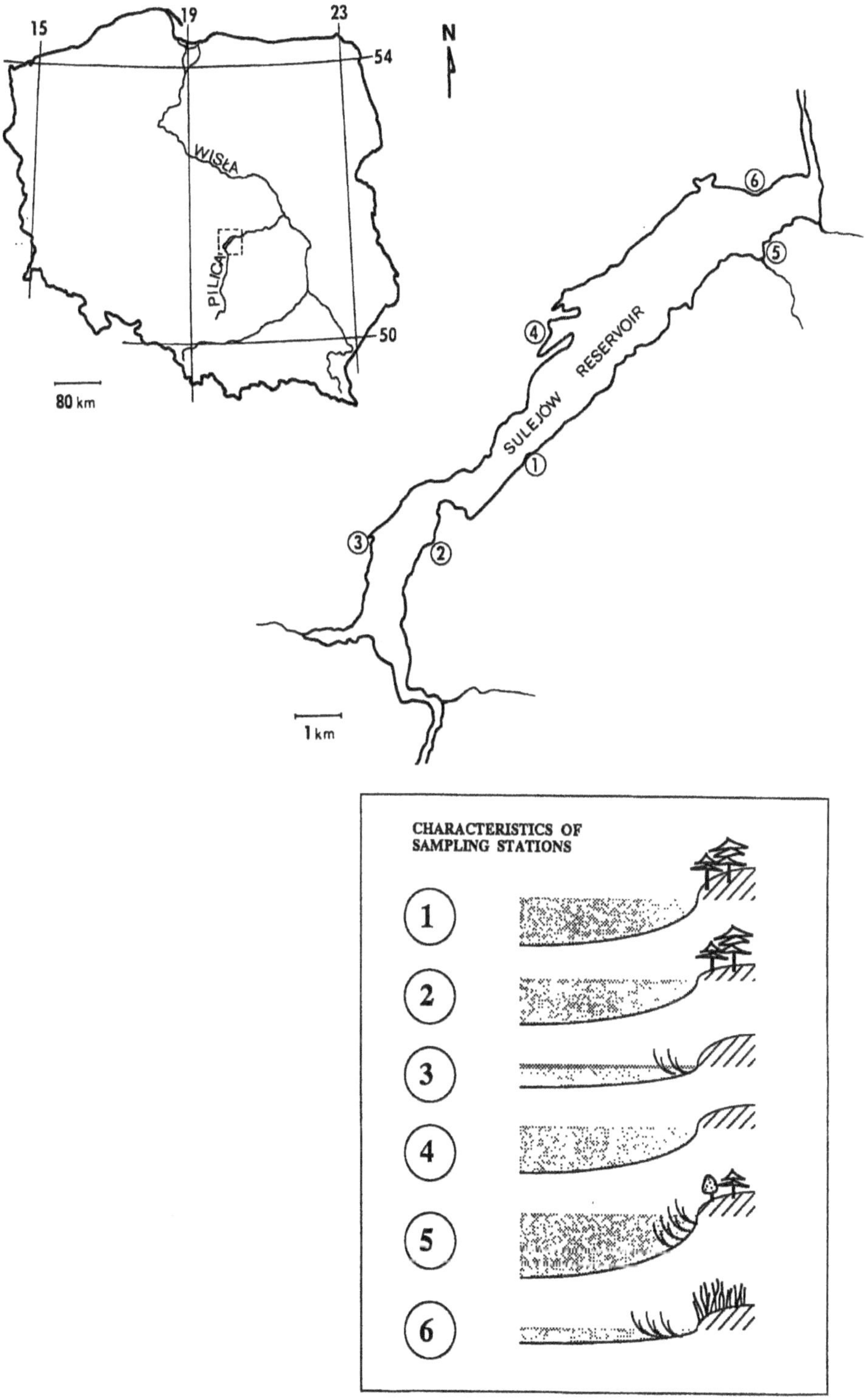

Fig. 1. The distribution and characteristics of sampling stations in the Sulejow Reservoir.

Table 1. Characteristics of sampling stations.

Stations	Characteristics				
	Depth (m)*	Sampling area (m^2)	Bottom substrate	Vegetation**	Surroundings
1	1.5	25	Gravel, sand (80%) (20%)	–	Coniferous forest
2	1.4	45	Sand, coarse detritus (20%) (80%)	–	Coniferous forest
3	0.6	150	Mud, fine detritus (70%) (30%)	++	Pastures, fields
4	1.4	200	Sand (100%)	–	Sand beach
5	2.1	100	Mud, sand (80%) (20%)	+++	Brushwood
6	0.5	60	Sand, pebble (70%) (30%)	+	Meadow, willows

* Depth was measured 10 m from the bank.
** In % of bottom cover: – 0%; + 0–10%; ++ 10–20%; +++ 20–30%.

our previous papers (Zalewski *et al.* 1990a, b), and by other authors (Mills & Forney, 1983; Post & McQueen, 1987) perch directly reduces large filtrating zooplankton (mainly daphnids), releasing pelagic algae from their control (Fig. 3). In consequence algae blooms are much more likely to appear and simultaneously perch fry growth rate will decreases due to the scarcity of large zooplankton filtrators which are their most profitable food (Figs 4, 5). Roach utilizes as food not only zooplankton but mainly periphytonic algae (especially at high temperatures) (Fig. 6). The low assimilation rate of algae (Prejs, 1984) causes a certain additional amount of nutrients from bottom algal assemblages to be released into the water after passing through intestinal tract of the roach. This may stimulate an increase in density of pelagic algae. Co-occurrence of these two species at high densities may thus amplify rapid development of pelagic algae.

Lack of zooplankton in roach food at stations where perch concentrated its foraging on this most profitable prey category and increases in the percentage of zooplankton eaten by roach at stations with very low perch density, together with the lowest growth rate of roach at the station with the highest contribution of algae to the gut content (Fig. 7), indicate competitive superiority of perch over roach in the reservoirs with little diversified littoral habitats and generally high YOY perch densities. This is contradictory to Persson's (1987a) findings, but differences might be due to much higher densities of perch in Sulejow Reservoir. Decreases in perch growth rate while in sympatry with roach were found in open water enclosures (Persson, 1987a) but the competitive effect may be modified with increases in habitat heterogeneity (Persson, 1987b). Moreover, enclosure experiments do not include aspects of diel spatial migrations of zooplankton (e.g. Lampert & Taylor, 1985), which may greatly influence food availability and may be followed by fish feeding migrations. In natural situations competitive influences of roach on perch may be negligible due to temporal separation in feeding activity which has been found during 24 hour investigations (Frankiewicz *et al.*, in press). Also the roach has a much higher opportunistic foraging behaviour than does the strictly predatory perch and can utilize detrital and plant material (e.g. Niederholzer & Hofer, 1980; Prejs, 1984). A long term study in Sulejow Reservoir has indicated that perch and roach try to separate spatially, especially in years when their densities are highest (Fig. 8). It seems that roach, due to their higher efficiency in zooplankton utilization (Persson,

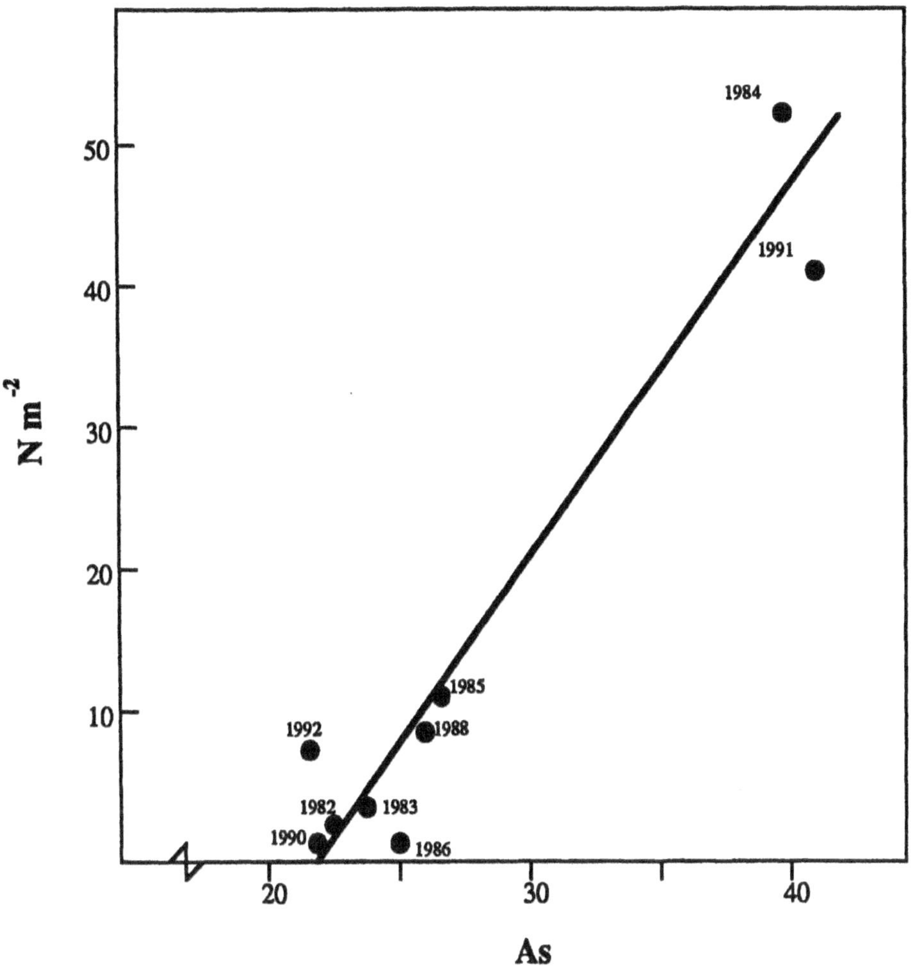

Fig. 2. The relationship between water level (Area Stability Index (As)) and perch reproductive success (fry density in the littoral zone). Numbers indicate years. For more information see Zalewski *et al.*, 1990b.

1987b) are superior to perch in exploitative competition, so when numerous they are able to suppress perch growth rate. But at low roach density, perch superiority in interference competition may cause the reverse competitive effect between these two species.

Hypothetical models of energy flow through the reservoir's ecosystem

Controlling the quality of the ecotonal shoreline zone from the point of view of spawning substrates together with enhancing the predatory fish population, density of fry may be kept at a level which will be in dynamic equilibrium with zooplankton resources and will maintain the energy flow in the reservoir characteristic for oligotrophic systems. It means that the largest portion of energy is diverted from primary producers (algae) to subsequent trophic levels – planktivorous and predatory fish (Fig. 9A). If the fish fry population is too dense, a reservoir with high primary production behaves as an eutrophic system: most of the phytoplankton production is not consumed by filtrators (Fig. 9B). This is because large daphnids are eliminated by dense populations of juvenile percid and cyprinid fishes in eutrophic waters (Zalewski *et al.*, 1990a, b). Nutrients from

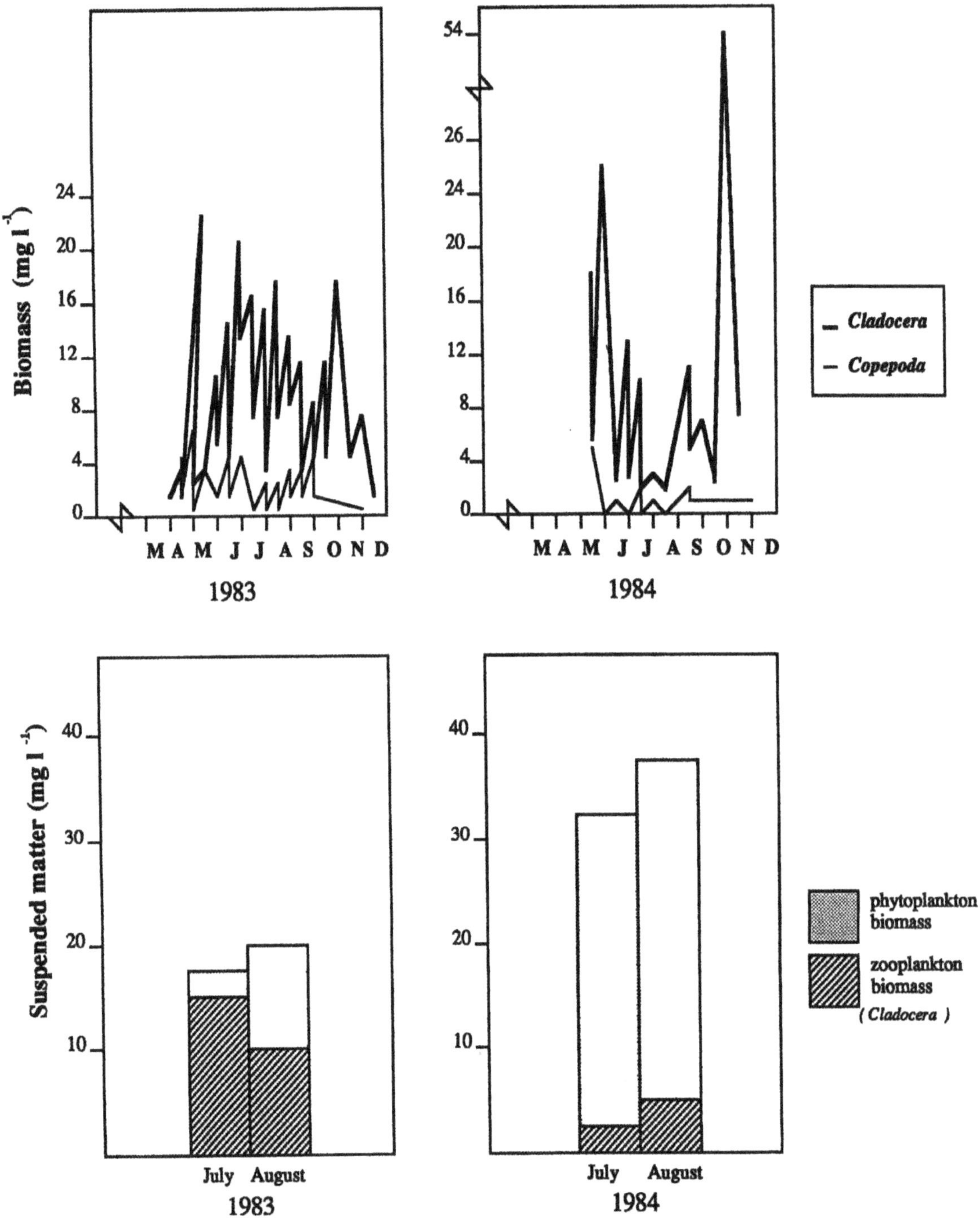

Fig. 3. Biomass of zooplankton (fresh weight) during years of low (1983) and high (1984) perch fry densities (upper panels), and summer biomass of zooplankton and phytoplankton (fresh weight) during those years (lower panels). Data from the Center of Environmental Research and Management in Łódź.

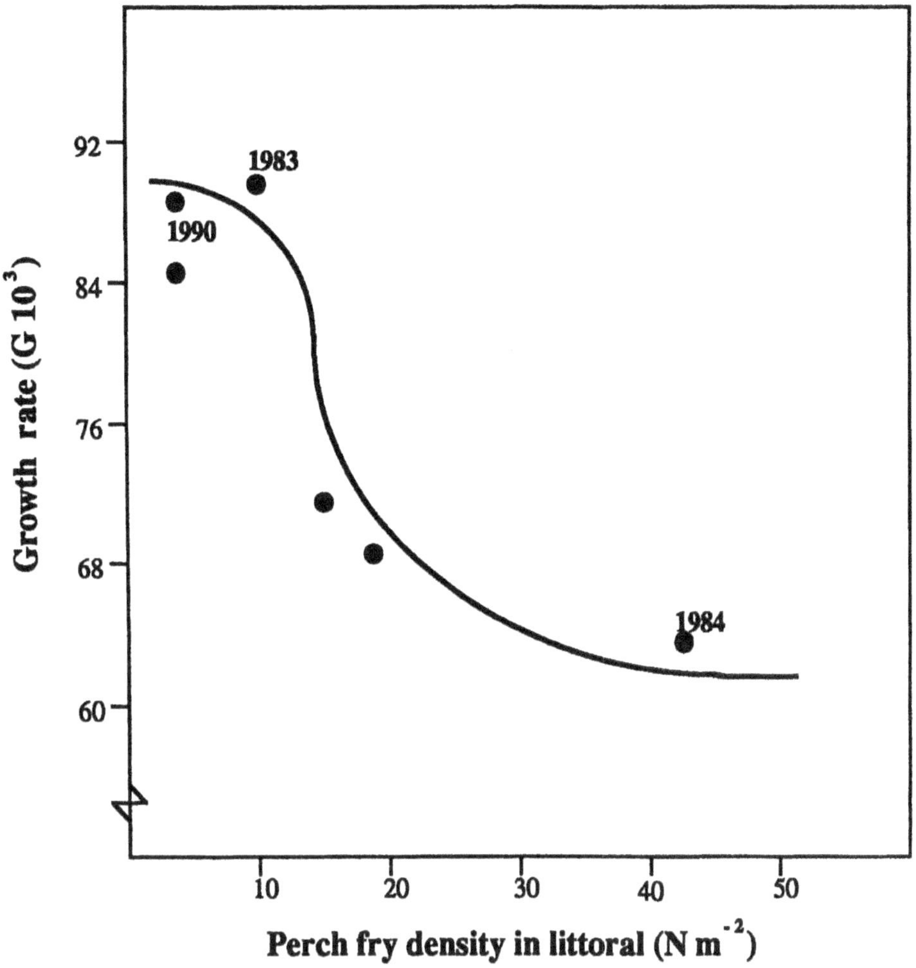

Fig. 4. The relation between instantaneous coefficient of perch fry growth (G) and their density during years of low (1983, 1990) and high (1984) fry densities. Adapted from Zalewski *et al.* 1990b.

dead algae are recirculated by the 'microbial loop' and a continuing high nutrient supply maintains the concentration of algae, resulting in deterioration of water quality. Benthic detritivores may be stimulated in the system as long as the bottom oxygen supply of the reservoir is sufficient. In such a system most of the energy is accumulated in bottom sediments leading to a decrease of water quality.

The total biomass (standing crop) of fish community was calculated on the basis of the chlorophyll a concentration and the phosphorus load (Oglesby, 1977; Hanson & Leggett, 1982; Lee & Jones, 1992), and was approximately 160–190 kg ha^{-1} in the years of low fry densities, and 350–380 kg ha^{-1} in the year of high fry density. Biomasses (standing crops) of given trophic levels presented in Fig. 9 were arrived at as follows: algae and zooplankton biomasses were calculated from data of Fig. 3; biomass of planktivorous fish (mainly juvenile roach and perch) was estimated using data from Zalewski *et al.*, 1990a. Those values are within the ranges noted by Kubecka (1989) for Rimov reservoir, which is at similar stage of ichthyofauna succession. Biomass of adult fish (benthivorous and piscivorous) was assumed to be similar in both years and was derived on the basis of anglers catches (which were approxi-

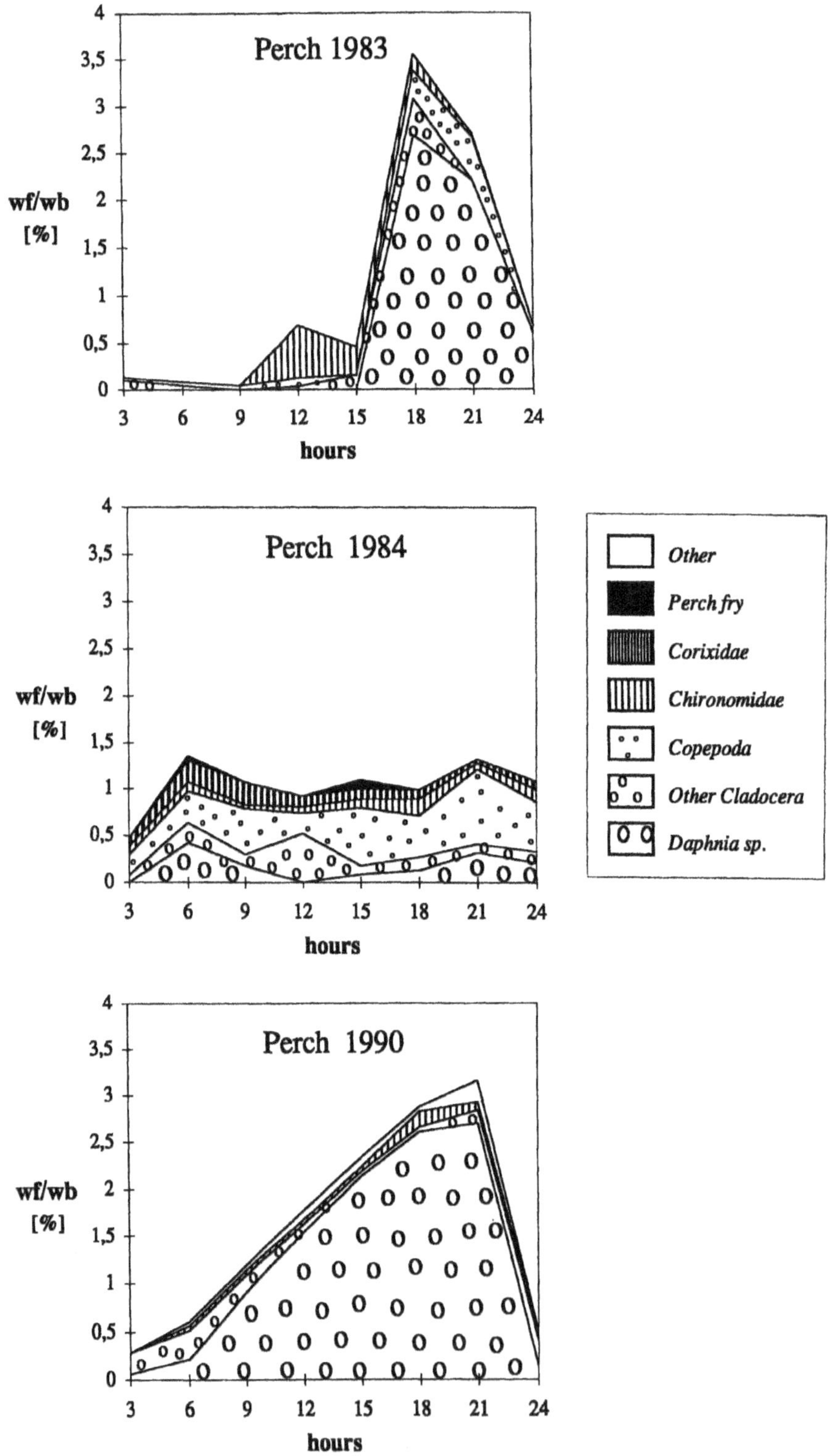

Fig. 5. Changes in perch fry fullness index (wf/wb) and amount of large cladocerans in their food in 24 hour cycle in years of low (1983, 1990) and high (1984) fry densities.

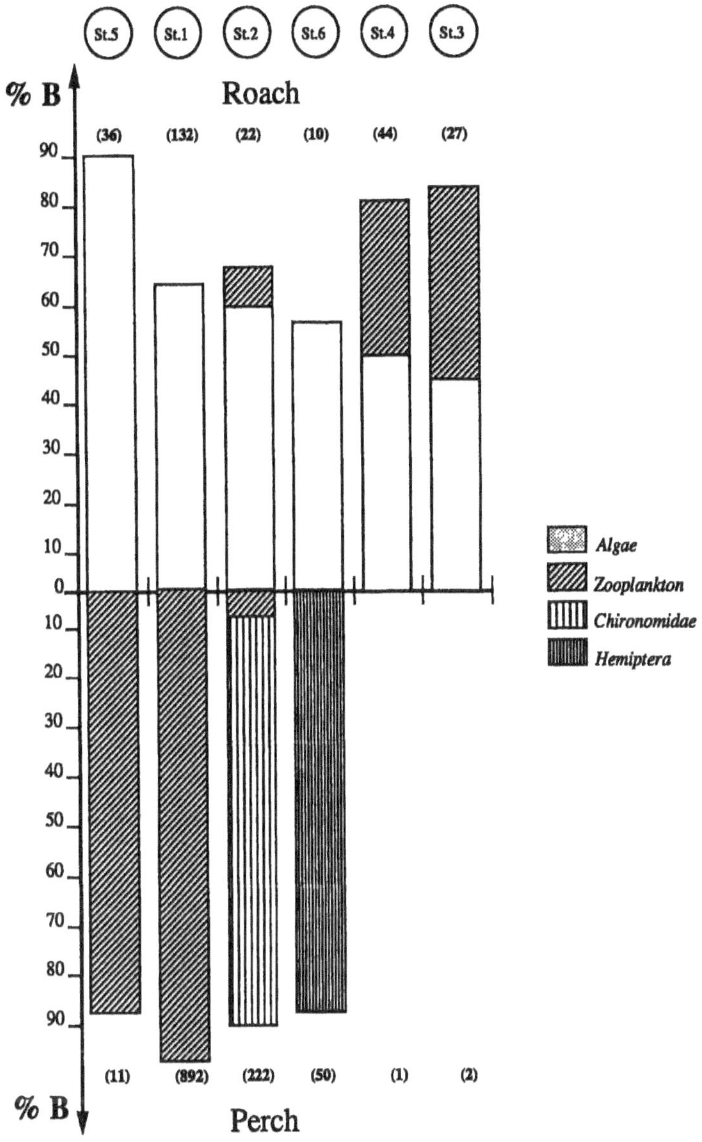

Fig. 6. Contribution of main prey categories to the food biomass of roach and perch collected in the littoral zone of the Sulejow Reservoir in July 1990. In parentheses – fry densities per 100 m^2.

mately 25 kg ha^{-1}) (Fiet, 1985), commercial gill-net catches (approx. 10 kg ha^{-1}) (unpublished data from Polish Anglers Association), and rate of bird predation (approx. 5 kg ha^{-1}) (Wojciechowski, unpublished manuscript). The magnitude of detritus and detritivore trophic levels were not investigated precisely, and present only gross approximations from monitoring samples.

Given that above data are imprecise, we believe that the general conclusion is correct. This holistic view of the dynamics of a lowland reservoir ecosystem indicates that the complexity and accessibility of the shoreline zone may regulate a cascading effect at all trophic levels, which modifies sharply the pattern of nutrient circulation and energy flow in the aquatic ecosystem. The shoreline ecotones are vulnerable to degradation (rec-

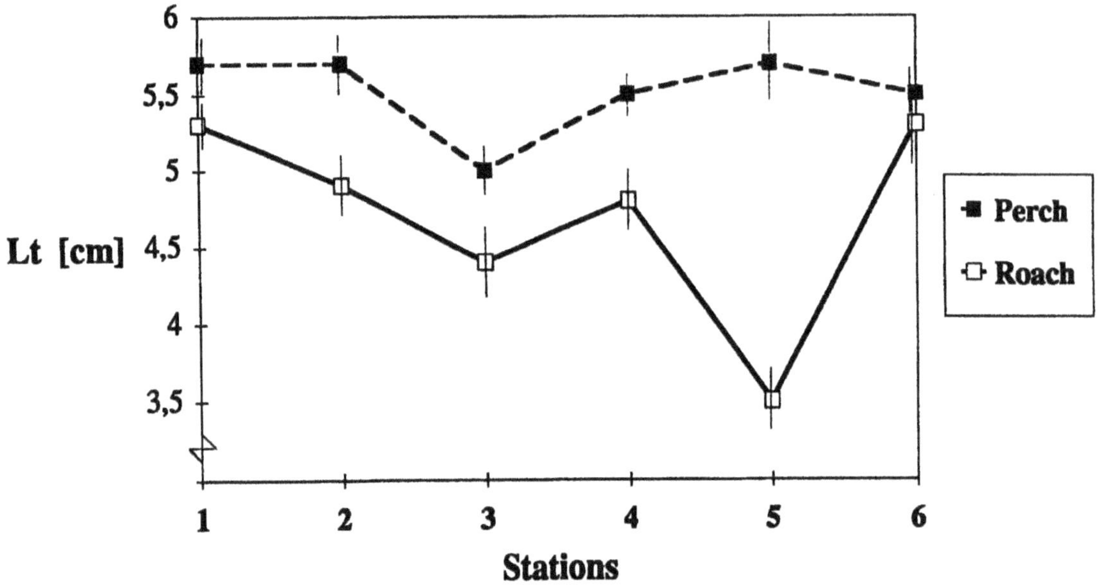

Fig. 7. Mean total length (+ SD) of perch and roach fry at different stations in the littoral zone of the Sulejow Reservoir in July 1990. The length of roach at station 5 was significantly lower from those at all other stations (one-way ANOVA; $p<0.01$).

reation), but on the other hand they are most easily manageable components of the aquatic ecosystem. That is why progress in understanding of feedback mechanisms, which start at land/water ecotones and determine fish communities and water quality, seems to be of a great importance for aquatic resource conservation and needs further research.

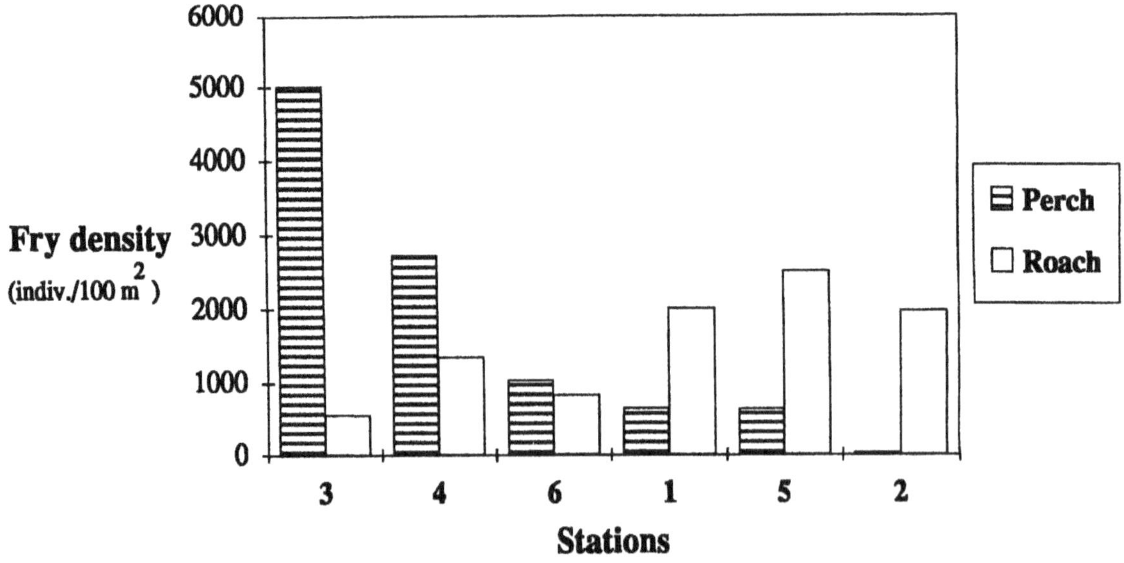

Fig. 8. Changes in perch and roach fry densities (N 100 m^2) in the littoral zone of the Sulejow Reservoir in July 1991.

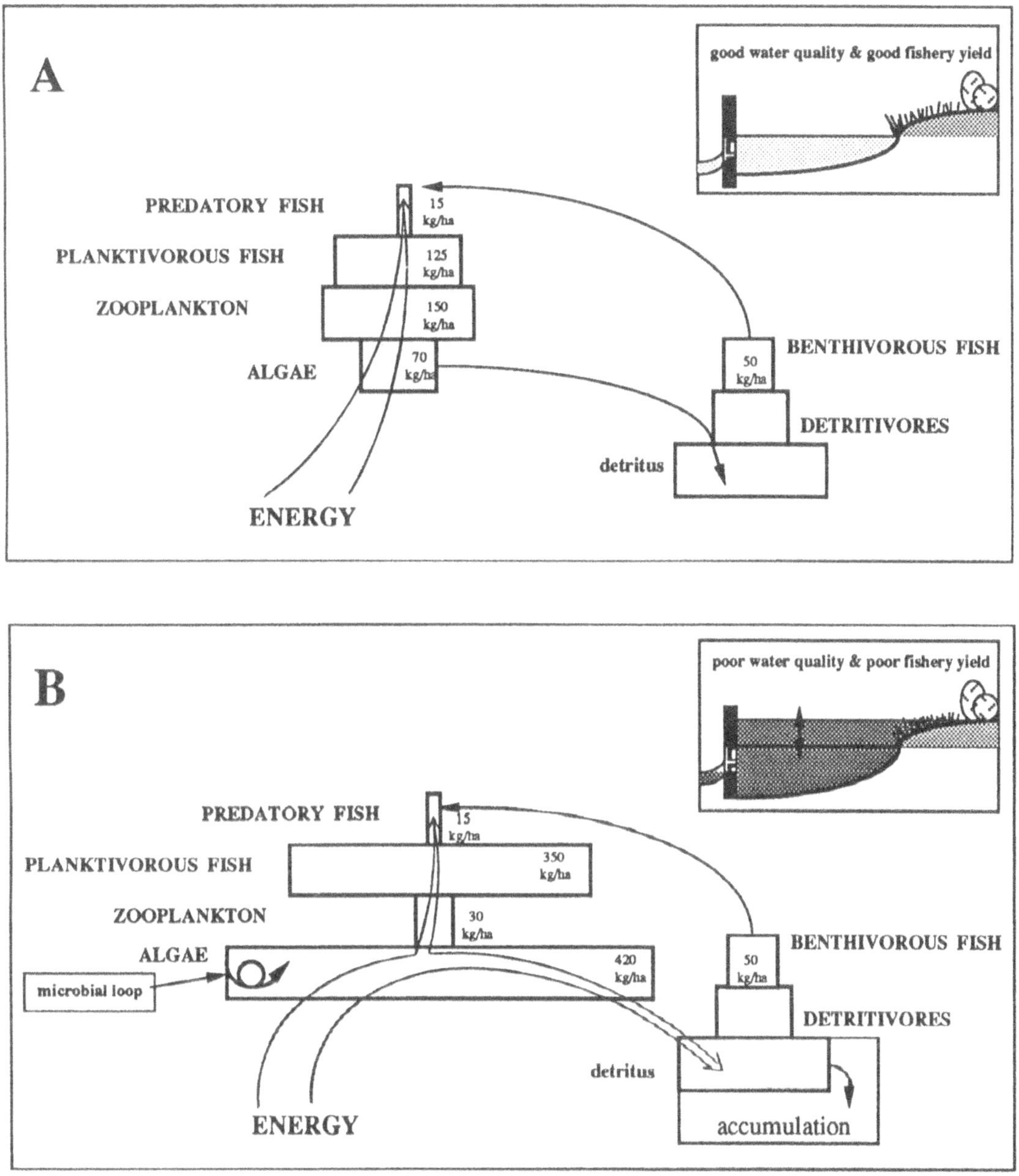

Fig. 9. Hypothetical models of energy flow through the reservoir ecosystem with poor (A) and well (B) developed riparian ecotone zone.

Acknowledgments

This work was financed by Agency for International Development (USA) Grant No. DHR-5600-G-00-1045-00 and supported by Central European University Foundation (CEU) Grant No. 884/1054.

References

Chapman, D. W., 1978: Production. In: T. B. Bagenal (ed.), Methods for assessment of fish production in fresh waters. 3rd edn. Blackwell Sci. Publ. Oxford: 202–217.

Fiet, W. T., 1985. Characteristic of anglers and fishermen catches in the Sulejow Reservoir. Msc. thesis, University of Łódź.

Frankiewicz, P., K. O'Hara & T. Penczak, 1986: Three small seine nets' method used for assessing the density of juvenile fishes in the Sulejow Reservoir. Ekol. pol. 34: 215–226.

Frankiewicz, P., M. Zalewski, A. Martyniak & U. Chylicka, in press. Trophic relations between dominant fish fry in lowland reservoir – consequences for biomanipulation. Proceedings of Second International Conference on Reservoir Limnology and Water Quality. Ceske Budejovice, Czechoslovakia 1992.

Hanson, J. M. & W. C. Leggett, 1982. Empirical prediction of fish biomass and yield. Can. J. Fish. aquat. Sci. 39: 257–263.

Hyslop, E. J., 1980. Stomach contents analysis – a review of methods and their application. J. Fish Biol. 17: 411–429.

Kubecka, J., 1989. Development of the ichthyofauna of the Rimov Reservoir and its management. Arch. Hydrobiol. Beih. Ergebn. Limnol. 33: 611–613.

Lampert, W. & B. E. Taylor, 1985. Zooplankton grazing in a eutrophic lake: implications of diel vertical migration. Ecology 66: 68–82.

Lee, G. F. & A. Jones, 1992. Effects of eutrophication on fisheries. Lake Line 12: 13–20.

Mills, E. L. & J. L. Forney, 1983. Impact on *Daphnia pulex* of predation by young yellow perch in Oneida Lake, New York. Trans. am. Fish. Soc. 112: 154–161.

Nelson, W. R. & C. H. Walburg, 1977. Population dynamics of yellow perch (*Perca flavescens*), sauger (*Stizostedion canadense*), and walleye (*S. vitreum vitreum*) in four main stem Missouri River reservoirs. J. Fish. Res. Bd Can. 34: 1748–1763.

Niederholzer, R. & R. Hofer, 1980. The feeding of roach (Rutilus rutilus L.) and rudd (Scardinius erythrophthalmus L.). I. Studies on natural populations. Ekologia Polska 28: 45–59.

Oglesby, R. T., 1977: Relationships of fish yield to lake phytoplankton standing crop, production, and morphoedaphic factors. J. Fish. Res. Bd Can. 34: 2271–2279.

Persson, L., 1987 a. Competition-induced switch in young of the year perch, *Perca fluviatilis*: an experimental test of resource limitation. Envir. Biol. Fish 19: 235–239.

Persson, L., 1987b. Effects of habitat and season on competitive interactions between roach (*Rutilus rutilus*) and perch (*Perca fluviatilis*). Oecologia 73: 170–177.

Ploskey, G. R., 1985. Impacts of terrestrial vegetation and preimpoundment clearing on reservoir ecology and fisheries in the United States and Canada. FAO Fisheries Technical Paper 258: 35 pp.

Post, J. R. & D. J. McQueen, 1987. The impact of planktivorous fish on the structure of a plankton community. Freshwat. Biol. 17: 79–89.

Prejs, A., 1984. Herbivory by temperate freshwater fishes and its consequences. Envir. Biol. Fishes 10: 281–296.

Zalewski, M., B. Brewińska-Zaraś & P. Frankiewicz, 1990a. Fry communities as a biomanipulating tool in a temperate lowland reservoir. Arch. Hydrobiol. Beih., Ergebn. Limnol. 33: 763–774.

Zalewski, M., B. Brewińska-Zaraś, P. Frankiewicz & S. Kalinowski, 1990b. The potential for biomanipulation using fry communities in a lowland reservoir: concordance between water quality and optimal recruitment. Hydrobiologia 200/201 (Dev. Hydrobiol. 61): 549–556.

Hydrobiologia **303**: 61–67, 1995.
F. Schiemer, M. Zalewski & J. E. Thorpe (eds), The Importance of Aquatic-Terrestrial Ecotones for Freshwater Fish.

Relationship between the spatial scale and biotic variability in a wetland ecotone

Jurek Kolasa & Lisa Weber
Department of Biology, McMaster University, Hamilton, Ontario L8S 4K1, Canada

Key words: wetland, spatial scale, carp impact, abundance variation, richness, algae, ecotone, recovery

Abstract

We studied potential effects of the expansion of a wetland ecotone on phytoplankton communities. The expansion of ecotone width and emergence of a mosaic of habitats are expected outcomes of a reduction of carp numbers and a consequent managed vegetation recovery. Specifically, we investigated the effects of decreasing size of open water cells on species richness, abundance and diversity. In a set of experimental enclosures of four different sizes we sampled phytoplankton weekly during 1991. The phytoplankton communities showed similar patterns of seasonal change but no clear differences in species abundance or richness that could be attributed to the system size. Notably, however, the magnitude of variation in the community structure metrics responded strongly to the enclosure size. This indicates that the importance of scale may not be fully appreciated when analyzed in the light of standard community measures. We tentatively conclude that the variation in species composition, individual responses, richness, and abundance may result in an increasing diversification of the wetland ecotone as the recovery of vegetation advances.

Introduction

An ecotone is a zone of transition between adjacent ecological systems such as a uniform macrophyte stand and open water (Holland, 1988). Wetland ecotones may vary from sharp and narrow to wide and diffused. In the case of a spreading vegetation front, the transition zone may take the form of an abrupt change in plant presence and density or may extend over a wider area with gradual and heterogeneous decrease in macrophyte densities. In the latter case, the ecotone appears as a mixed zone of vegetation and open water patches. Characteristics of such patches and their role in wetland function are unknown (Holland *et al.*, 1990). Cootes Paradise, a coastal wetland of Lake Ontario, presents an opportunity for postulating and testing hypotheses addressing the properties of ecotone internal patches. We will articulate a specific hypothesis later on.

In Cootes Paradise, both emergent and submergent macrophyte vegetation is currently at its lowest compared to the past patterns (Painter *et al.*, 1989). Several factors are implicated in the demise of macrophytes, and they appear to interact. Carp reproduction and feeding, wind driven sediment resuspension resulting in low light penetration, long-term water level fluctuations, and excessive algal densities were postulated. Fish role in the demise of the wetland is of particular interest because of its complexity and amenability to remedial manipulation. Carp has a direct or

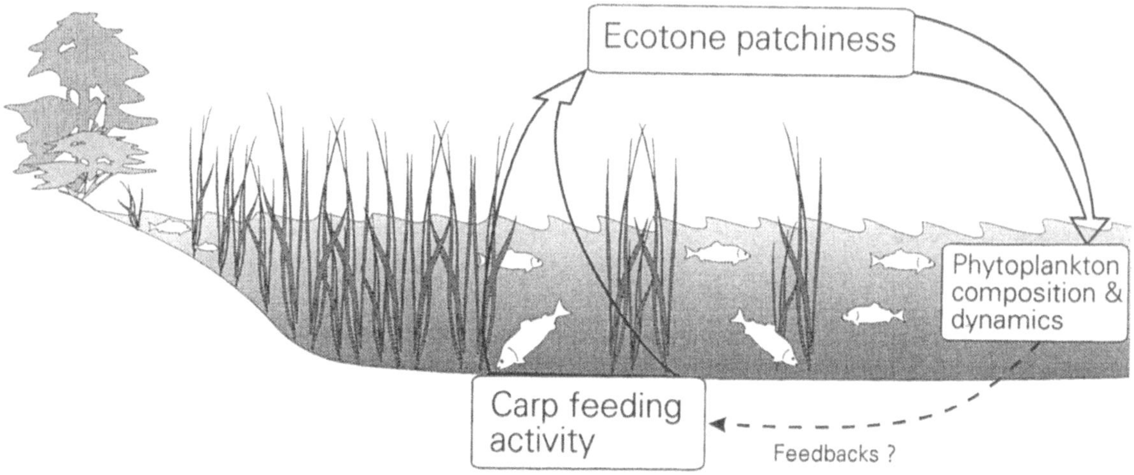

Fig. 1. Major links between carp, *Cyprinus carpio*, emergent vegetation, habitat heterogeneity, and other components of the wetland ecosystem. Carp is a 'keystone' species.

indirect impact on macrophyte development and spread, amount of suspended sediments, light availability, nutrient dynamics, benthos composition and abundance. Many of these factors influence algal ecology (Fig. 1). A restoration plan under development aims at the reduction of carp densities and activities. Preliminary experiments indicate that the reduction of carp density will allow plant recovery (unpublished). Based on the past states of Cootes Paradise documented by aerial photographs, we expect to observe the following trends in the wetland characteristics to be associated with the carp control (Fig. 2): (a) a reduction in open water areas, (b) increase in plant, fish and wildlife diversity, (c) improvement in water clarity, (d) increase in the amount of 'edge' habitats and (e) a decline in the size of remaining open water patches or cells.

The last two expected trends represent changes in spatial scale and patterns and can be addressed experimentally. Furthermore, biological trends associated with these changes can be predicted and tested within the framework of hierarchy theory (cf. Allen & Hoekstra, 1992). Specifically, this theory predicts that population and community parameters should be more variable at smaller spatial scales (Kolasa & Waltho, 1994).

The goal of the study was to test this prediction and provide a quantitative evaluation of the vari-

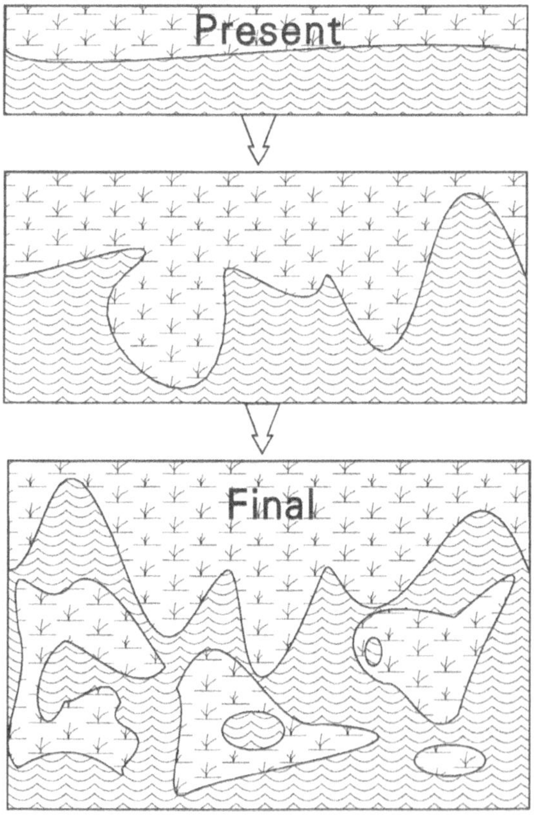

Fig. 2. Schematic expected trends in wetland development based on the past known conditions in Cootes Paradise.

ability of selected community parameters. Our approach to achieve this goal was experimental.

Study site and methods

Emergent macrophytes persist in the shallowest littoral zone only. They consist primarily of cattails (*Typha angustifolia*) and occasional patches of manna grass (*Glyceria aquatica*), and bur reed (*Sparganium sp.*). Submersed plants are absent throughout most of the season but may appear in small numbers towards the end of summer.

We established a total of 12 enclosures of four different sizes (three replicates of each 1 m^2, 4 m^2, 9 m^2, 16 m^2) and 3 control plots at a site near the western end of Cootes Paradise Fig. 3), about 15 m away from the nearest emergent vegetation. The enclosures were constructed of plastic foil reinforced externally by chicken wire. The placement of enclosures and controls was randomized on a grid of 5 m by 5 m squares, in four rows

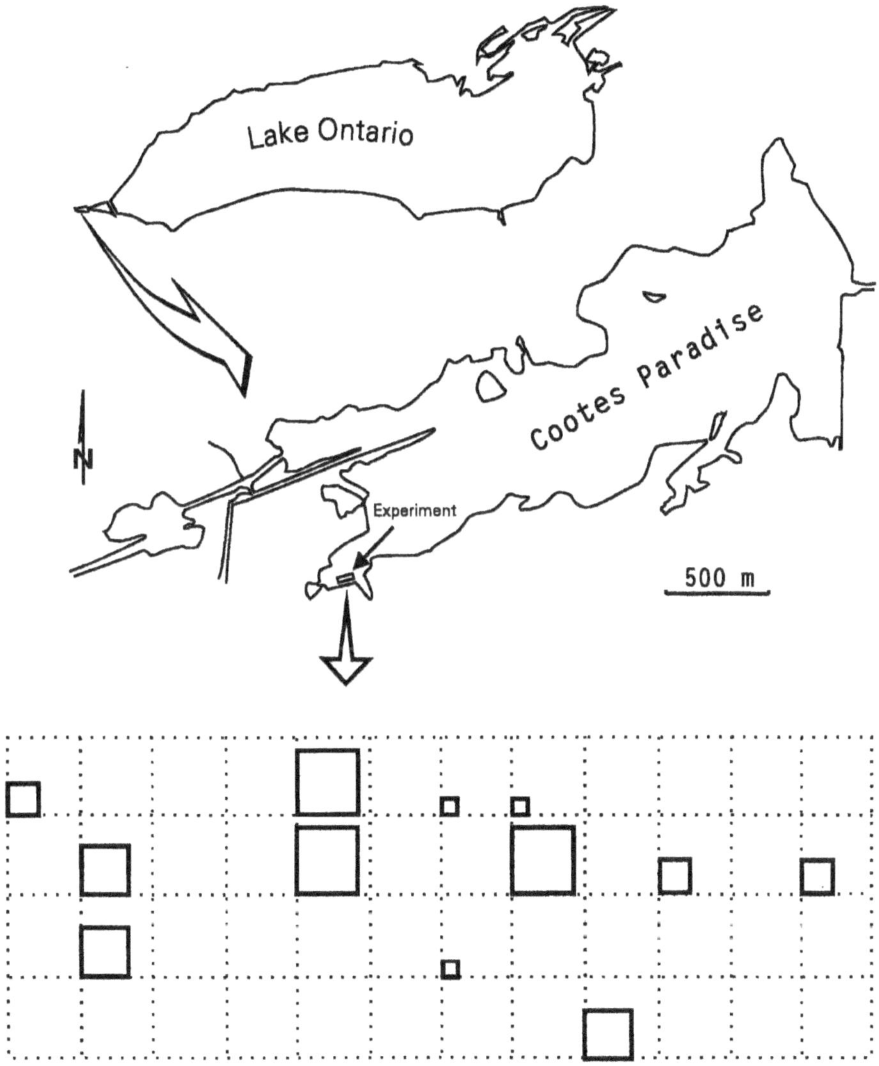

Fig. 3. Geographical location and placement of treatment enclosures in the field. Bold squares represent plastic enclosures of different sizes. Other enclosures and controls (not shown) were randomly spread in the remaining cells of the grid.

parallel to the shore (Fig. 3). We measured basic physical parameters within the enclosures and in controls. Light conditions were assessed using a Secchi disk. Oxygen and pH were measured using portable meters.

Sampling commenced June 21, 1991 and ended August 15, 1991. One sample was taken from each enclosure and control site, at a sampling frequency of every third day. Samples were taken at approximately the same time each day, 10:00 am $\pm$ 1 hour.

A sampling device was applied to ensure that the composite sample contained plankton from the whole water column. The device consisted of a 1 l NALGENE® bottle fitted between 2 square plates of plexiglass, with a hole drilled in the top plate to accommodate the neck of the bottle. The plates were maintained in position by the use of four metal posts screwed tightly to each corner. The apparatus was weighted by a brick wedged between the plexiglass at the bottom of the bottle and another piece placed below it. The entire apparatus was attached to a rope and lowered and lifted twice before the 1 l bottle was filled. A portion of the integrated sample was then transferred to a labelled 25 ml glass scintillation vial. After the fifteen scintillation vials were filled with their respective samples, they were taken to the lab and preserved with 0.2 ml of LUGOL's iodine within an hour of finishing sampling. Samples were counted using a ZEISS inverted microscope at magnification of 40 × with phase contrast. Ten random viewing fields in which all algae were identified and counted were chosen along a straight (transect) line. Organisms were identified using a variety of sources (Findlay & Kling, 1979; Prescott, 1970; Skuja, 1964; Skuja, 1956; Taft & Taft, 1971; U.S. Federal Water Pollution Control Administration, 1966).

Results

The enclosures showed almost no consistent differences among sizes and did not differ from controls except for the water transparency. The water transparency was higher in the enclosures (15 cm on average) than in controls (6 cm on average; ANOVA, $p < 0.0000$). The variability of oxygen, pH, light, and temperature showed no significant differences among treatments and controls (one way ANOVA). We have not conducted post-hoc analyses with contrasts that might reveal less prominent patterns. We have found forty-eight algal taxa to occur at least on one occasion in either treatments or controls. These taxa responded in a variety of ways which we briefly characterize below. The most important aspects of this response include (1) the number of taxa involved in community changes and (2) the magnitude of variability. We concentrate on these two results because of their significance for the interpretation of the test outlined earlier.

Species abundance and richness increased steadily within the enclosures over the season and were correlated (adjusted $r^2 = 0.641$). By contrast,

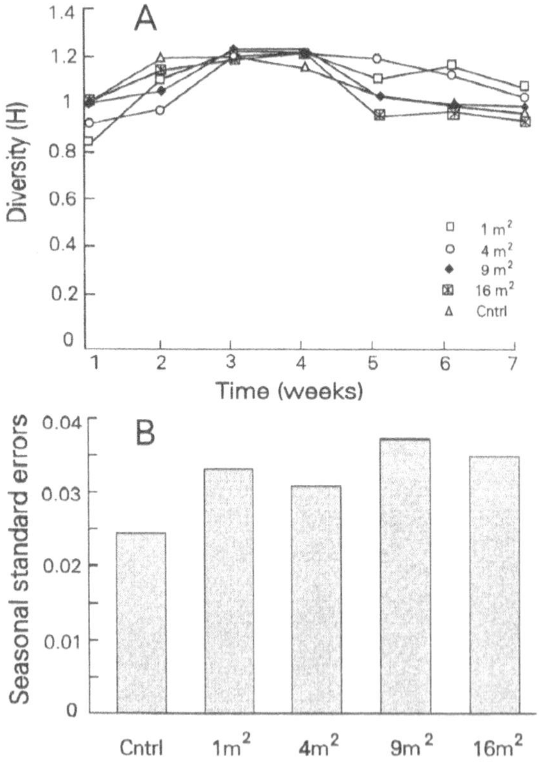

Fig. 4. Richness (A) and abundance (B) of algae in the experimental enclosures and control areas during the summer of 1991.

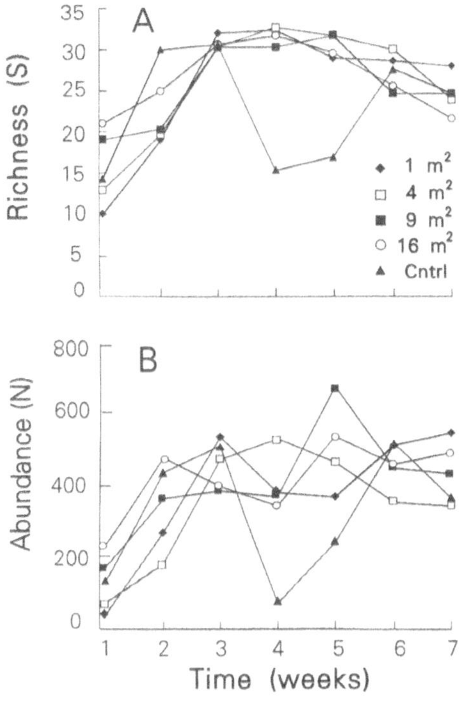

Fig. 5. Diversity means (A) and seasonal variation (B) in enclosures of various sizes and in control areas.

species richness and abundance outside the enclosures showed much greater variation over the season (Fig. 4A, B). The community structure was evaluated by the Shannon-Weaver index of diversity. This index summarizes the pattern of abundance distribution across species. Diversity showed some seasonal variation (Fig. 5A) but, once again, we failed to find significant differences between the controls and treatments (Fig. 5B). In the next step, we examined the variation at the species level. First, we found that different sets of species were responsible for seasonal variation in different enclosures. Second, and more importantly, we found that more species contributed significantly to that variation in small enclosures than in large enclosures. We observed a striking difference between the number of species responsible for temporal density variation in large enclosures (11 and 14 species for 9 m^2 and 16 m^2 enclosures, respectively) and controls (22 species). Furthermore, as many as 32 species showed significant temporal variation in all enclosures taken together than in the control system outside. The question emerges then as to

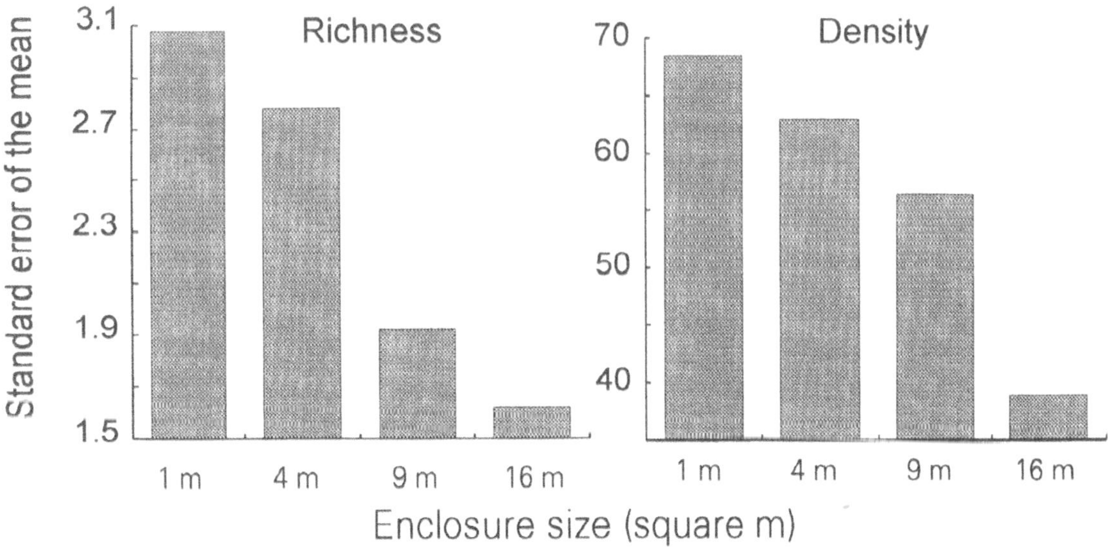

Fig. 6. Variability of algal richness and abundance (density) represented as seasonal standard errors in enclosures of various sizes. Variability in the control areas was substantially greater. Trends are statistically significant (see text).

the nature of this variability. An ANOVA showed that the species richness, abundance, and diversity responded in a highly variable manner but that differences among the means were not significant. In order to assess what was the nature of this variability, we compared standard errors of the seasonal means for both the abundance and richness of algal taxa (Fig. 6). We found that there was a clear declining trend with the increasing enclosure size (Regression fitted to standard errors, $r^2 = 0.9177$, $p < 0.042$).

Discussion and conclusions

The quality of our results might have been influenced by the taxonomic level of identification even though a generic name identifies a single species in most cases. Generally, a taxonomic detail might be expected to strengthen some of our quantitative analyses without, however, having a major impact on the conclusions (e.g. Rahel, 1990). Other sources of error potentially include changes in periphyton to phytoplankton ratios associated with the decline in the amount of enclosure material available for attached algae in larger enclosures relative to smaller ones. We have no data that would allow to explain a major decline in abundance and richness in control areas observed in the middle of the summer (Fig. 4). One possible cause of the difference between controls and treatments is wind driven water movement from other portions of the marsh. Such incoming water would not affect the community structure within the enclosures.

We are not aware of other studies varying the size of enclosures in wetlands. Various scales of variation in aquatic systems were however investigated using other organisms or measures. For example, Carpenter & Kitchell (1987) used model simulations and found that the variance of primary production and its covariance with other dynamic variables are a function of temporal sampling scales. Their observations agree with our conclusion that scale affects variability but differ in the specific pattern observed and the type of scale investigated. Specifically, Carpenter & Kitchell (1987) found that coefficients of variation stabilized at a certain level when temporal scales larger than 7 days were treated as samples. We found variation in species abundance and richness to be less at larger spatial scales.

While this study aims at identification of the patterns of variability in the phytoplanktonic community structure, we feel obliged to provide some plausible hypotheses to address the more puzzling results. The general pattern of smaller variation in larger enclosures could be tentatively explained by cancelling effects of patch heterogeneity on the population dynamics of individual species. This idea has been first proposed by Hutchinson (1961) and further resumed by Hebert & Crease (1980) with respect to zooplankton. Such a heterogeneity would presumably increase in larger enclosures and thus could be responsible for the reduction of variation in those enclosures. The observed larger variation in controls as opposed to treatments could be explained by wind driven water movements in the marsh. Such movements are likely to bring about different communities of algae from various parts of the marsh that is known to be limnologically diverse (Painter *et al.*, 1989). Given the isolation of aquatic communities in the enclosures, water movements could not result in species substitutions directly. The difference between the number of significantly variable species in enclosures taken jointly and the control areas is of considerable interest because it indicates the importance of microhabitat isolation for the generation and maintenance of diversity. While the variation in small isolated microhabitats may threaten the persistence of some species, the interplay between the diversity of those microhabitats these habitats offer and the differential species responses to individual microhabitats appears to provide an overall enhancement of ecosystem diversity at the phytoplankton level.

In conclusion we may identify the following processes and trends:

1. Variation of phytoplankton communities declines with an increase in the spatial scale the community exists at. This variation is ex-

pressed at various levels of ecological resolution. It was observed in agglomerative parameters such as species richness or total abundance and at the compositional level.

2. Microhabitat isolation and fragmentation, as exemplified by the collection of enclosures, appears to increase overall diversity of ecological components and behaviors. This general observation is illustrated by the increasing number of species in small enclosures that show significant variation through time.
3. Diversity remains constant suggesting that the phytoplanktonic communities quickly compensated for major abundance changes at the species level.

All the above trends are likely to intensify with the increasing width and complexity of the ecotone postulated by the restoration plan (Fig. 2). A comparison of patterns exhibited by zooplankton and benthos with those of phytoplankton is likely to shed more light on the generality of the observations made here and on the possible interactions among all those ecotone components.

Acknowledgement

This study was financially supported by NSERC grant to JK. Additional funds were made through the Canada Department of Fisheries and Oceans, Environmental Youth Corps, and the Ontario Government Campus Employment for Aboriginal and Native Youth (grants to B.N. White). We express our sincere thanks to Len Simser and his colleagues from Royal Botanical Gardens for generous help in setting up the experiment and to our students involved in various stages of the work: K. Barnes-Steele, A. Chaves, S. Jaffer, P. Kuzmich, P. Miller, E. Mussel, T. Mussel, B. Bullough, S. Ward & J. Wilson. Dr D. Coxson and the editors and reviewers have provided helpful comments on an earlier draft of the manuscript.

References

Allen, T. F. H. & T. W. Hoekstra, 1992. Toward a unified ecology. Columbia University, New York, 384 pp.

Carpenter, S. R. & J. F. Kitchell, 1987. The temporal scale of variance in limnetic primary production. Am. Nat. 129: 417–433.

Findlay, D. L. & H. J. Kling, 1979. A species list and pictorial reference to the phytoplankton of central and northern Canada – parts I and II. Fisheries & Marine Services Manuscript Report No. 1503. Department of Fisheries and the Environment, Winnipeg, Manitoba, Canada.

Hebert, P. D. N. & T. J. Crease, 1980. Clonal coexistence in Daphnia pulex (Leydig): another planktonic paradox. Science 207: 1363–1365.

Holland, M. M. (compiler), 1988. SCOPE/MAB technical consultations on landscape boundaries: report of a SCOPE/MAB workshop on ecotones. Biology International, Special Issue 17: 47–106.

Holland, M. M., D. F. Whigham & B. Gopal, 1990. The characteristics of wetland ecotones. In R. J. Naiman & H. Décamps (eds), The ecology and management of aquatic-terrestrial ecotones. UNESCO and Parthenon Publishing Group, Paris and Park Ridge: 171–198.

Hutchinson, G. E., 1961. The paradox of the plankton. Am. Nat. 95: 137–145.

Kolasa, J. & N. D. Waltho, 1994. Predictions of the hierarchy theory (in manuscript).

Painter, D. S., K. J. McCabe & W. L. Simser, 1989. Past and present limnological conditions in Cootes Paradise affecting aquatic vegetation. Royal Botanical Gardens Technical Bulletin No. 13, Burlington, Canada, 34 pp.

Rahel, F. J., 1990. The hierarchial nature of community persistence: a problem of scale Am. Nat. 136: 328–344.

Skuja, H., 1956. Taxonomische and biologische Studien uber das Phytoplankton schwedischer Binnengewasser. Nova Acta Regiae Soc. Sci. Upsal. 16: 445–461.

Skuja, H., 1964. Grundzuge der Algenflora und Algenvegetation der Fjeldgegenden um Abisko in schwedisch-Lappland. Nova Acta Regiae Soc. Sci. Upsal. 18: 391–400.

Taft, C. E. & C. W. Taft, 1971. The algae of Western Lake Erie. Ohio Biol. Survey Bull. 4: 1–189.

U.S. Department of the Interior, Federal Water Poll. Admin. 1966. A guide to the common diatoms at water pollution surveillance system stations, 98 pp.

Streams

Hydrobiologia **303**: 71–81, 1995.
F. Schiemer, M. Zalewski & J. E. Thorpe (eds), The Importance of Aquatic-Terrestrial Ecotones for Freshwater Fish.

Critical landscape attributes that influence fish population dynamics in headwater streams

Isaac J. Schlosser
Department of Biology, University of North Dakota, Grand Forks, ND 58202–9019, USA

Key words: fish, habitat, landscape ecology, refugin, riparian, streams

Abstract

Three landscape attributes are likely to have strong effects on the rate-dependent processes determining fish population dynamics in headwater streams: (1) functional interactions at terrestrial-aquatic ecotones and their influence on temporal and spatial variation in resource supply and predator-prey interactions, (2) large-scale spatial habitat relationships and their effect on resource use and fish movement, and (3) presence of refugia from harsh environmental conditions and their influence on fish survival and emigration/immigration rates. Elucidating how these factors interact over a range of temporal and spatial scales should be a major goal of lotic fish ecologists.

Introduction

Stream fish ecologists are increasingly asked to predict the effect of landscape disturbances due to agriculture, deforestation, and grazing on the diversity and population dynamics of stream fish. Unfortunately, these disturbances alter ecological processes that operate over large spatial scales, while most studies of fish ecology in headwater streams have been conducted at small spatial scales (Matthews & Heins, 1987). Consequently, we lack a conceptual framework as to how large-scale ecological processes influence stream fish, which limits our ability to either predict the impact of future landscape disturbances on these organisms or design effective restoration efforts once such disturbances have occurred.

The emerging discipline of landscape ecology (Risser *et al.*, 1984; Naveh & Lieberman, 1984; Forman & Godron, 1986) offers a useful context for the development of this conceptual framework. Three subareas within landscape ecology are particularly relevant, including: (1) the influence of terrestrial-aquatic ecotones on trophic processes in streams (Karr & Schlosser, 1978; Naiman *et al.*, 1988; Schiemer & Zalewski, 1992), (2) the role of large-scale spatial habitat relationships in regulating population dynamics (Forman & Godron, 1986; Pringle *et al.*, 1988; Dunning *et al.*., 1992), and (3) the influence of the presence or absence of refugia on the response of organisms to harsh environmental conditions (Pickett & White, 1985; Sedell *et al.*, 1990). My objective in this paper is to briefly review our current knowledge about the influence of these three factors on fish in headwater streams. I have not provided a general discussion of the role terrestrial landscape attributes, other than the riparian environment, play in influencing stream ecosystems since others have already published succinct descriptions of these linkages (Swanston, 1991, Gregory *et al.*, 1991).

Life cycles and the population dynamics of stream fish

To understand the potential influence of landscape attributes on stream fish requires a brief review of their life cycle and the rate-dependent processes determining their population dynamics. Stream fish vary dramatically in size from embryo to adult. Associated with this variation they exhibit complex life cycles and habitat-use patterns, with these habitat use patterns being mediated by migratory processes (Jones, 1968; Northcote, 1978; Fig. 1). Life begins at spawning, after which there is an incubation period lasting anywhere from a few days to several months. Hatched fish usually then move to 'feeding habitats' where most growth and development occurs. These 'feeding habitats' normally consist of a complex mosaic (Fig. 1) of several habitat types (e.g., pools, riffles, stream margins etc.), which serve as nursery and rearing habitats for larval stages and juveniles. Depending on the life cycle of the fish and local climatic conditions, the fish then normally experience a series of seasonally favorable periods with rapid growth and seasonably unfavorable periods with reduced growth until sexual maturity is reached. In north-temperate streams these favorable and unfavorable periods frequently involve movement between summer feeding habitats and winter refugia (Fig. 1; Cunjak, 1988). In tropical streams, favorable and unfavorable periods frequently involve movement between wet season feeding habitats and dry season refugia (Lowe-McConnell, 1987). Once sexual maturity is reached, the fish then undergo spawning movements to appropriate sites for egg deposition and reinitiation of the cycle.

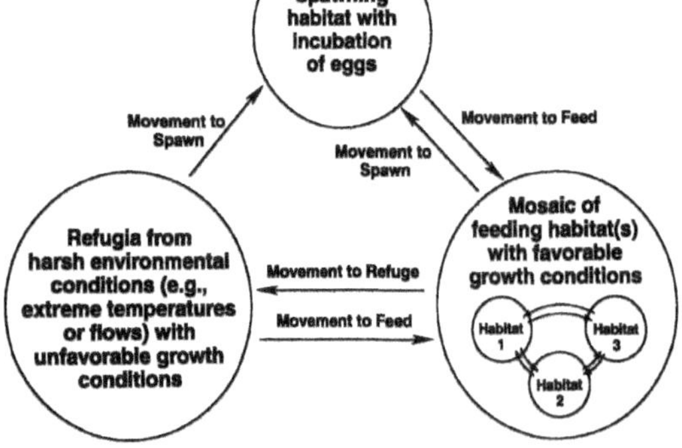

Fig. 1. The basic life cycle of stream fish with emphasis on patterns of habitat use and movement.
[Based on Jones (1968), Northcote (1978), and Schlosser (1991).]

Based on this general life cycle, three things should be apparent. First, because various life stages of fish require different physical habitats, spatial heterogeneity and the connectivity of habitat patches will be critical for completion of their life cycles (Bisson *et al.*, 1982; Schiemer & Spindler, 1989). Second, the life cycle of stream fish can potentially extend over long temporal and large spatial scales, with the maximum time and length of these scales being defined by the life history characteristics of the species considered. Third, because fish complete their life cycles across potentially long temporal and large spatial scales, their population dynamics will be a function of numerous rate-dependent processes, including their immigration/emigration, natality, and mortality rates, with mortality rates during early life history stages being strongly influenced by the growth rates of fish during this period (Andrewartha & Birch, 1954; Werner & Gilliam, 1984; Elliott, 1989).

The role of functional interactions

Functional interactions involving the flow of nutrients and energy at terrestrial-aquatic ecotones have numerous potential effects on the rate-dependent processes determining fish population dynamics. Two functional interactions are likely to be particularly critical: (1) the influence of nutrient and energy exchanges between terrestrial and aquatic environments on the 'bottom-up' supply of resources and (2) the 'top-down' effect of predator-prey interactions on habitat use and resource availability (Fig. 2).

Current conceptual models of nutrient and energy exchanges in headwater streams emphasize the critical role of lateral transport processes (Vannote *et al.*, 1980). Since headwater channels

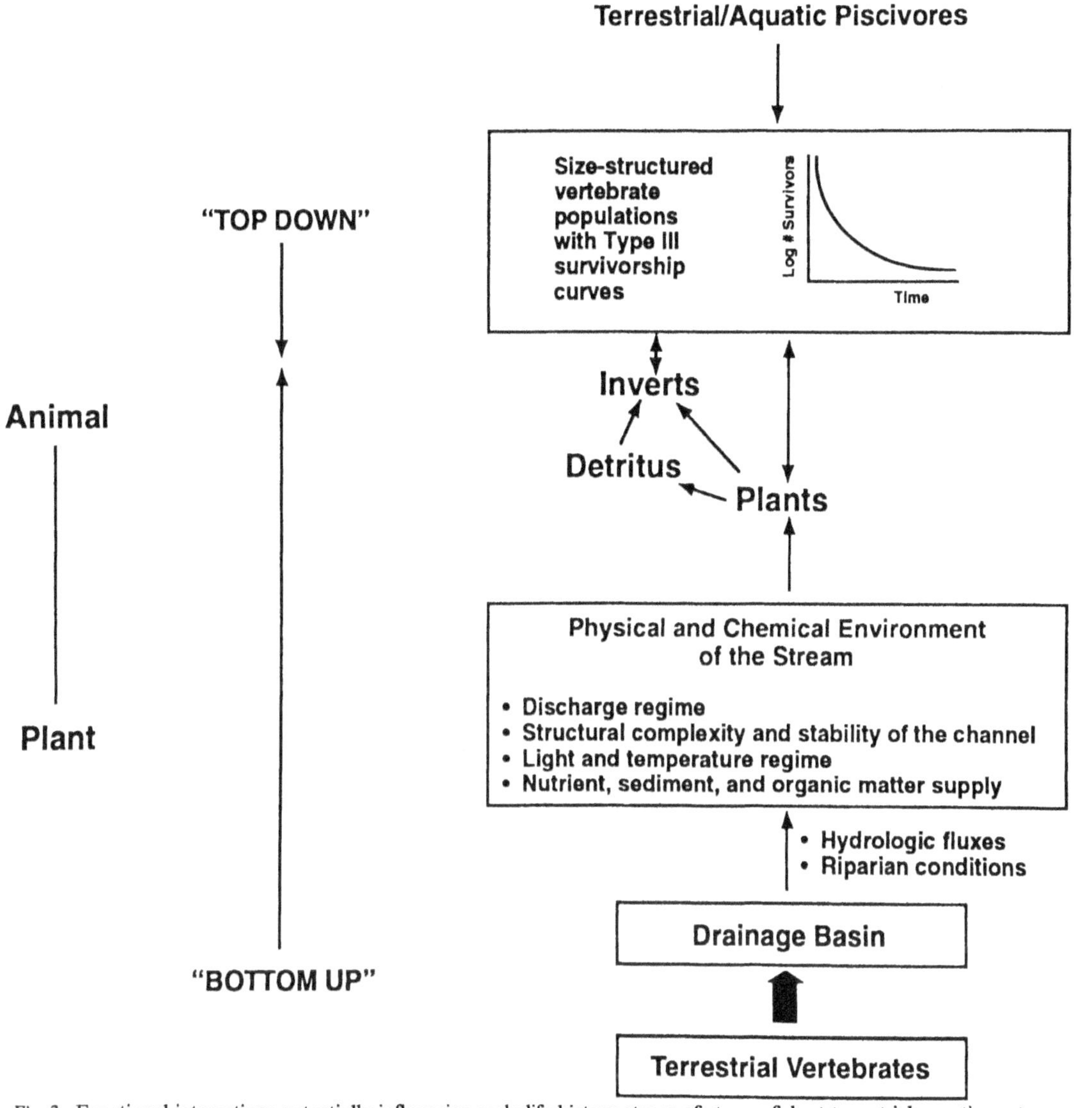

Fig. 2. Functional interactions potentially influencing early life history stages of stream fish at terrestrial-aquatic ecotones.

are narrow, floodplains poorly developed, and absolute peak discharges small and of short duration, the active capture of nutrients and energy from the floodplain is thought to be of minor importance. Instead, current models of nutrient and energy exchanges in headwater streams emphasize the downhill flux of water, nutrients, and organic matter to the lotic environment (Vannote *et al.*, 1980), with the riparian vegetation playing a critical role in regulating these exchanges (Fig. 2; Karr & Schlosser, 1978; Peterjohn & Correll, 1984; Pinay & Decamps, 1988; Correll, 1991).

With respect to the fish, one of the most critical aspects of these nutrient and energy exchanges is their influence on the temporal distribution of 'bottom-up' resource supply to higher trophic levels (Fig. 2). In most natural, north temperate headwater streams the supply of organic matter to organisms on the surface of the streambed is highly seasonal. Leaf inputs tend to occur in autumn, which are then processed throughout the winter and spring, with an associated large pulse in invertebrate production in late spring and early summer (Hynes, 1970; Schlosser & Toth, 1984).

During the spring, through summer and into autumn radiant energy reaching the stream is filtered through the riparian canopy and is limited. Carbon fixation in the stream is reduced and invertebrate biomass falls as aquatic insects emerge to complete their life cycle (Hynes, 1970; McIntire & Colby, 1978, but see Minshall, 1978). Descriptive studies of seasonal variation in number of food items, breadth of foods utilized, and growth rates of early life stages of stream fish suggest the 'bottom-up' supply of invertebrate food resources is limiting to fish during summer in natural headwater streams (Angermeier, 1982; Schlosser, 1982; Schlosser & Angermeier, 1990). Even stronger evidence for resource limitation during summer comes from experimental studies of resource supplementation that reveal decreased emigration, increased growth, and increased lipid reserves of fish in streams with supplemental feeding (Mason, 1976).

'Top-down' predator-prey interactions also have potentially strong effects on the growth and survival of fish at terrestrial-aquatic ecotones in headwater streams (Fig. 2). Two major groups of piscivores, differing in their spatial distribution and size-selectivity of prey, are likely to be important. Terrestrial piscivores (e.g., herons) are most efficient in shallow habitats and, since they are usually not gape limited, they tend to select the largest prey (Power, 1987). In contrast, aquatic piscivores tend to be gape limited and select the smallest prey (Schlosser, 1988) and because aquatic piscivores tend to be large fish, they are restricted to deeper habitats by terrestrial piscivores. The combination of terrestrial and aquatic predator-prey interactions results in a complementary distribution of small fish in shallow habitats and large fish in deep habitats (Power, 1987; Schlosser, 1987a), which is further accentuated by the fact that small fish also select stream edge and backwater habitats to avoid the swiftest water currents (Moore & Gregory, 1988a, 1988b). The combination of terrestrial and aquatic predator-prey interactions also increases interspecific and intraspecific overlap in habitat use among early life history stages of the fish (Schlosser, 1987a), which enhances the likelihood of competitive interactions and subsequent constraints on their growth and survival (Werner & Gilliam, 1984; Elliott, 1989).

The role of spatial habitat relationships

The energy flux created by the downhill movement of water and sediment out of the basin, as modified by obstructions due to the presence of boulders or large organic debris (LOD), is the primary factor controlling spatial heterogeneity in headwater streams (Keller & Swanson, 1979; Richards, 1982). LOD derived from riparian areas is particularly important in upstream reaches because of the reduced stream energy available for transporting the wood once it enters the lotic environment (Keller & Swanson, 1979). The type of habitat created by the LOD depends on the size and placement of wood in the stream (Bisson *et al.*, 1987).

The interaction between the energy flux associated with the downhill movement of water/sediment and obstructions in the channel creates substantial longitudinal and lateral heterogeneity in the stream (Frissell *et al.*, 1986). Longitudinal habitat heterogeneity is primarily associated with differences in depth, substrate, and current velocity related to pool-riffle development (Yang, 1971). Lateral heterogeneity is primarily associated with the proximity of the habitat to the main stream channel and the amount of woody debris contributed by the riparian environment. In upstream areas, these lateral habitats normally exist over small spatial scales (e.g., meters) and are closely associated with the main stream channel. They include stream margins, backwaters, and small isolated pools (Moore & Gregory, 1988a, 1988b).

Numerous studies have revealed associations between the structural characteristics of the stream environment and ontogenetic changes in the static distribution of fish species and size classes. Normally, spawning habitats differ from juvenile feeding habitats, juvenile feeding habitats differ from adult habitats, and habitat use by adults and juveniles varies among species (Bisson *et al.*, 1982; Lobb & Orth, 1991). In addition, several studies have revealed considerable complementarity in the distribution of small and large fish along both longitudinal and lateral dimensions. Small fish and/or earlier life stages tend to be found predominantly in shallow riffle or stream margin habitats, while large fish and/or later life stages are more abundant in deeper pool or midchannel areas (Schlosser, 1987a; Moore & Gregory, 1988a; Lobb & Orth, 1991). Consequently, it is widely recognized by fish biologists that, when historical factors concerning the evolutionary histories of drainage basins are taken into consideration, spatial habitat heterogeneity is a fundamental factor influencing the diversity of fish species and life stages in headwater streams (Gorman & Karr, 1978; Schlosser, 1982).

Few studies by lotic fish ecologists, however, have addressed the more important question of how large-scale spatial habitat relationships influences the dynamic, or rate-dependent, processes regulating population size. Four large-scale spatial habitat relationships are likely to be important: habitat complementation, habitat supplementation, source *vs* sink interactions, and neighborhood effects (Dunning *et al.*, 1992).

Habitat complementation

Habitat complementation refers to the spatial proximity of different nonsubstitutable resources or habitat types required by a particular species (Dunning *et al.*, 1992). For example, if a stream consists of two habitat types and each habitat contains a different 'resource' required by a particular species, the stream segments can differ in the spatial proximity or complementation of the habitats (Fig. 3.1). Furthermore, since these 'resources' are found in different habitats, the organism must travel between the habitats to be successful. For stream fish these 'resources' might be spawning *vs* feeding habitats, feeding *vs* refugia habitats, etc. (Fig. 1).

Portions of the landscape where different habitat types are in close proximity (Fig. 3.1A) will supposedly support higher populations of fish than portions of the landscape where they are farther apart (Fig. 3.1B; Dunning *et al.*, 1992). The mechanism(s) by which habitat complementation increases fish population size is due to its effect on the rate-dependent processes regulating population dynamics. In particular, habitat complementation would potentially: (1) decrease the energetic costs of migration between different habitat types, with an ensuing increase in growth and survival rates and (2) decrease the travel time between habitat patches, with an ensuing decrease in their exposure to predation or other risks of mortality during transition periods.

If the spatial arrangement among habitats is temporally stable, increased habitat complementation can also result in increased stability or persistence of populations (Dunning *et al.*, 1992). Complementation will increase persistence of populations by increasing the probability fish will find an appropriate habitat during critical transition or migratory periods (e.g., from summer feeding habitats to overwintering habitats), which enhances their survival and persistence in that region of the landscape. If, however, the spatial arrangement of habitats is temporally unstable, then the stability of the fish population becomes a more complex function of both the temporal and spatial persistence of habitats and the ability of fish to find those habitats.

Recent studies in large rivers and headwater streams suggest habitat complementation is likely to be a critical factor determining population size and persistence of fish. For example, based on detailed studies of habitat use and recruitment by fish in the Danube River, Schiemer *et al.* (1991) concluded that a fundamental factor influencing juvenile recruitment and population size was not just the presence of both spawning and larval feeding habitats but also the spatial proximity of the two areas. Similarly, in headwater streams in the central United States, Schlosser (1987b) concluded that the spatial proximity of deep pool refugia to shallow juvenile rearing habitats was a fundamental factor determining the temporal persistence of juvenile fish, particularly during harsh winter periods.

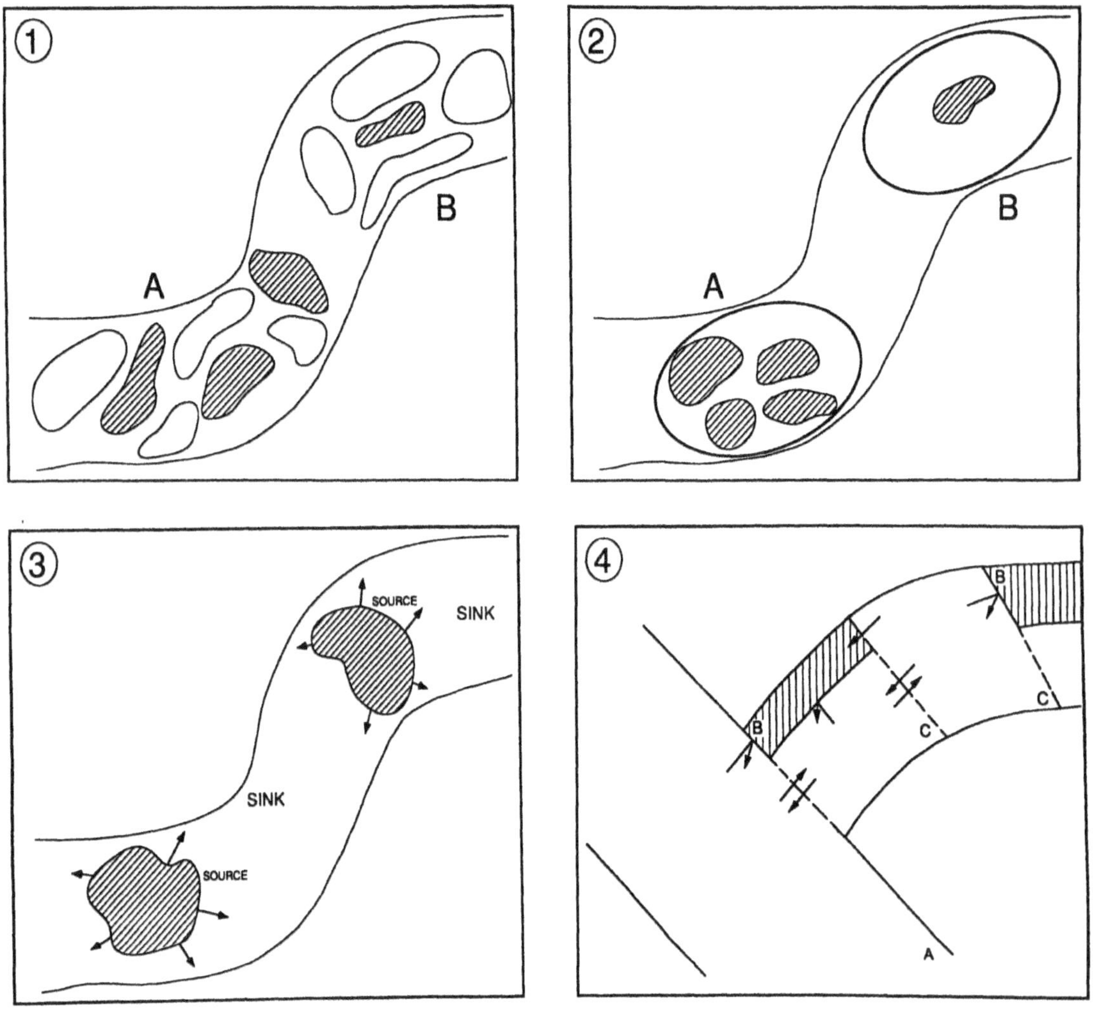

Fig. 3. Four large-scale spatial habitat relationships potentially influencing the population dynamics of stream fish: (1) habitat complementation: a species requires non-substitutable resources found in two different habitats (cross-hatched and open); (2) habitat supplementation: a species requires substitutable resources found in the cross-hatched habitat patches and all patches within the ovals are accessible; (3) source *vs* sink relationships: a species occupies two habitat types, with one (sources: shaded area) producing excess individuals or dispersers (arrows) and the other (sink: open areas) not producing enough individuals to maintain a local population; (4) neighborhood effects: dispersers (arrows) are more likely to move into adjacent than distant habitat patches and habitat patches with permeable (dotted lines) versus impermeable (solid lines) boundaries.
[Adapted from Dunning et al. (1992.)]

Habitat supplementation

In contrast to habitat complementation, habitat supplementation occurs when 'resources' are substitutable and these 'resources' can be either food or physical habitat. Since fish in headwater streams frequently exhibit flexible foraging behavior (Angermeier, 1982, 1987), they can expand or supplement their 'resource' supply by using different resources in the same habitat or the same resource in different habitats, particularly when local densities are high. Population size will supposedly be higher in regions of the landscape where supplementation occurs (Fig. 3.2A) than where it does not occur (Fig. 3.2B) because it: (1) allows individuals to increase their food intake and subsequently their growth, survival, and fecundity rates and/or

(2) increases the range of feeding habitats available for use by the organism that are within range of suitable refugia from predation (Dunning *et al.*, 1992). The end result is that supplementation will allow a stock of fish to increase and dampen density-dependent mortality rates following high levels of recruitment.

I know of no studies explicitly testing the effect of habitat supplementation on recruitments dynamics or the population size of stream fish. Recent studies, however, have demonstrated that habitat supplementation does affect critical rate-dependent processes influencing fish population dynamics. For example, Power and her colleagues (Power, 1984ab; Power *et al.*, 1989) assessed the response of armored catfish (Loricariidae) to spatial heterogeneity in food resources and predation intensity. They documented that: (1) periphyton productivity in the stream was patchily distributed, with the highest rates in shallow, open sunny reaches and the lowest rates in deep, densely shaded areas, (2) individual loricariids continually redistributed themselves in response to changes in food availability in pool habitats, to the extent that fish in sunny crowded pools had food intake and growth rates similar to fish in dark, sparsely populated pools, and (3) although periphyton resources were most abundant in shallow habitats, they were largely unexploited by loricariids because fish in these areas experienced higher mortality rates from terrestrial predators. These results strongly suggest individual fish in streams supplement resource intake by using a number of different habitat patches, that the selection of these habitat patches is finely balanced by its influence on the food intake and growth rate of individual fish, and that predators place major constraints on the ability of fish to engage in habitat shifts for the purpose of resource supplementation.

Source vs *sink relationships*

In contrast to habitat complementation and habitat supplementation, which deals with the spatial distribution of nonsubstitutable or substitutable resources for individual organisms, source *vs* sink relationships deal with spatial variation in the production of juveniles (Pulliam, 1988). Sources are defined as those areas that for a variety of reasons (e.g., high organic matter supply and low piscivory) exhibit high juvenile recruitment. These source areas in turn provide immigrants to less productive habitats on the landscape, which are termed sinks (Fig. 3.3). Since local recruitment in sinks is not sufficient to maintain the population, immigration from source habitats is the primary rate-dependent process determining population dynamics in sinks (Fig. 3.3). Similarly, since individuals in sinks do not exhibit strong recruitment, the relative amount of source/sink habitat on the landscape can have strong effects on overall recruitment in the population (Pulliam, 1988).

Do source *vs* sink relationships play a critical role in determining the population dynamics of stream fish? I know of no study in the current literature that directly addresses this question. This is partly due to the difficulty in collecting the appropriate data to document such a relationship, which would include measurements of reproductive success, dispersal, and persistence of dispersers in adjacent habitats (Dunning *et al.*, 1992). However, data recently collected in beaver (*Castor canadensis*) ponds and associated stream reaches suggest source *vs* sink relationships play a critical role in determining the population dynamics of fish in these areas (I. J. Schlosser, unpublished data). Specifically, new beaver ponds are organically rich relative to lotic environments (Naiman *et al.*, 1988), juvenile fish recruitment tends to be higher in ponds than associated lotic habitats, and pond species disperse along streams in search of new pond habitats but usually do not reproduce successfully in intervening lotic reaches. Consequently, immigration from pond (source) habitats is a primary factor maintaining many of the pond species in associated lotic (sink) habitats.

Neighborhood effects

Since movement between different habitats is a fundamental part of the life cycle of stream fish (Fig. 1), the abundance of a species in a particular patch of habitat can be strongly influenced by the type of adjoining habitats and the nature of the boundaries between them (Dunning *et al.*, 1992). For example, since species are more likely to disperse by chance into adjoining rather than distant habitat patches, the species composition and temporal dynamics of fish would probably be very different in the two, type C habitats in Fig. 3.4 because of differences in the nature of their adjacent habitats. Similarly, the ability of fish to disperse into a neighboring habitat depends on the size and permeability of the boundary between them (Fig. 3.4).

Is there any evidence that spatial proximity of different habitat patches influences immigration processes and, consequently, the population and community dynamics of stream fish? Osborne & Wiley (1992) examined the influence of stream channel position within the drainage network on species richness in the midwestern United States. They observed that the proximity of the tributary sampling location to the downstream main channel was more important than local habitat structure in determining species richness in tributary streams. Osborne & Wiley (1992) concluded that dispersal-mediated interchanges between the main channel and tributary were more critical than local biological interactions, such as habitat mediated competition, in determining species richness in these areas. Gorman (1986) reached a similar conclusion regarding the importance of the spatial proximity of large rivers to tributary sites in determining fish species richness in tributaries.

The permeability of habitat boundaries also appears to be important in some natural streams. In many mountain streams, for example, natural geological barriers such as waterfalls strongly influence species colonization (Gilliam *et al.*, 1993). Species composition across these barriers depends on the height of the barrier, stream discharge, and the colonizing ability of different species.

The role of temporal environmental variability and refugia

Because headwater streams are at the periphery of the drainage network, fluctuations in temperature and discharge are potentially extreme (Hynes, 1970). Consequently, the presence of refugia from harsh environmental conditions is likely to be a critical factor regulating the population and community dynamics of fish in these areas (Sedell *et al.*, 1990).

Stream temperatures associated with either extreme summer or winter conditions have been documented as having strong effects on fish survival and habitat use. The importance of high summer temperatures and cool water refugia have been widely documented, for example, among salmonids in cool or coldwater streams (Beschta *et al.*, 1987; Bilby, 1984). These cool temperature refugia are usually associated with subsurface springs entering the stream (Bilby, 1984). Similarly, cold winter temperatures (0–4 °C) in conjunction with elevated stream discharge are especially stressful to small fish because of their reduced metabolic reserves and weak swimming capabilities (Cunjak, 1988). As a result, winter conditions normally result in significant mortality of age 0 fish (Schlosser, 1987b; Cunjak, 1988) and large shifts in habitat use from summer feeding to overwintering refugia. Overwintering refugia frequently exhibit increased habitat volume, reduced flow velocity, and increased structural complexity (Moore & Gregory, 1988a, 1988b, Power & Power, in press) all of which reduce the rate of change in water temperature and decrease the exposure of fish to high water velocities.

Fluctuations in stream discharge also have direct and strong effects on spatial habitat heterogeneity and large-scale patterns of habitat use. For example, extreme high or low discharges result in decreased length of stream edge and decreased perimeter of lateral (flow $<4\ \mathrm{cm\ s^{-1}}$) habitats (Fig. 4), which are likely to be critical juvenile rearing habitats (Moore & Gregory, 1988a, 1988b). In addition, very low discharges have strong effects on large-scale patterns of habitat use. Fish normally respond to low dis-

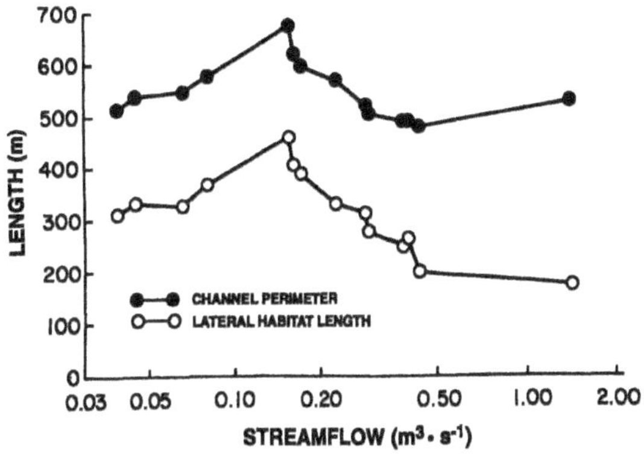

Fig. 4. Effect of streamflow on the total length of the channel perimeter and the perimeter of lateral habitats (stream margins, backwaters, and isolated pools) in a 100-m reference reach of Mack Creek, Oregon, USA.
[Adapted from Moore & Gregory (1988b).]

charge by shifting their habitat use to deeper pool refugia, either in nearby reaches or downstream areas (Ross *et al.*, 1985).

Conclusions

Fish in headwater streams exhibit complex life cycles that potentially extend over long temporal and large spatial scales. Conceptual models describing the influence of landscape attributes on fish population dynamics will therefore need to integrate the effect of multiple landscape attributes on different life-history stages of these fish in a scale-dependent manner (sensu Frissell *et al.*., 1986).

Three landscape attributes are likely to be particularly critical: functional interactions at terrestrial-aquatic ecotones, large-scale spatial habitat relationships, and the distribution of refugia from harsh environmental conditions. Functional interactions at terrestrial-aquatic ecotones can have strong effects on the temporal distribution of food resources, risk of predation, and patterns of habitat use. Spatial habitat relationships are likely to have strong effects on fish population dynamics by influencing the ability of fish to find appropriate habitat types during critical transition periods, the energetic costs and benefits of foraging in these habitats, and migration rates among adjacent habitat units. Finally, the spatial distribution of refugia from harsh environmental conditions is likely to be a fundamental factor controlling fish population dynamics by either directly influencing fish survival or by controlling immigration/emigration rates between feeding and refugia habitats.

Unfortunately, our current understanding of how these three landscape attributes interact to influence fish population and community dynamics in streams is qualitative at best. A more quantitative understanding of how these factors interact will require more refined descriptive, experimental, and modelling approaches. Since landscape disturbances alter all three of these attributes (Schiemer & Spindler, 1989; Hicks *et al.*, 1991; Schlosser, 1991), development of a more rigorous conceptual framework describing how these factors interact will be important from both a theoretical and applied perspective.

Acknowledgements

I am indebted to our Austrian hosts, especially F. Schiemer, G. Bretschko, and M. Leichtfried, for their gracious hospitality during the mid-term meeting of the Fish and Land/Inland Water Ecotone Group in Lunz and Vienna. Support for attendance at the meeting was partially provided by UNESCO and by the Department of Biology at the University of North Dakota. Manuscript preparation was supported by National Science Foundation grants to IJS (BSR 8804926) and the state of North Dakota (STIAA-RII EPSCoR). Two anonymous reviewers provided numerous useful comments on an earlier version of the manuscript.

References

Andrewartha, H. G. & L. C. Birch., 1954. The Distribution and Abundance of Animals. University of Chicago Press, Chicago.

Angermeier, P. L., 1982. Resource seasonality and fish diets in an Illinois stream. Envir. Biol. Fishes 7: 251–264.

Angermeier, P. L., 1987. Spatiotemporal variation in habitat selection by fishes in small Illinois streams. In W. J. Matthews & D. C. Heins (eds), The Ecology and Evolution of North American Stream Fishes. University of Oklahoma Press, Norman (O.K.): 52–60.

Beschta, R. L., R. E. Bilby, G. W. Brown, L. B. Holtby & T. D. Hoffstra, 1987. Stream temperature and aquatic habitat: Fisheries and forestry interactions. In E. O. Salo & T. W. Cundy (eds), Streamside Management: Forestry and Fishery Interactions. University of Washington, Institute of Forest Resources Contribution No. 57; Seattle (W.A.): 191–232.

Bibly, R. E., 1984. Characteristics and frequency of cool-water areas in a western Washington stream. J. Freshwat. Ecol. 2: 593–602.

Bisson, P. A, J. A. Nielsen, R. A. Palmason & E. L. Grove, 1982. A system of naming habitat types in small streams, with examples of habitat utilization by salmonids during low. In N. B. Armantrout (ed.), Acquisition and Utilization of Aquatic Habitat Inventory Information. Western Division of the American Fisheries Society, Portland (O.R.): 62–73.

Bisson, P. A., R. E. Bilby, M. D. Bryant, C. A. Dolloff, G. B. Grette, R. A. House, M. L. Murphy, K. V. Koski & J. R. Sedell, 1987. Large woody debris in forested streams of the Pacific northwest: Past, present, and future. In E. O. Salo & T. W. Cundy (eds), Streamside Management: Forestry and Fishery Interactions. University of Washington, Institute of Forest Resources Contribution No. 57, Seattle (W.A.): 143–190.

Correll, D. L., 1991. Human impact on the functioning of landscape boundaries. In M. M. Holland, P. G. Risser & R. J. Naiman (eds), Ecotones: The Role of Landscape Boundaries in the Management and Restoration of Changing Environments. Chapman and Hall, New York: 90–109.

Cunjak, R. A., 1988. Physiological consequences of over wintering in streams: the cost of acclimatization. Can. J. Fish. aquat. Sci. 45: 443–452.

Dunning, J. B., B. J. Danielson & H. R. Pulliam, 1992. Ecological processes that affect populations in complex landscapes. Oikos 65: 169–175.

Elliott, J. M., 1989. The natural regulation of numbers and growth in contrasting populations of brown trout, *Salmo trutta*, in two lake district streams. Freshwat. Biol. 21: 7–19.

Forman, R. T. & M. Godron, 1986. Landscape Ecology. John Wiley & Sons, New York, 619 pp.

Frissell, C. A., W. J. Liss, C. E. Warren & M. D. Hurley, 1986. A hierarchical framework for stream habitat classification: Viewing streams in a watershed context. Envir. Mgmt 10: 199–214.

Gilliam, J. F., D. F. Fraser & M. Alkins-Koo, 1993. Structure of a tropical fish community: A role for biotic interactions. Ecology. In press.

Gorman, O. T., 1986: Assemblage organization of stream fishes: the effect of rivers on adventitious streams. Am. Nat. 128: 611–616.

Gorman, O. T. & J. R. Karr, 1978. Habitat structure and stream fish communities. Ecology 59: 507–515.

Gregory, S. V., F. J. Swanson, W. A. McKee & K. W. Cummins, 1991. An ecosystem perspective of riparian zones. BioScience 41: 540–551.

Hicks, B. J., J. D. Hall, P. A. Bisson & J. R. Sedell, 1991. In W. R. Meehan (ed.), Influences of Forest and Rangeland Management on Salmonid Fishes and Their Habitats. American Fisheries Society Special Publication 19, Bethesda (M.D.): 483–518.

Hynes, H. B. N., 1970. Ecology of Running Waters. University of Toronto Press, Toronto, 555 pp.

Jones, F. R., 1968. Fish Migration. Edward Arnold, London, 325 pp.

Karr, J. R. & I. J. Schlosser, 1978. Water resources and the land-water interface. Science 201: 229–234.

Keller, E. A. & F. J. Swanson, 1979. Effects of large organic material on channel form and fluvial processes. Earth Surface Processes 4: 361–380.

Lobb, M. D. & D. J. Orth, 1991. Habitat use by an assemblage of fish in a large warmwater stream. Trans. am. Fish. Soc. 120: 65–78.

Lowe-McConnell, R. H., 1987. Ecological Studies in Tropical Fish Communities. Cambridge University Press, London, 382 pp.

Mason, J. C., 1976. Response of underyearling coho salmon to supplemental feeding in a natural stream. Wildl. Manag. 40: 775–788.

Matthews, W. J. & D. C. Heins (eds), 1987. Community and Evolutionary Ecology of North American Stream Fishes. University of Oklahoma Press, Norman, 310 pp.

McIntire, C. D. & J. A. Colby, 1978. A hierarchical model of lotic ecosystems. Ecol. Monogr. 31: 61–104.

Minshall, G. W., 1978. Autotrophy in stream ecosystems. BioScience 28: 767–771.

Moore, K. M. S. & S. V. Gregory, 1988a. Summer habitat utilization and ecology of cutthroat trout fry (*Salmo clarki*) in Cascade Mountain streams. Can. J. Fish. aquat. Sci. 45: 1921–1930.

Moore, K. M. S. & S. V. Gregory, 1988b. Response of young-of-the-year cutthroat trout to manipulation of habitat structure in small stream. Trans. am. Fish. Soc. 117: 162–170.

Naiman, R. J., H. DeCamps, J. Pastor & C. A. Johnson, 1988. The potential importance of boundaries to fluvial ecosystems. J. North Am. Benthol. Soc. 7: 289–306.

Naveh, Z. & A. S. Lieberman, 1984. Landscape Ecology. Springer-Verlag, New York, 356 pp.

Northcote, T. G., 1978. Migratory strategies and production in freshwater fishes. In S. D. Gerking (ed.), Ecology of Freshwater Fish Production. Blackwell Scientific Publications, Oxford: 326–359.

Osborne, L. L. & M. J. Wiley, 1992. Influence of tributary spatial position on the structure of warmwater fish communities. Can. J. Fish. aquat. Sci. 49: 671–681.

Peterjohn, W. T. & D. L. Correll, 1984. Nutrient dynamics in an agricultural watershed: Observations on the role of a riparian forest. Ecology 65: 1466–1475.

Pickett, S. T. A. & P. S. White (eds), 1985. The Ecology of Natural Disturbance and Patch Dynamics. Academic Press, New York.

Pinay, G. & H. Decamps, 1988. The role of riparian woods in regulating nitrogen fluxes between the alluvial aquifer and surface water: a conceptual model. Regul. Rivers 2: 507–516.

Power, G. & M. Power, 1995. Ecotones and fluvial regimes in arctic lotic environments. Hydrobiologia 303 (Dev. Hydrobiol. 105): 111–124.

Power, M. E., 1984a: Habitat quality and the distribution of algae-grazing catfish in a Panamanian stream. J. anim. Ecol. 53: 357–374.

Power, M. E., 1984b: Depth distributions of armored catfish: predator-induced resource avoidance? Ecology 65: 523–528.

Power, M. E., 1987. Predator avoidance by grazing fishes in temperate and tropical streams: Importance of stream depth and prey size. In W. C. Kerfoot & A. Sih (eds), Predation: Direct and Indirect Impacts on Aquatic Communities. University Press of New England, Hanover, (N.H.): 333–351.

Power, M. E., T. L. Dudley & S. D. Cooper, 1989. Grazing catfish, fishing birds, and attached algae in a Panamanian stream. Envir. Biol. Fishes. 26: 285–294.

Pringle, C. M., R. J. Naiman, G. Bretschko, J. R. Karr, M. W. Oswood, J. R. Webster, P. B. Moyle, B. Statzner & I. R. Wais de Bagden, 1988. Patch dynamics in lotic ecosystems: the stream as a mosaic. J. N. Amer. Benth. Soc. 7: 503–524.

Pulliam, H. R., 1988. Sources, sinks, and population regulation. Am. Nat. 132: 652–661.

Richards, K., 1982. Rivers: Form and Process in Alluvial Channels. Methuen, New York.

Risser, P. G., J. R. Karr & R. T. T. Forman, 1984. Landscape ecology: directions and approaches. Illinois Natural History Special Publication No. 2. Champaign (I.L.).

Ross, S. T., W. J. Matthews & A. A. Echelle, 1985. Persistence of stream fish assemblages: effects of environmental change. Am. Nat. 126: 24–40.

Schiemer, F. & T. Spindler, 1989. Endangered fish species of the Danube River in Austria. Regul. Rivers 4: 397–407.

Schiemer, F. & M. Zalewski, 1992. The importance of riparian ecotones for diversity and productivity of riverine fish communities. Neth. J. Zool. 42: 323–335.

Schiemer, F., T. Spindler, H. Wintersberger, A. Schneider & A. Chovanec, 1991. Fish fry associations: important indicators for the ecological status of large rivers. Verh. int. Ver. Limnol. 24: 2497–2500.

Schlosser, I. J., 1982. Fish community structure and function along two habitat gradients in a headwater stream. Ecol. Monogr. 52: 395–414.

Schlosser, I. J., 1987a. The role of predation in age and size related habitat use by stream fishes. Ecology 68: 651–659.

Schlosser, I. J., 1987b. A conceptual framework for fish communities in small warmwater streams. In W. J. Matthews & D. C. Heins (eds), The Ecology and Evolution of North American Stream Fishes. University of Oklahoma Press, Norman (O.K.): 17–24

Schlosser, I. J., 1988. Predation risk and habitat use by two size classes of a stream cyprinid: Experimental test of a hypothesis. Oikos 52: 36–40.

Schlosser, I. J., 1991. Stream fish ecology: a landscape perspective. BioScience 41: 704–712.

Schlosser, I. J. & L. A. Toth, 1984. Niche relationships and population ecology of rainbow (*Etheostoma caeruleum*) and fantail (*E. flabellare*) darters in a temporally variable environment. Oikos 42: 229–238.

Schlosser, I. J. & P. L. Angermeier, 1990. The influence of environmental variability, resource abundance and predation on juvenile cyprinid and centrachid fishes. Pol. Arch. Hydrobiol. 37: 267–286.

Sedell, J. R., G. R. Reeves, F. R. Hauer, J. A. Stanford & C. P. Hawkins, 1990. Role of refugia in recovery from disturbances: modern fragmented and disconnected river systems. Envir. Mgmt 14: 711–724.

Swanston, D. N., 1991. Natural processes. In W. R. Meehan (ed.), Influences of Forest and Rangeland Management on Salmonid Fishes and Their Habitats. American Fisheries Society Special Publication 19, Bethesda (M.D.): 139–179.

Vannote, R. L., G. W. Minshall, K. W. Cummins, J. R. Sedell & C. E. Cushing, 1980. The river continuum concept. Can. J. Fish. aquat. Sci. 37: 130–137.

Werner, E. E. & J. F. Gilliam, 1984. The ontogenetic niche and species interactions in size structured populations. Annu. Rev. Ecol. Syst. 15: 395–425.

Yang, C. T., 1971. Formation of riffles and pools. Wat. Resour. Res. 7: 1567–1574.

Hydrobiologia **303**: 83–91, 1995.
F. Schiemer, M. Zalewski & J. E. Thorpe (eds), The Importance of Aquatic-Terrestrial Ecotones for Freshwater Fish.

River/land ecotones: scales and patterns

Dedicated to Professor Franz Berger on the occasion of his 90th birthday

G. Bretschko
Institute for Limnology, Austrian Academy of Sciences, Biological Station Lunz, A-3293 Lunz/See, Austria

Key words: river/land ecotones, River Continuum Concept (RCC), Patch/RCC

Abstract

On a continental scale a river system itself is an ecotone, mediating between the terrestrial system and the sea. On the landscape scale, ecotones appear between terrestrial and river systems. At decreasing scales, the number and diversity of ecotones increases. Ecotone processes are discussed in relation to the entire 'river' ecosystem, on scales ranging from the activity range of fish fry up to that of adult fish.

The introduction of patch theory into the River Continuum Concept (RCC) allows for the proper consideration of hydraulics and time. The relationships between stream order and patch distribution, patch size and lifetime and age diversity of patches are described. This 'combination model' is much better adjusted to the situations occurring in nature than the original, very abstract RCC-model.

Introduction

'An ecotone is a zone of transition between adjacent ecological systems, having a set of characteristics uniquely defined by space and time scales and by the strength of interactions between adjacent ecological systems (Holland, 1988).' Although this definition is widely used and agreed on (Decamps & Naiman, 1990), further differentiation is necessary between ecotones and simple boundaries, as well as from ecosystems, in order to justify retention of the concept (Jeffers *et al.*, 1989):

- The interactions between ecosystems result in mechanisms that do not exist within the component ecosystems but in the ecotone alone.
- The fundamental principles of energy, material and species dynamics are valid for both ecotones and ecosystems, but ecotones are characterized by discontinuities and by the variability of flows, in amounts as well as directions.
- Ecosystem dynamics are stabilized by feedback mechanisms; ecotones are basically unstable, depending on the interactions between the component ecosystems (Table 1; Jeffers *et al.*, 1989).

According to these ecotone definitions, the total stream system can be seen as an ecotone on a large, continental scale, mediating between land

Table 1. Basic interactions differentiating between ecotones and ecosystems (from Jeffers *et al.*, 1989).

	Within system interactions	Between system interactions
Ecotones	Variable	Strong and direct
Ecosystems	Strong	Weak

and sea. Consequently, anadromous and catadromous fish would spend the freshwater part of their lives entirely in an ecotone. This point of view is valid only on this very large and therefore rough scale. On a smaller, more detailed, landscape scale, streams acquire the properties of ecosystems, as might be expected: considering the size and complexity of streams, the observed system stability is based on positive feedback mechanisms, typical ecosystem properties (Table 1). River/land ecotones form whenever the terrestrial and the lotic systems overlap, in space or time or in a combination of both (Fig. 1).

The stream ecosystem

Most studies on the level of a river system aim at a characterisation of stretches of streams as the basis for the river stream zonations (for compilations see Illies & Botosaneanu, 1963). But previously Hynes (1970) had pointed out, that 'the general theme of the change in biotic associations from source to mouth in a river will tend to be one not of zonation but of transition.' The 'River Continuum Concept' (RCC, Vannote *et al.*, 1980) considers the transition properties of streams only slightly by the downstream dependence on particulate organic matter (POM) processed upstream. The main, driving variable of this concept is the quality and origin of organic matter as the

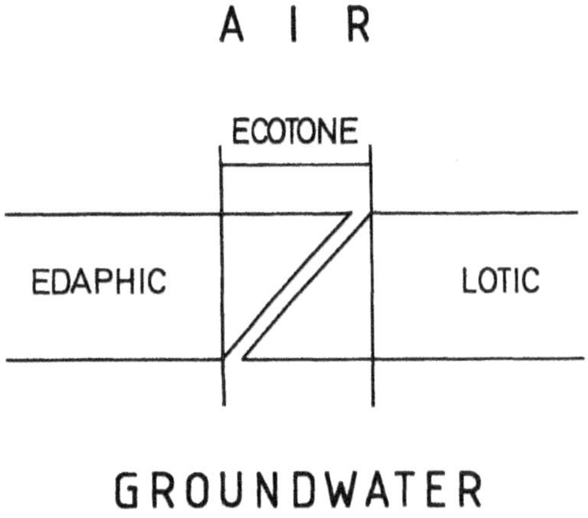

Fig. 1. River/land ecotone.

energy source for the consumer community. On this basis, three system constructs are developed and longitudinally arranged (Cummins, 1979; Fig. 2, A). The main characteristics are in short:

- low order streams: Autochthonous primary production is small because of closed canopies and substrate instabilities. The main energy source is allochthonous organic matter.
  P/R = < 1.
- medium order streams: Autochthonous primary production is high because of open canopies and shallow water. The main energy sources are macrophytes and periphyton.
  P/R = > 1.
- high order streams: Autochthonous primary production is small because of deep and turbid water. The main energy source is FPOM derived from upstream regions.
  P/R = < 1.

Although the RCC is widely accepted because of its explanatory power, it is also open to obvious criticisms, such as:

- The longitudinal zonation, describing three typical reaches with increasing interdependence, is often critized because nature rarely matches the concept (e.g. Winterbourne *et al.*, 1981).
- For stream ecology the importance of stream hydraulics and connected parameters is beyond discussion but not covered adequately by the RCC (e.g. Statzner & Higler, 1985, 1986).

Patch dynamics and the 'River Continuum Concept'

The RCC-critique, summarized above, can be met by supplementing the basic idea of the RCC with the patch concept (Kotliar & Wiens, 1990; Pickett & White, 1985; Pringle *et al.*, 1988). The incorporation of the patch concept changes the longitudinal array of the three basic system constructs (Fig. 2, A) into a patchy distribution of all three constructs over the entire course of the stream (Fig. 2, B). This arrangement results in a much more realistic situation: for instance, in headwater streams moss patches with dense epiphytic

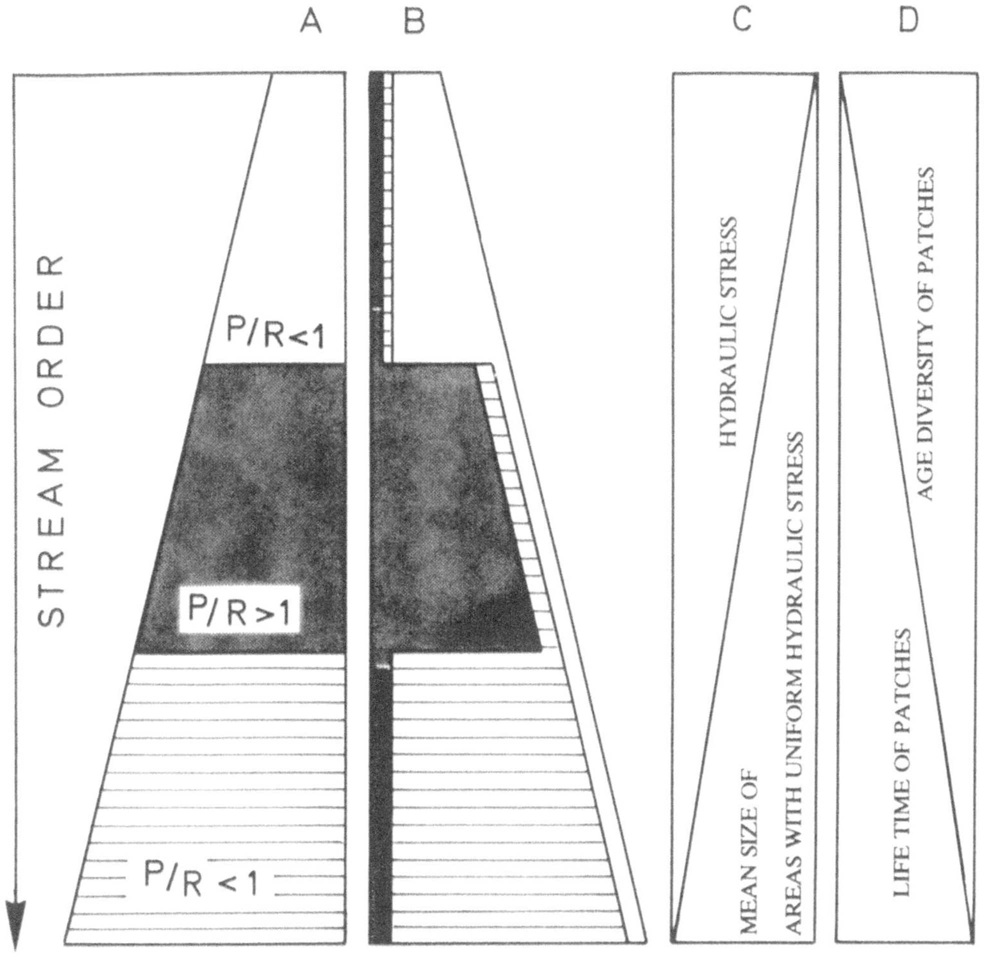

Fig. 2. A: The classical RCC configuration. B: Modification of the classical RCC to include habitat heterogeneity. C: Hydraulic regulation and size of physical habitats. D: Temporal responses to hydraulics and streamsize.

growth and mud patches do occur in conjunction with the typical biocoenoses described by the RCC for low order streams. The character of a specific reach is then given by the dominance of one of the three system constructs (Fig. 2, B).

Patch dynamic processes involve hydrology as well as stream hydraulics. Generally, stream size and discharge increase whereas the slope decreases, with increasing stream order. Consequently, the hydraulic stress (as a summarizing heading, after Statzner & Highler, 1985) acting on the biocoenosis decreases with increasing stream order. The combination of decreasing hydraulic stress and increasing stream size results in an increase in the mean size of areas with uniform hydraulic stress (Fig. 2, C). The mosaic-like distribution of different patches in headwater streams and the large and dominating areas with uniform biocoenoses in lowland streams are well known (e.g. Illies & Botosaneanu, 1963). It always takes larger floods to thoroughly disturb or destroy larger patches. Since the frequency of occurrence of floods decreases with their increasing sizes, the lifetime of patches increases with increasing stream order (Fig. 2, D).

Even more important is the age diversity of

patches of the same kind. Because the hydraulic stress is never uniform in its distribution over the crossectional area, a flood of a certain size may destroy some of the patches but not all of them. This is most accentuated in headwater streams with the mosaic-like distribution of many small patches. Therefore, age diversity of patches of the same kind decreases with increasing stream order (Fig. 2, D). But a greater age diversity of patches accommodates many more species of similar ecological demands in a limited space, based on the idea of disturbance mediated coexistence of species (Caswell, 1978; Townsend, 1989).

River/land ecotones

When the terrestrial and lotic system overlap in space, in time or in a combination of both, a river/land ecotone forms (Fig. 1). Overlaps do occur on both sides of the bankfull line (Leopold *et al.*., 1964): Instream ecotones are mostly the result of spatio/temporal overlaps, because vertically streams are segmented into three layers. These three interdependent, functional units are (Bretschko, 1992; Bretschko & Moser, 1993):

- The surface water.
- The wet sediment surface down to the depth of the light discontinuity. This layer is characterized by the presence of green, assimilating algae.
- The 'bedsediments' characterized by the quantitative dominance of epigeic faunal elements (Bretschko & Klemens, 1986; Bretschko & Leichtfried, 1988). The upper limit of this layer is the light discontinuity, the lower limit is variable and depends on the throughflow pattern (Bretschko, 1992). In the case of a discontinuity near the sediment surface, bedsediments may be missing at all. Horizontally, bedsediments do not extend far beyond the bank full lines, but a few exceptional cases are known (Stanford & Gaufin, 1974).

Whenever bedsediments are present, periodically flooded channel areas develop ecotones during

Fig. 3. The RITRODAT study area in May, 1978, after levelling was completed in 1977.

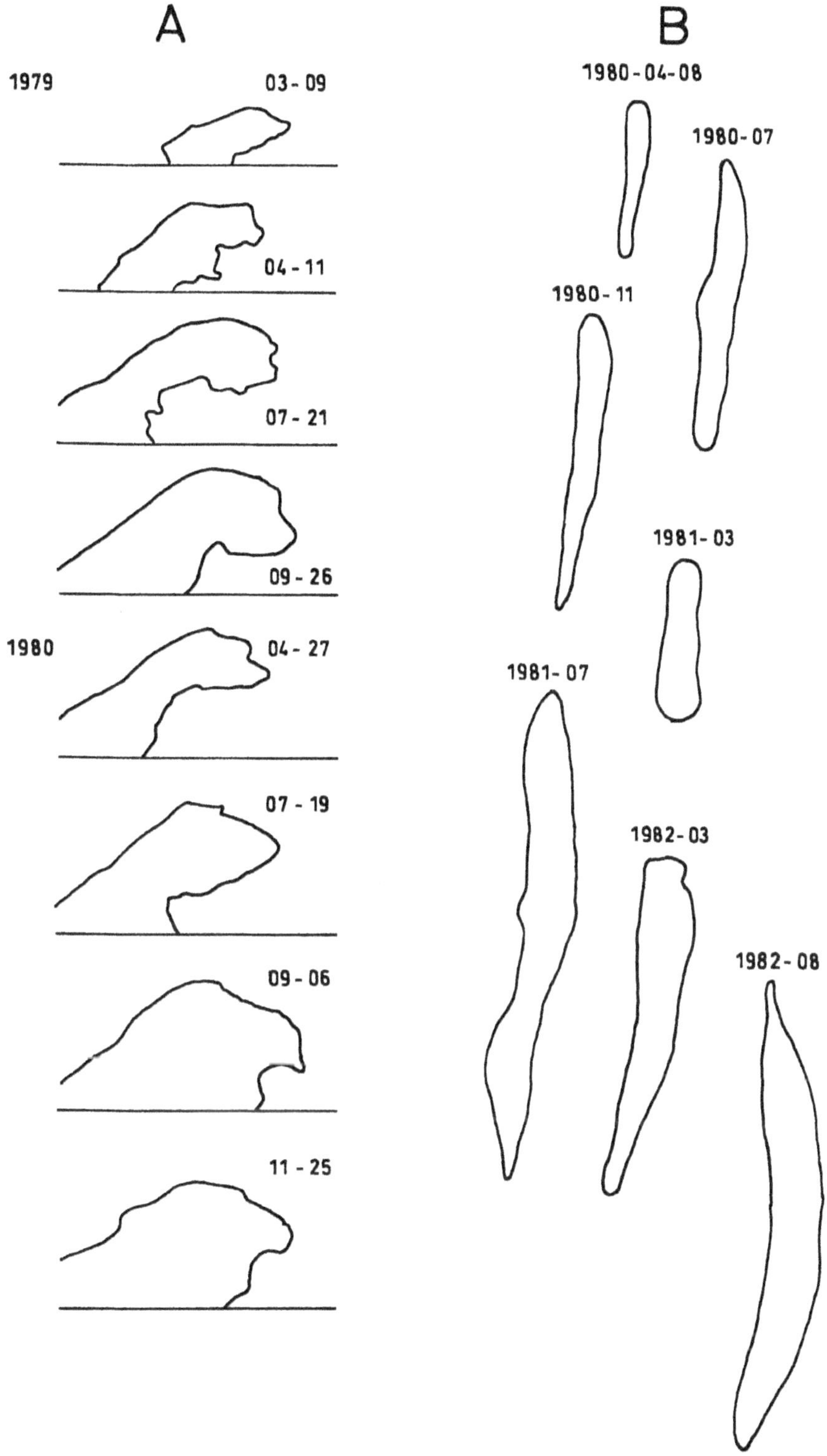

Fig. 4. Seebach Lunz, RITRODAT study area. A: The beginning of the gravel bar seen at mean discharge downstream from the bridge in Fig. 3. B: Areas higher than 25 cm in relation to the rod gauge 'RITRODAT-BRIDGE'. Interpolation based on level measurements along a 5 × 10 m grid.

their dry phases, because the lotic bedsediments persist below the edaphic developments on and in the dry sediment surface above. A very typical example of these processes is the evolution of a gravel bar: during formation, the gravel bar increases in size and height, resulting in ever larger areas above the low water mark for increasingly longer periods of time. Edaphic developments start. As soon as larger and stronger plants, such as willows, grow, the edaphic development itself becomes enhanced, because of the increasing retention capacity. Longer and longer periods of inundation can be tolerated. Without major disturbances gravel bars will grow above the bankfull line, becoming islands. But generally, floods are frequently strong enough to destroy at least part of the edaphic achievements. Therefore, the evolution of a bar towards an island is frequently set back to a former stage of development. If a flood occurs which is strong enough to destroy the entire bar, the evolution is set back to zero. An appropriate example is the development of a gravel bar in the second order limestone gravel stream 'Oberer Seebach', Lunz, RITRODAT-study-area (Bretschko, 1981, 1990). In 1977 the entire study area was levelled with a caterpillar. The material was used for reinforcing the right banks (Fig. 3). Already in March, 1979, a gravel bar started to form in the central part of the stream channel. It increased in size and moved downstream and towards the left bank. Until 1983 the structure was changing its shape and position rapidly (Fig. 4). During this period, willows and other terrestric plants are washed away shortly after they have started to grow. In winter, 1983, the structure is large but still devoid of woody plants (Fig. 5). In spite of this barrenness the gravel bar contributes already to the metabolism of the stream (Bretschko, 1990). Very few and relatively modest floods allowed a lush growth of plants over the period 1982 to 1986 (Bretschko, 1985). In summer, 1984, willows grew strong enough to withstand flooding. In 1986 the bushes are already wellseized and nearly the entire gravel bar is covered by dense willow stands five years later (Fig. 6). These edaphic developments in-

Fig. 5. The gravel bar at mean discharge. 1983-01-10.

crease the retention capacity tremendously, stabilize the center part of the gravel bar, are a mid-stream source of organic matter and provide feeding- and shelter-places for the stream fauna.

Fig. 6. The growth of willows on the gravel bank. Top: September, 1986. Bottom: July, 1991.

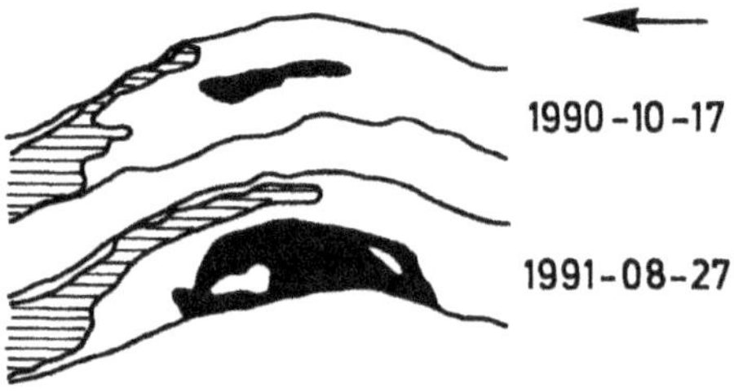

Fig. 7. RITRODAT study area. Reference: rod gauge RITRODAT-BRIDGE. Shore line at bankfull discharge (70 cm). Hatched areas: deeper than −20 cm. Black areas: higher than +25 cm.

Besides of minor but permanent changes in size and shape, the gravel bar developed towards a structure projecting above the bankfull line. A great flood (flood frequency: 30–50 years) stopped this smooth development in July/August 1991: The bar increased greatly in size, vertically as well as horizontally. Furthermore, it became connected with the left bank (Fig. 7). But some of the willows survived inspite of this rather dramatic impact. The preflood willow density was regained in about one year. Ecotone processes in bank and debris dam areas are quite comparable, despite site specific pecularities.

The existence of instream river/land ecotones is of great importance for fish:

- The combination of relatively high primary production and a high retention capacity makes instream ecotones important energy sources for the stream community. Fish participate in the increased production across the general foodweb.
- Specific feeding and shelter places are provided in close proximity for various fish species of various ages around and, when inundated, over the whole ecotone area (Schlosser, 1995).
- In harsh conditions during large floods, plant stands of instream ecotones are preferred shelter places, even for large fish.

Floodplains and the belt of riparian vegetation are river/land ecotones beyond the bankfull line. Their direct contact with the stream ecosystem is restricted to flood events and to side arms and small tributaries at lower discharges. Apart from this restricted accessibility the functional connections between ecotone and stream are quite similar to those described for the instream ecotones. In contrast to instream ecotones, floodplain-specific backwaters and estuaries are essential spawning sites for various fish species as well as feeding and shelter places for many kinds of fish fry (Schiemer, 1985; Schiemer & Zalewski, 1992).

With increasing stream order, instream ecotones become more and more restricted to the immediate bank areas. At the same time, ecotones beyond the bankfull line increase in size and complexity. Floodplains begin to replace the decreasing importance of instream ecotones, in spite of their limited accessibility. The increasing stabilization of the general hydrological cycle enables fish to synchronize their biological demands and to use the entire floodplain. Exotic, but well known examples are the forest fruit eating Myleus spp. in the Amazon.

Summarizing the facts we can conclude, that river/land ecotones influence the stream ecosystem greatly, irrespective of stream order. Fish participate in the overall system productivity and use ecotones directly in various ways. For the latter, the accessibility of the proper ecotone and the life history of the fish have to match in time to ensure the existence of the fish species in the system at all.

References

Bretschko, G., 1981. Vertical distribution of zoobenthos in an alpine brook of the RITRODAT – Lunz study area. Verh. int. Ver. Limnol. 21: 873–876.

Bretschko, G., 1985. Das hydrographische Geschehen im RITRODAT – Areal in der Zeit von 1980 bis 1984. Jber. Biol.Stn. Lunz 8: 12–16.

Bretschko, G., 1990. The dynamic aspect of coarse particulate organic matter (CPOM) on the sediment surface of a second order stream free of debris dams (RITRODAT-Lunz study area). Hydrobiologia 203: 15–28.

Bretschko, G., 1992. Differentiation between epigeic and hypogeic fauna. Regulated Rivers 7: 17–22.

Bretschko, G. & W. E. Klemens, 1986. Quantitative methods and aspects in the study of the interstitial fauna of running waters. Stycologia 2: 279–316.

Bretschko, G. & M. Leichtfried, 1988. Distribution of organic matter and fauna in a second order, alpine gravel stream

(Ritrodat. Lunz study area, Austria). Verh. int. Ver. Limnol. 23: 1333–1339.

Bretschko, G. & H. Mosert, 1993. Transport and retention of matter in riparian ecotones. Hydrobiologia 251: 95–101.

Caswell, H., 1978. Predator-mediated coexistence: a non-equilibrium model. Am. Nat. 112: 127–154.

Cummins, K. W., 1979. The natural stream ecosystem. In: J. V. Ward & J. A. Stanford (eds), The Ecology of Regulated Streams. Plenum Press, New York and London: 7–24.

Decamps, H. & R. J. Naiman, 1990. Aquatic terrestrial ecotones: summary and recommendations. In R. J. Naiman & H. Decamps (eds), The Ecology and Management of Aquatic-Terrestrial Ecotones. Man and the Biosphere Series, Vol. 4. Pathenon Publ. Group, Paris: 295–301.

Holland, M. M., 1988. SCOPE/MAB technical consultations on landscape boundaries: report of a SCOPE/MAB workshop on ecotones, 5–7 January 1987, France. Biology International, Special Issue 17: 47–106.

Hynes, H. B. N., 1970. The Ecology of Running Waters. Liverpool Univ. Press, 555 pp.

Illies, J. & L. Botosaneanu, 1963. Problèmes et méthodes de la classification et de la zonation écologique des eaux courantes, considerées surtout du point de vue faunistique. Mitt. int. Verein. theor. angew. Limnol. 12: 57 pp.

Jeffers, J., G. Bretschko, F. El Daoushi, E. Fernald, M. Hadley, A. Hansen & R. C. Wissmar, 1989. Theoretical considerations of the ecotone concept. In: R. J. Nainam, H. Decamps & F. Fournier (eds), The Role of Land/Inland Water Ecotones in Landscape Management and Restoration: a Proposal for Collaborative Research. MAB Digest 4, Unesco, Paris: 55–59.

Kotliar, N. B. & J. A. Wiens, 1990. Multiple scales of patchiness and patch structure: a hierarchical framework for the study of heterogeneity. Oikos 59: 253–260.

Leopold, L. B., M. G. Wolman & J. P. Miller, 1964. Fluvial processes in geomorphology. W. H. Freeman & Comp., San Francisco, 522 pp.

Moog, O. & R. Wimmer, 1990. Grundlagen zur typologischen Charakteristik österreichischer Flie-gewässer. Wasser & Abwasser 34: 55–211.

Pickett, S. T. A. & P. S. White (eds), 1985. The Ecology of Natural Disturbance and Patch Dynamics. Academic Press, New York, 472 pp.

Pringle, C. M., R. J. Naiman, G. Bretschko, J. Karr, M. Oswood, J. Webster, R. Welcomme & M. J. Winterbourn, 1988. Patch dynamics in lotic systems: the stream as a mosaic. J. N. Am. Benthol. Soc. 7: 503–524.

Schiemer, F., 1985. Bedeutung von Augewässern als Schutzzonen für die Fischfauna. Österr. Wasserwirtschaft 37: 239–245.

Schiemer, F. & M. Zalewski 1992. The importance of riparian ecotones for diversity and productivity of riverine fish communities. Neth. J. Zool. 42: 323–335.

Schlosser, I. J., 1995. Critical landscape attributes that influence fish population dynamics in headwater streams. Hydrobiologia 303 (Dev. Hydrobiol. 105): 71–81.

Stanford, J. A. & A. R. Gaufin, 1974. Hyporheic communities of two Montana rivers. Science 185: 700–702.

Statzner, B. & B. Higler., 1985. Questions and comments on the River Continuum concept. Can. J. Fish. aquat. Sci. 42: 1038–1044.

Statzner, B. & B. Higler, 1986. Stream hydraulics as a major determinant of benthic invertebrate zonation patterns. Freshwat. Biol. 16: 127–139.

Townsend, C. R., 1989. The patch dynamics concept of stream community ecology. J. N. Am. Benthol. Soc. 8: 36–50.

Vannote, R. L., G. W. Minshall, K. W. Cummins, J. R. Sedell & C. E. Cushing, 1980. The River Continuum Concept. Can. J. Fish. aquat. Sci. 37: 130–137.

Winterbourn, M. J., J. S. Rounick & B. Cowie, 1981. Are New Zealand stream ecosystems really different? N. Z. J. Mar. Freshwat. Res. 15: 321–328.

Hydrobiologia **303**: 93–101, 1995.
F. Schiemer, M. Zalewski & J. E. Thorpe (eds), The Importance of Aquatic-Terrestrial Ecotones for Freshwater Fish.

Riparian ecotones and spatial variation of fish assemblages in Portuguese lowland streams

M. J. Collares-Pereira, M. F. Magalhães, A. M. Geraldes & M. M. Coelho
Departamento de Zoologia e Antropologia, Faculdade de Ciências, Campo Grande, Bloco C2, 3º Piso, 1700 Lisboa, Portugal

Key words: riparian ecotones, fish assemblages, spatial variation, Iberian streams

Abstract

The first results of a long-term study on the role of riparian ecotones on the population and community dynamics of Iberian stream fish are presented and discussed. Riparian and macrophyte cover, bank slope and depth were among the most important variables affecting fish distribution. In general small fish favoured shallow areas with high macrophyte cover, whereas large fish dominated in deep areas with a high riparian cover. Slight spatial changes in terrestrial prey use were found suggesting a minor role for this resource during autumn. Finally, no significant spatial differences were found for linear growth, although some differences were obtained for the condition factor.

Introduction

Studies of stream fish assemblages have shown that abiotic factors such as temperature, current velocity, depth and substratum, can determine the distribution and abundance of individual species as well as influence community-level properties such as species diversity, guild-composition and production (Rahel & Hubert, 1991).

In recent years it has been progressively emphasized that the population and community dynamics of stream fish may also greatly depend on the structure and complexity of riparian zones (Zalewski, 1990, 1991; Schiemer & Zalewski, 1992). In fact, the aquatic-terrestrial interface is a critical area on the landscape, combining spatial heterogeneity, availability of refugia from harsh environmental conditions and from predators, and a large supply of nutrients and organic matter, including terrestrial invertebrates (Schlosser, 1991).

Southern streams of the Iberian Peninsula have marked seasonal variations from occasional floods to extensive droughts, which may enhance the importance of riparian habitats. If so, management options for the conservation/restoration of riparian ecotones should be urgently defined before human influences will irreversibly affect fish assemblages, particularly the endemic cyprinid fauna.

In the Sorraia system (the largest left tributary of the R. Tejo), some zones are under a strong human pressure (sand/silt and water extraction, river bed and banks linearization, eutrophication and chemical pollution, dam construction), whereas others are still under a low level of disturbance. Thus this system was considered suitable to further analyse the buffer role played by riparian zones in regulating the spatial and temporal dynamics of fish populations and communities. Besides, base line studies on the spatial and temporal structure of aquatic and riparian vegetation were already available for this hydrographic system (Ferreira & Moreira, 1987; Ferreira, 1992).

Included in a long-term study, some data are now presented focusing on the spatial variation of (1) the distribution and size-structure of individual species as well as of fish assemblage composition, (2) the patterns of terrestrial prey use by non-benthic fish, and (3) the growth and condition of fish.

Study area

The 12 surveyed sites listed in Fig. 1, belong to the Sor stream (sites A and B) and to the Sorraia river (sites C to M). These range from four to six order streams, and are characterised by a mosaic of habitat conditions and highly diversified biocenosis. Banks have a moist deep soil layer allowing the occurrence of dense riparian formations. Sand is the dominant substratum, although in Sor upper reaches calco-alkaline granite deposits may be found.

In the Sorraia R., the three zones defined by Ferreira (1992) for the alluvial corridor Sta Justa-Coruche, were sampled:

Zone I – 'semi-natural' zone (7.5 km), partially-regularized, owing to the existence of river bed and banks linearization (constant apparent river bed 60–65 m), with regular wall-barriers and agricultural discharges; the river bed is regular, only rarely with wet sand flats; the relative surface of the distinct habitats and associate vegetation are very constant. Dominant trees are the willows (*Salix alba* spp. *vitelina, Salix salvifolia* spp. *salvifolia*), and the water and bank herbaceous dominants are *Erodium moschatum, Cynodon dactylon, Poa annua, Phragmites australis* and *Arundo donax* (site C).

Zone II – the most 'natural' zone (12.5 km), in which the apparent river bed width is highly variable (from 40 to 100 m) and the human influence is much less marked; river water branches and ondulates through wet sand flats creating open river areas, with distinct depths. Dominant trees are willows, poplars (*Populus nigra*) and ashe trees (*Fraxinus angustifolia*) (some of these trees grow down to the water level), and the major water and bank herbaceous are *Potamogetom fluitans, Myriophyllum aquaticum, M. spicatum* and *Ceratophyllum demersum* (sites D to L).

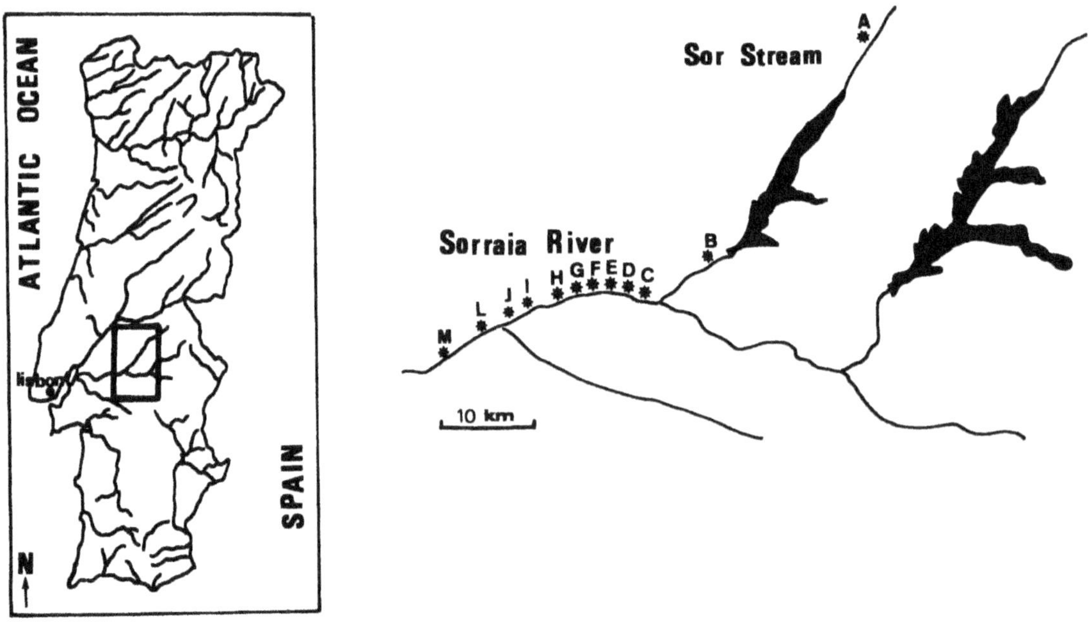

Fig. 1. Study sites along the Sorraia system,the main left tributary of Tagus river in Portugal.

Zone III – the most 'artificial' zone (2.5 km), annually regularised and prepared for fish game competition with the use of flood-gates, river bed and banks linearization and with willows and helophytic vegetation control. There are no sand habitats and the riparian zone is relatively reduced, presenting almost exclusively willows as lotic vegetation cover and a patchy distribution of *Myriophyllum aquaticum* (site M).

Two sites in the Sor stream were also included because a previous analysis (Magalhaes *et al.*, 1991) suggested distinct spatial patterns for the distribution of the dominant cyprinid species in the R. Sorraia and its tributaries. A sand substratum predominates in the study sites isolated since 1958 by the Montargil reservoir. Below the reservoir (site B) the typical water and bank herbaceous species are *Mentha aquatica*, *Licopus*, *europeus*, *Juncus articulatus* and *Myriophyllum aquaticum*, and above (site A) species like *Paspallum paspalodes*, *Frimbristillis bisumbellata*, *Lotus pedunculatum* and *Illecebrum verticillatum* dominate (Ferreira & Moreira, 1987). Willows, poplars and alders (*Alnus glutinosa*) not found in site B are dominant trees in site A.

The Sorraia system has a rich-species fish community, clearly dominated by the family Cyprinidae (Table 1). The most abundant taxa are the endemic roach '*Rutilus alburnoides* complex', the Iberian barbel *Barbus bocagei* and the endemic chub *Leuciscus pyrenaicus*.

The roach represents an important fraction of the fish community and constitutes a well-established diploid-triploid complex of forms involved in bisexual-unisexual modes of reproduction, but it is still waiting for more reliable morphological and genetic characterisations (see Collares-Pereira, 1989). The second species is the barbel, which has a quite distinct migratory behaviour, at least during the reproductive period (Rustarazo *et al.*, 1989).

Table 1. Fish fauna of R. Sorraia system. First figure = number of fish captured; figure in parenthesis = % number of captures.

Family	Species	Number
Cyprinidae	– *Rutilus alburnoides* complex (Steindachner, 1866)	2298 (57.88)
	Barbus bocagei Steindachner, 1865	1184 (29.82)
	Leuciscus pyrenaicus (Gunther, 1868)	185 (4.66)
	Chondrostoma polylepis Steindachner, 1865	156 (3.93)
	Gobio gobio (Linnaeus, 1758)	22 (0.55)
	Cyprinus carpio Linnaeus, 1758	18 (0.45)
	Carassius auratus (Linnaeus, 1758)	18 (0.45)
	Chondrostoma lusitanicum Collares-Pereira, 1980	26 (0.66)
	Barbus comiza Steindachner, 1865	1 (0.03)
Cobitidae	– *Cobitis maroccana* Pellegrin, 1921	54 (1.36)
Centrarchidae	– *Lepomis gibbosus* (Linnaeus, 1758)	1 (0.03)
	Micropterus salmoides Lacépède, 1802	–
Poeciliidae	– *Gambusia holbrooki* Girard, 1859	3 (0.07)
Petromyzontidae	– *Lampetra fluviatilis* (Linnaeus, 1758)	–
	Petromyzon marinus Linnaeus, 1758	–
Anguillidae	– *Anguilla anguilla* (Linnaeus, 1758)	2 (0.05)
Mugilidae	– *Liza ramada* (Rizzo, 1826)	2 (0.05)
Atherinidae	– *Atherina boyeri* Risso, 1810	–

Material and methods

Fish assemblages and environmental variables

Altogether eleven sites, each 50–70 m long, were selected (sites B to M). The number and length of sites were chosen to represent the range of habitats and plant communities occurring in both streams and all sites were sampled in October 1991.

Fishes were captured by electrofishing along both banks, identified, measured for total and standard length (to within 1 mm) and weight (to within 0.01 g).

Eleven environmental variables were measured for each sampling date and site:(1) channel width (m); (2) water depth (m); (3) current velocity ($m\ s^{-1}$); (4) temperature (°C); (5) conductivity (μS); (6) oxygen ($mg\ l^{-1}$), (7) dominant substratum; (8) bank slope; (9) macrophytes cover; (10) riparian cover (bushes and/or trees); (11) percent covered by canopy. Variables 7 to 10 were measured using qualitative categories: *7)* substratum – (1) organic cover; (2) silty sand (1–2 mm); (3) sand (2–5 mm); (4) gravel (5–25 mm); (5) pebble (25–50 mm); (6) rock (50–100 mm); (7) cobble (100–250 mm); (8) boulder (> 250 mm); *8)* bank slope – (1) very steep; (2) steep; (3) gentle slope; (4) convex slope; *9)* and *10)* macrophytes and riparian cover – (1) absent; (2) light presence; (2) median presence; (3) dense; (4) very dense.

Sampling protocol for variables 1 to 7 followed Gorman & Karr (1978) with slight modifications. Beginning 10–20 cm from the banks, point samples were taken at four meters intervals along the stream. Repeated sets of points were taken across the stream at ten meters intervals, moving upstream. Variables 8 to 10 were measured at five to seven points taken along each bank at ten meter intervals. Habitat volume (m^3) and habitat area (m^2) for each sampling site were calculated according to Schlosser (1987).

To detect patterns in fish assemblages and to relate these patterns to environmental factors, the general procedure recommended by Green & Vascotto (1978) was used. Fish assemblage patterns were identified using cluster analysis (Pearson correlation coefficient, UPGMA) for all the species individually, except those with a percentage occurrence lower than 0.1%, were pooled together. A Kruskall-Wallis test was carried out for each environmental variable in order to determine whether the species assemblage clusters differed significantly from each other in their median value of that variable.

Terrestrial prey utilization

Previous studies on the cyprinid diets and patterns of food resource partitioning in the Sorraia system (Magalhães & Collares-Pereira, 1991; Magalhães, 1992; 1993), suggested that terrestrial invertebrates were minor components of the diet of benthic-cyprinids over all the year but they were major components of the roach and chub diets during the dry season.

So, in the present analysis, roach and chub were sampled at sites B, C, E and M. Whenever available, twenty fish of each species, spanning the entire size range in the collection, were selected. These fish were placed in an ice-bath, and deep frozen within three to six hours.

The entire gut contents were analysed, and the percentage occurrence of plant material and the numerical percentage of animal prey, were determined. Percentage occurrence was calculated omitting empty guts and numerical percentage was based on total numbers of prey in a sample (Hyslop, 1980).

Overlaps in animal prey use were computed using the Schoener formula and were regarded as high when values exceeded 0.6 (Wallace, 1981). To assess the importance of terrestrial prey in relation to species interactions, animal prey overlaps were calculated considering only the aquatic prey and then both aquatic and terrestrial prey. Prey were considered terrestrial if they were the result of terrestrial secondary production, not aerial adults of aquatic forms (*sensu* Garman, 1991).

Linear growth and condition

For the linear growth and condition analyses, the chub was the species selected first. Fish were captured at sites A, B, C, E and M, measured for standard lengths and four to six scales were removed. Condition factor was calculated over fish total length and linear growth over fish standard length. Back-calculations were performed by the SPH (Scale Proportional Hypothesis) method (*in* Francis, 1990). One-way analysis of variance (ANOVA) and the Kruskall-Wallis test were used to test for spatial variations on the back-calculated length and condition factor of fish, respectively.

Results

Of the 18 species reported for the Sorraia system, only 14 species were collected in this study (Table 1). Among the 3970 fish sampled, the most common species were the roach (57.9%) and the barbel (29.8%). The other species captured were much less represented and those with a numerical percentage less than 0.1% were jointly considered for the cluster analysis of relative abundance data matrix.

Fish relative abundances clustered into three assemblage groups (Fig. 2). In the first group (sites B, C, and M), fish assemblages were dominated by roach. Sites in the second group (D, E, F and I) presented an equal proportion of roach and barbel. In the third group (sites G, H, J and L) the barbel was the predominant species.

Means for the environmental variables grouped according to fish assemblage type (1, 2 or 3) were positioned along an environmental gradient. Variables describing width, depth, oxygen, riparian cover and substratum were significantly different among groups (Table 2). Sites dominated by roach (group 1) were narrow and shallow, presented the lowest oxygen concentration and riparian cover and the most coarse substratum. Sites dominated by barbel (group 3) were deepest, more oxygenated and presented a smooth substratum. Sites with both species (group 2) were wider and presented more riparian cover.

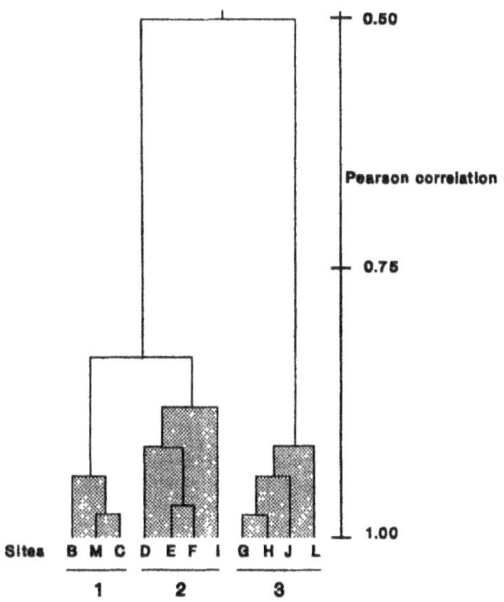

Fig. 2. Similarity of fish assemblages in the sampling sites. Dendogram showing three groups based on cluster analysis of relative fish number data (UPGMA method, cophenetic correlation 0.870). Sampling site codes as in Fig. 1.

A preliminary insight into the spatial variation of fish populations structure was obtained by comparing the size-compositions of roach, barbel and chub from the sampling sites where the species were abundant (Fig. 3). Length-frequency

Table 2. Means of the 12 environmental variables calculated for clusters of sampling sites. Results of Kruskall-Wallis test are also presented. * $p<0.05$; ** $p<0.01$. Cluster codes as in Fig. 2.

Variables	Sampling site clusters			H
	1	2	3	
Width (m)	8.7	25.9	17.6	21.9**
Depth (m)	0.3	0.4	0.5	29.5**
Current (ms^{-1})	14.8	12.2	12.0	4.0
Temperature (°C)	17.1	17.6	17.5	6.2
Conductivity (μS)	337.2	341.5	344.5	5.4
Oxygen ($mg\ l^{-1}$)	7.1	7.8	8.0	11.5**
Substratum	3.4	3.1	3.0	7.5*
Bank slope	2.5	2.6	2.7	2.6
Macrophytes	1.6	2.3	1.9	0.7
Riparian cover	0.3	1.6	0.8	6.0*
Canopy (%)	5.0	8.3	8.4	3.2
Volume	243.7	451.8	472.6	3.4

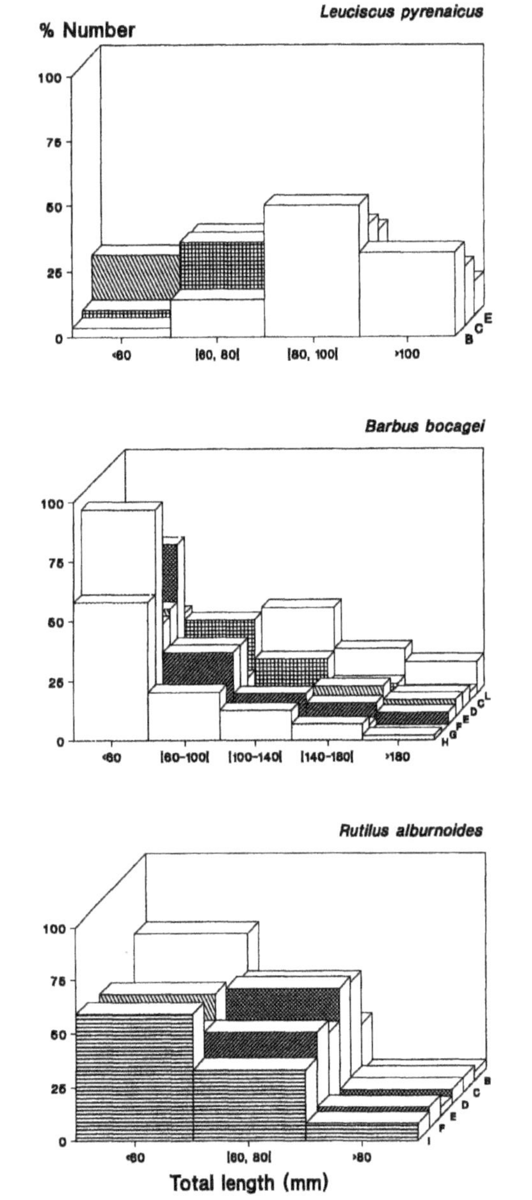

Fig. 3. Spatial changes in length-frequency distribution of roach, barbel and chub. Sampling site codes as in Fig. 1.

distributions of roach and chub were similar among all sampling sites. Most roach were < 80 mm, whereas most chub were between 60 and 100 mm. Most barbel were less than 100 mm at all sites, except site L, where most specimens were between 100 and 140 mm and the highest oxygen values were observed (9.2 mg l^{-1} versus 6.5–7.7 mg l^{-1}. The low frequency of small fish at site L is probably related to the high current velocity (22.6 m s^{-1} versus 6.5–15.8 m s^{-1} in the remaining sites). Figure 4, shows the percentage captures for the different size-classes of roach, barbel and chub along the river banks at site E. Significant differences between river banks were found for depth ($H = 5.769$, $p < 0.05$), macrophytes ($H = 5.771$, $p < 0.05$), riparian cover

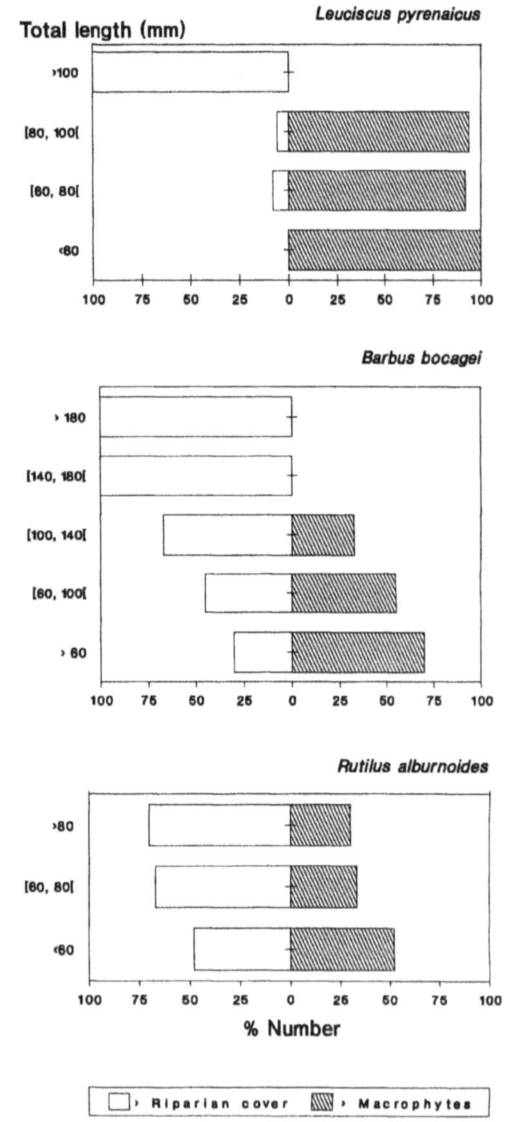

Fig. 4. Distribution of roach, barbel and chub along river banks, in site E.

($H = 3.938$, $p < 0.05$) and bank slope ($H = 3.938$, $p < 0.05$). Barbel and chub showed a trend for size-class segregation. Small fish (< 100 mm) of both species were found mainly in the shallowest bank, presenting the highest slope and macrophyte cover. Large fish were mainly found in the deepest bank, with a high riparian cover. Small and large roach were found in both banks.

Altogether the gut contents of 105 roach and 103 chub were analysed, yielding a total of 12 and 15 food categories (Table 3). Plant material, including both vascular plant parts, algae and seeds occurred more often in chub diet (35.9%) than in roach diet (23.8%). Among animal prey, opposite patterns of consumption were found for ephemeropteran nymphs and dipteran larvae, the former being the major prey for chub (56.8%) whereas the latter were the major prey for roach (71.2%). Terrestrial insects (dipteran adults and Formicidae) accounted for 3.7% and 7.4% of total animal prey for roach and chub respectively.

The percentage occurrence of plant material in the diets of fish from distinct sites is presented in Fig. 5. Plant material had an overall major importance in fish diets in site B and M. However, opposite patterns of plant consumption by chub and roach were found at these sites. At site B, plant material was more frequent in the diet of

Table 3. Percentage occurrence of plant material and numeric percentage of animal prey in the diets of roach and chub, over all sampling sites.

	Roach	Chub
Sample size	105	103
Prey number	646	416
Plant material	23.8	35.9
Mollusca	–	0.5
Athyaephira sp.	–	1.4
Baetidae nymphs	8.0	22.4
Other Ephemeroptera	15.8	34.4
Odonata nymphs	0.5	1.4
Chironomidae larvae	50.5	17.8
Simulidae larvae	14.7	6.2
Dipteran nymphs	1.4	–
Dipteran adults	1.5	4.8
Heteroptera	3.9	3.8
Coleopteran adults	0.6	1.4
Hydropsychidae larvae	0.9	0.3
Formicidae	2.2	2.4
Lepidoptera larvae	–	0.3
Teleostei	–	2.9

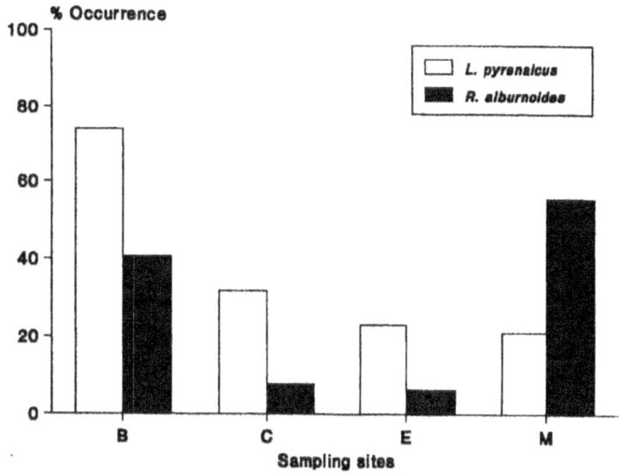

Fig. 5. Spatial variation of the percentage occurrence of plant material in the diets of roach and chub. Sampling site codes as in Fig. 1.

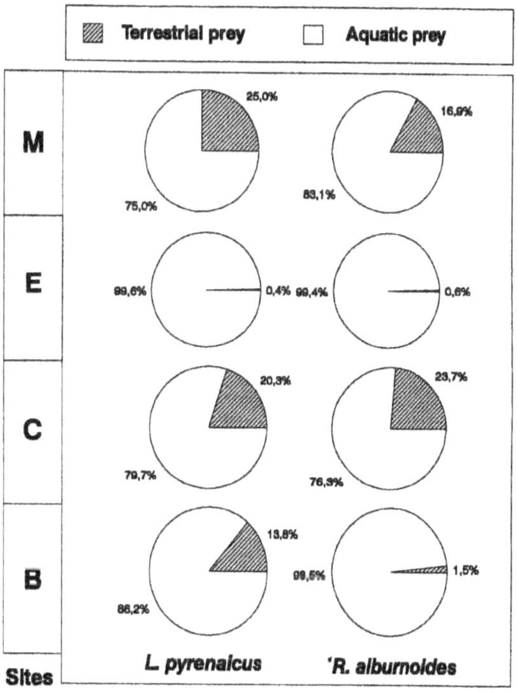

Fig. 6. Spatial variation of the percent composition of terrestrial and aquatic prey in the diets of roach and chub. Sampling site codes as in Fig. 1.

Table 4. Prey overlap between roach and chub at the different sampling sites, considering only aquatic prey and both aquatic and terrestrial prey. Sampling site codes as in Fig. 1.

	Sampling sites			
	B	C	E	M
Total prey	0.664	0.658	0.602	0.369
Aquatic prey	0.714	0.686	0.602	0.408

chub, whereas in site M it was more frequent in roach diet.

Spatial fluctuations in the importance of animal prey were assessed by combining prey into two categories, terrestrial and aquatic (Fig. 6). Terrestrial prey did not occur in fish diets at site E. Conversely, this category represented more than 10.0% of total numbers of prey for both chub and roach, at sites M and C. At site B, terrestrial prey was consumed mainly by chub, but was of low importance in roach diet.

Prey overlaps between roach and chub were high at sites B to E, but dropped considerably at site M (Table 4). Irrespective of the sampling site, prey-use overlaps were not significantly increased by the exclusion of terrestrial prey from the calculations.

The analysis of chub linear growth revealed no significant spatial differences for the back-calculated lengths of fish at any age ($F = 0.655$ for age 1^+, $F = 0.331$ for age 2^+ and $F = 1.149$ for age 3^+). However, it should be emphasized that for all age groups the smallest fish were found at site A. Conversely, significant spatial differences were obtained for the condition factor for fish between 60 to 80 mm ($H = 8.855$, $p < 0.05$) and for fish between 80 to 100 mm ($H = 10.300$, $p < 0.01$). Both size-classes had the highest condition at site B.

Discussion

Habitat structure greatly affected the composition and size-structure of fish assemblages. Riparian and macrophyte cover, bank slope and depth were among the most important variables affecting fish distribution. In general, small barbel, chub and roach favoured shallow areas with high macrophyte cover and bank slope, whereas large barbel and chub dominated in deep areas with a high riparian cover.

The factors leading to such habitat preferences were not investigated but would likely include physical stresses, food availability, predation-risk or competitive interactions (e.g. Schlosser, 1991). Previous studies addressing the relationships among fish and plants have shown that vegetation increases structural habitat complexity, improves invertebrate production and affects fish predator-prey interactions (*in* Blain & Boltz, 1992).

Terrestrial insects were minor components of the diet of both roach and chub. Irrespective of the sampling sites, fish fed mainly on plant material and aquatic prey. As no direct estimates of terrestrial invertebrates are available it is unclear whether the low consumption of terrestrial insects by fish was related to low inputs from the riparian zone or to high availability of alternative prey. Inputs of allochthonous material to temperate streams are highly seasonal, with maxima in spring and autumn (Mason & MacDonald, 1982; Garman, 1991). However, terrestrial prey is consumed by fish mainly in the summer when aquatic invertebrates are in comparatively short supply (Garman, 1991). Spatial variations in the intake of terrestrial insects by fish may also be related to changes in their availability due to differences in the type of tree canopies among sampling sites. According to Mason & MacDonald (1982), ash trees are poor in insects, in spite of their light canopy, whereas alders provide much larger inputs of invertebrates to streams. Moreover, other factors such as stream hydraulics may also be responsible for heterogeneous spatial distribution of terrestrial invertebrates (*in* Garman, 1991).

The present study is only an early step towards the assessment of the role played by riparian zones over the population and community dynamics of Iberian stream fish. Future studies should focus on the effects of temporal environmental heterogeneity on fish assemblages. It is expected that in a long-term perspective, the functional interactions of landscape elements on fish and the ecological consequences of riparian zone management will be fully evaluated. Thus, the definition of correct options for their conservation/restoration will allow the maintenance of the biodiversity of Iberian aquatic systems.

Acknowledgements

Thanks are specially due to J. A. Rodrigues, M. J. Alves, R. M. Brito, M. F. Fino and J. L. Costa for their collaboration in field work. This research project (PEA/115/92) is funded by the Junta Nacional da Investigaçao Cientifica (Portugal).

References

Bain, M. B. & S. E. Boltz, 1992. Effect of aquatic plant control on the microdistribution and population characteristics of largemouth bass. Trans. am. Fish. Soc. 121: 94–103.

Collares-Pereira, M. J., 1989. Hybridization in european cyprinids: evolutionary potential of unisexual populations. In R. M. Dawley & J. P. Bogart (eds), Evolution and ecology of unisexual vertebrates, Bull. 466, New York State Museum, Albany, New York: 281–288.

Ferreira, M. T., 1992. Estrutura e dinâmica das comunidades de macrofitos loticos da Bacia Hidrográfica do Sorraia. PhD Thesis, ISA, Univ. Técnica de Lisboa, Lisboa.

Ferreira, M. T. & I. Moreira, 1987. Distribution patterns of aquatic vegetation in Sor river system, central Portugal. Simp. Intern. Botanica, Lérida.

Francis, R. I., 1990. Back-calculations of fish length: a critical review. J. Fish Biol. 36: 883–902.

Garman, G. C., 1991. Use of terrestrial arthropod prey by a stream-dwelling cyprinid fish. Envir. Biol. Fishes 30: 325–331.

Gorman, O. T. & J. R. Karr, 1978. Habitat structure and stream fish communities. Ecology 59: 507–515.

Green, R. H. & G. L. Vascotto, 1978. A method for the analysis of environmental factors controlling patterns of species composition in aquatic communities. Wat. Res. 12: 583–590.

Hyslop, E. J., 1980. Stomach content analysis – a review of methods and their application. J. Fish Biol. 17: 411–429.

Magalhães, M. F., 1992. Feeding ecology of the iberian cyprinid *Barbus bocagei* Steindachner, 1865 in a lowland river. J. Fish Biol. 40: 123–133.

Magalhães, M. F., 1993. Effects of season and body-size in the distribution and diet of the Iberian chub *Leuciscus pyrenaicus* (Gunther) in a lowland catchment. J. Fish Biol. 42: 875–888.

Magalhães, M. F. & M. J. Collares-Pereira, 1991. Food resources partitioning within an iberian stream cyprinid assemblage. 7th Int. Ichth. Congr., Abstract: 45.

Magalhães, M. F., M. J. Collares-Pereira & M. M. Coelho, 1991. Spatial and temporal patterns of distribution and abundance of cyprinids in a portuguese lowland system. 7th Int. Ichth. Congr., Abstract: 45.

Mason, C. F. & S. M. Macdonald, 1982. The input of terrestrial invertebrates from tree canopies to a stream. Freshwat. Biol. 12: 305–311.

Rahel, F. J. & W. A. Hubert 1991. Fish assemblages and habitat gradients in a Rocky-Mountain-Great Plains stream: biotic zonation and additive patterns of community change. Trans. am. Fish. Soc. 120: 319–332.

Rustarazo, M. M., E. B. Colebras & D. Garcia de Jalon, 1989. Estudio de las poblaciones del barbo *Barbus bocagei* Stein., 1865) del Rio Jarama. Actas Col. Luso-esp. Ecol. Bacias Hidrog. Rec. Zool., 277–286.

Schiemer, F. & M. Zalewski, 1992. The importance of riparian ecotones for diversity and productivity of riverine fish communities. Neth. J. of Zool. 42: 323–335.

Schlosser, I. J., 1987. A conceptual framework for fish communities in small warmwater streams. In W. J. Matthews & D. C. Heins (eds), Ecology and Evolution of North American stream fishes. Univ. Oklahoma Press, Norman: 17–24.

Schlosser, I. J., 1991. Stream fish ecology: a landscape perspective. BioScience 41: 704–712.

Wallace, R. K., 1980. An assessment of diet overlap indexes. Trans. am. Fish. Soc. 110: 72–76.

Zalewski, M., 1990. The ecotone concept in conservation and fisheries management of riverine ecosystems. Proc. Unesco/Mab Int. Seminar: 80–94.

Zalewski, M., 1991. Review of the session: rivers. In M. Zalewski, J. E. Thorpe & P. Gaudin (eds), Role of land/inland water ecotones in landscape management and restoration, Lodz: Unesco, 17–20.

Hydrobiologia **303**: 103–110, 1995.
F. Schiemer, M. Zalewski & J. E. Thorpe (eds). The Importance of Aquatic-Terrestrial Ecotones for Freshwater Fish

Morphological variability in the ecotone – an important factor for the conservation of fish species richness in Swiss rivers

Arthur Kirchhofer
Zoological Institute University of Bern, Baltzerstr. 3, CH–3012 Bern, Switzerland

Key words: river morphology, ecotone, variability, fish fauna, threatened species, conservation

Abstract

Data from a survey on the distribution of fish species in Switzerland were used to test the hypothesis that species richness of the fish fauna is closely related with the structure, heterogeneity and complexity of the ecotonal zone in rivers. Species richness increased with increasing structural heterogeneity and variability of the ecotone in rivers wider than 2.5 m at altitudes below 500 m. In lowland brooks and in rivers at higher altitudes other factors seemed to be more important than morphological variability: alpine rivers are more often left natural than lowland rivers and salmonid stocking is more important in these naturally species poor fish communities. The proportion of threatened species was higher if riverbed and -banks were natural, the slope was flat, the variability in depths and sediment composition was high and if the ecotonal zone was well structured. Specialized species such as nase and riffle minnow only occurred in larger rivers, whereas in small and medium-sized rivers with low morphological variability these threatened species were missing. On the other hand, ubiquitous species such as chub, roach and perch, occurred in all sizes of rivers, even if morphological variability was small. Low morphological variability and its impact on the composition of the fish community is mainly a consequence of past human interventions. Therefore river management has to take into account the ecological requirements of specialized species and increasing the variability of riverbed and banks would considerably contribute to the conservation of endangered fish species.

Introduction

With the increasing impact of human activities on the aquatic environment, one of our main concerns must be the conservation of aquatic communities. The fish fauna occupies a prominent place in this context, not only because of its commercial importance, but also because of its character as a long time indicator for the condition of aquatic habitats.

The still on-going process of simplification and monotonization of habitat diversity in Swiss rivers evoked the implementation of revised laws of fishery, water protection and hydraulic engineering in the past years, requiring ecologically-oriented management in rivers and lakes, in order to preserve natural habitats, improve badly disturbed sites and protect biodiversity, especially threatened fish species. To provide the information necessary for the achievement of this goal, detailed studies about the condition of aquatic habitats and the present distribution and ecological requirements of fish species have to be undertaken. Many studies have shown that channel morphology, stream bed parameters and adjacent structures are the most important factors in determining community composition (Gorman & Karr, 1978; Barila *et al.*, 1981; Jungwirth & Winkler, 1983), and that human disturbances, such as channelization, damming and impoundments affect fish diversity to a great extent (Jungwirth, 1984; Schiemer & Waidbacher, 1992).

In the last years a nation-wide survey was undertaken to study the current fish distribution and to provide the basic data for an assessment of the vulnerability of fish taxa in Switzerland (Kirchhofer *et al.*, 1990). This study documents the detailed distribution of all 58 taxa occurring in Switzerland, with indications on the population development (Pedroli *et al.*, 1991). Besides these results, the database contains information on the condition of the sampled rivers and lakes and provides

Table 1. Number of sampling sites in rivers according to width-class, altitude and degree of modification of the riverbed and banks (NAT = 1: heavily modified, NAT = 2: moderately modified, NAT = 3: natural); Out of 1900 rivers, 186 with missing data were excluded.

Width-class	Altitude < 501 m		Altitude 501–750 m		Altitude > 750 m		Total	
	n	%	n	%	n	%	n	%
<1.1 m	**125**		**233**		**569**		**927**	
NAT = 1		20.0		9.0		1.4		5.8
NAT = 2		15.2		15.0		3.2		7.8
NAT = 3		64.8		76.0		95.4		86.4
1.1–2.5 m	**54**		**88**		**271**		**413**	
NAT = 1		27.8		17.0		3.0		9.2
NAT = 2		29.6		23.9		4.4		11.9
NAT = 3		42.6		59.1		92.6		78.9
2.6–5 m	**51**		**44**		**88**		**183**	
NAT = 1		25.5		20.5		0		12.0
NAT = 2		37.3		20.5		14.8		22.4
NAT = 3		37.2		59.0		85.2		65.6
5.1–10 m	**30**		**27**		**27**		**84**	
NAT = 1		46.7		7.4		0		19.0
NAT = 2		23.3		44.4		22.2		29.8
NAT = 3		30.0		48.1		77.8		51.2
10.1–25 m	**18**		**24**		**15**		**57**	
NAT = 1		16.6		4.2		6.7		8.8
NAT = 2		56.6		50.0		33.3		47.3
NAT = 3		27.8		45.8		60.0		43.9
>25 m	**43**		**6**		**1**		**50**	
NAT = 1		14.0		0		0		12.0
NAT = 2		62.8		50.0		0		60.0
NAT = 3		23.2		50.0		100		28.0
total	**321**		**422**		**971**		**1714**	
NAT = 1		23.7		11.4		1.8		8.3
NAT = 2		30.5		21.8		5.6		14.2
NAT = 3		45.8		66.8		92.6		77.5

a basis for further work on more detailed questions. According to a hypothesis in Naiman *et al.* (1989), fish diversity depends to a great extent on the condition of the ecotonal zone between aquatic and terrestrial habitats. In the present contribution this hypothesis is tested on a large scale, the data from the distribution study being evaluated with special respect to the morphological variability of riverbed and banks.

Methods

Description of study sites

The study area comprised the whole territory of Switzerland with 50,000 km of rivers and over 2000 lakes at altitudes between 195 and 4000 m. Of these a statistically representative sample was selected: On the map 1: 50,000 one river and/or one lake out of a

Table 2. Examined structures of the ecotonal zone and corresponding ratings used to describe morphological variability.

Morphological character	Rating 1	2	3	4
Modification of riverbed and banks	heavily modified	moderately modified	natural	–
Slope of banks	steep (>1:1)	moderate (1:3–1:1)	flat (<1:3)	–
Variability of depths	uniform	irregular	–	–
Variability of sediments	homogenous	variable	–	–
Structural heterogeneity of the ecotonal zone	very low	low	medium	high
"Global variability"	Sum of ratings of modifications + slope + depths + sediments + structural heterogeneity (min. = 5, max = 14)			

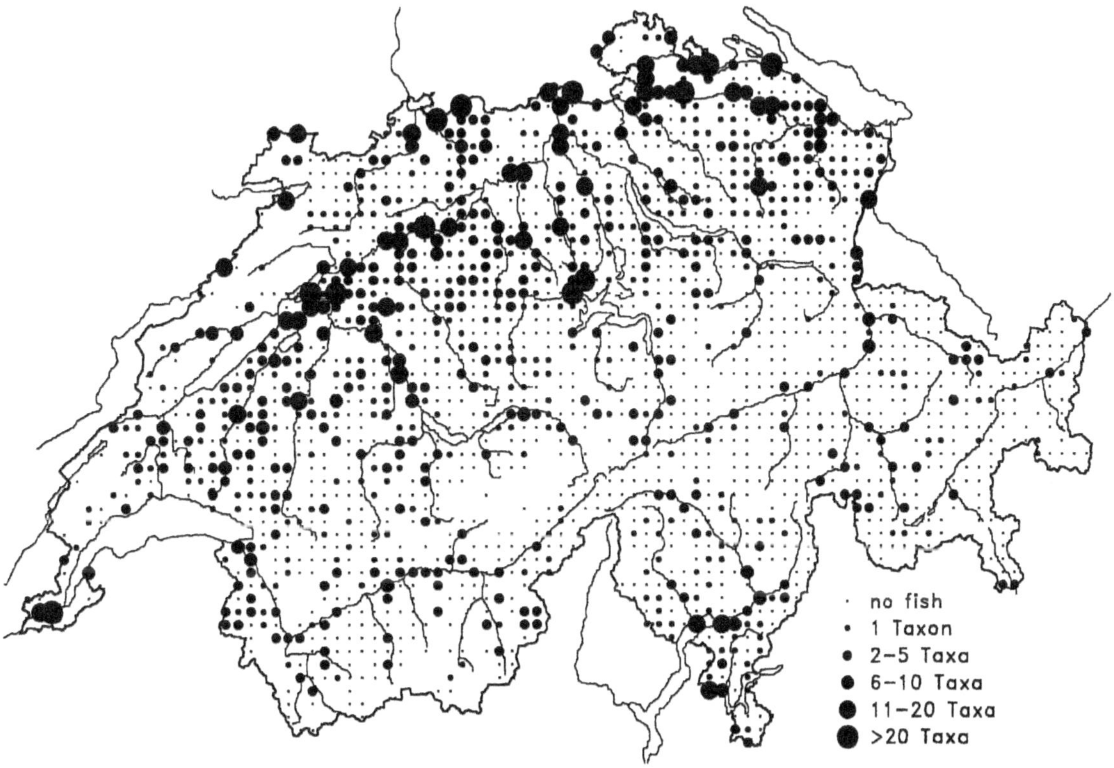

Fig. 1. Species richness of the fish fauna in 1900 riverine sampling sites in Switzerland; data from Pedroli *et al.* (1991).

grid with 4 km edge length (= 16 km^2) was chosen, and in the rivers a stretch of 200 m was designated as the sampling site. With this method 1900 rivers and 249 lakes (= 2149 sampling sites) were selected, comprising all altitudes and sizes of rivers and lakes (Table 1).

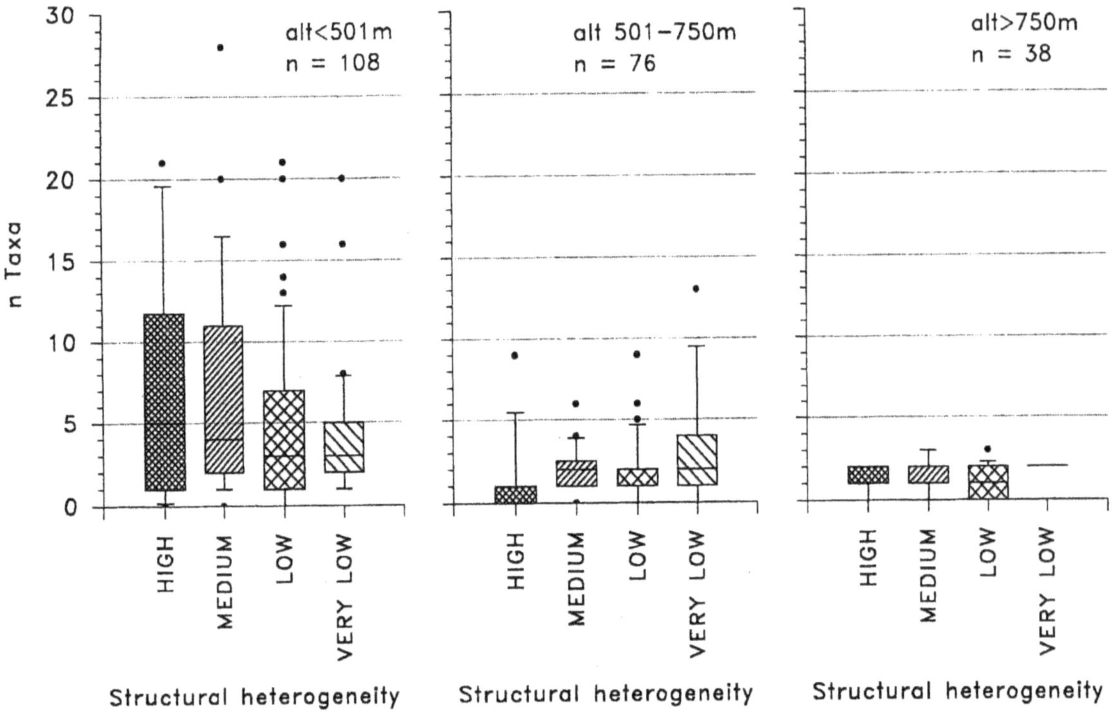

Fig. 2. Number of taxa versus structural heterogeneity of the riparian zone in 222 river sections classified by altitudes. The boxes enclose 25th and 75th percentiles, the horizontal lines indicate the medians and wispers assign 10th and 90th percentiles, filled circles mark points outside the 10th and 90th percentile.

Sampling procedure

The data on the distribution of the fish fauna were obtained in two steps:

1. Between 1984–86 the existing knowledge about the sampling sites was gathered by a nation-wide inquiry among the 250 official fishery attendants and staff of the fishery administrations. Besides the fish taxa present or having disappeared during the last 20 years at this site, information on hydrology and morphology of the sampling site was collected such as width, depth, pollutions, water diversions, fishing intensity and the degree of modification of riverbed and banks (Table 1).
2. After a preliminary evaluation of the results from step one, 222 rivers and 42 lakes were chosen at random as a representative sub-sample for further investigations, again comprising all sizes and altitudes. In 1987–88 these sampling-sites were electro-fished and photographed in summer, and the information on the habitat from the first step was controlled and completed. In small rivers the whole water course could be sampled, in large rivers and lakes only the accessible border was fished and further information gained by direct observations with scuba-diving. The data were summarized with distribution maps and numerical tables for each taxon in Pedroli *et al.* (1991).

Examination of river morphology

The present contribution only comprises the 222 rivers sampled during the second step of the study. The morphology of riverbed and banks was examined on the photographs in attributing ratings to different variables, which were considered as essential for habitat quality in the ecotonal zone (Table 2): Naturalness of riverbed and banks, slope of the banks, variability of depths, variability of sediments and structural heterogeneity. The latter is defined as high, if aquatic or overhanging riparian vegetation, and woody debris, stones of different sizes and boulders were present, very low if neither of theses structures could be detected. In order to compare species richness to the combined effects of these variables, a "global variability" was computed, comprising the sum of ratings of the individual param-

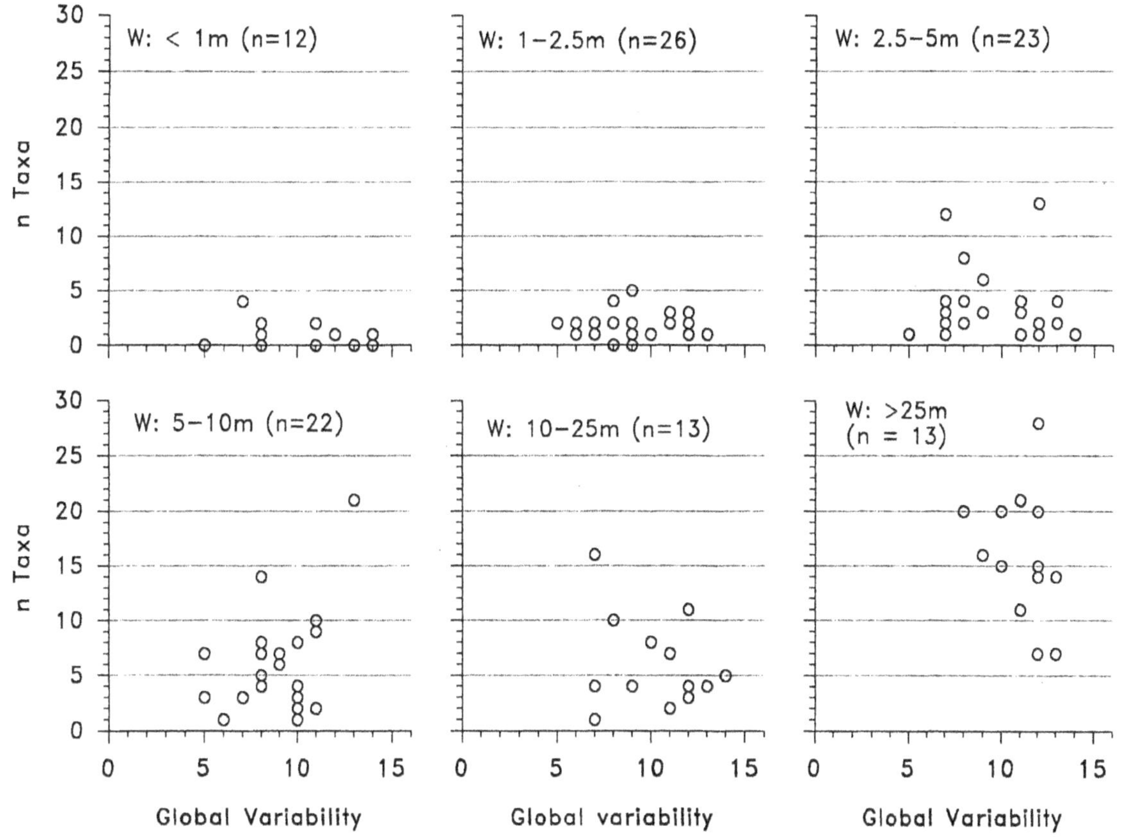

Fig. 3. Number of taxa versus global variability as sum of ratings for the 5 morphological characters of sampling sections in 108 rivers up to an altitude of 500 m; W = width class.

eters (global variability = ratings of modification of riverbed and banks + slope of banks + depth variability + sediment variability + structural heterogeneity). The higher this value, the higher the variability of all the features in the ecotone and the more natural the river morphology. The rivers were classified in three altitudinal and six width classes and the ratings of structural heterogeneity and the combination of all variables were correlated with species richness (= number of taxa present today). Furthermore the condition of the ecotonal zone was correlated with the number of threatened taxa present, according to the Red List of Fish and Cyclostomes of Switzerland (Kirchhofer *et al.*, 1990), and with the presence of selected specialized or ubiquitious species.

Results

The data on fish distribution provided the basis for an examination of species richness of the fish fauna in the rivers of Switzerland (Fig. 1). Species richness is greatest in the lowlands and least in the higher regions of the Prealps, Alps and Jura. Species-rich communities with 5 or more taxa were only found below 750 m in rivers wider than 2.5 m, more than 20 species lived exclusively below 500 m in rivers wider than 10 m. This result is consistent with the concept of fish-regions in European rivers (Huet, 1949), and implies a dependence of fish species richness mainly on two physical factors: the gradient which is related to altitude, and the width of the river. Analysis of species richness and habitat structures has to take into account these differences in fish distribution and therefore the studied rivers were classified according to altitude and river-width at the sampling station. In rivers up to 500 m species number increased with increasing heterogeneity in the bound-

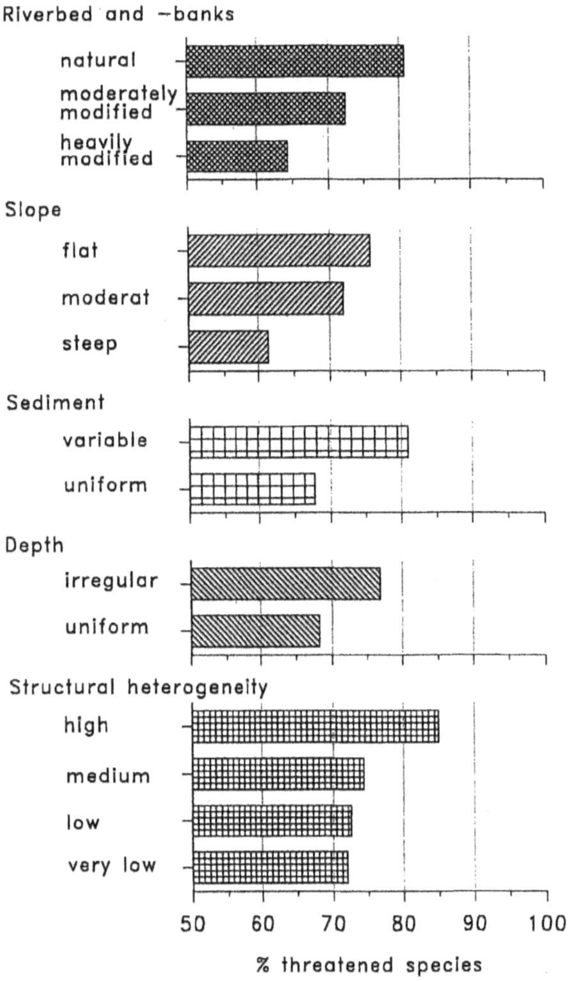

Fig. 4. Proportion of threatened species versus morphological characters in 195 sampled river sections containing fish. Threatened species are taxa in categories 1–4 in the Red List of Fish and Cyclostomes of Switzerland (Kirchhofer *et al.*, 1990).

ary zones (Fig. 2) and the numbers of taxa in the 4 categories are significantly different (one-way ANOVA, $p < 0.05$) with high and medium heterogeneous rivers containing significantly more taxa than river sections with low and very low structural heterogeneity (multiple t-test, $p < 0.0005$). Above 500 m no significant difference in number of taxa could be confirmed for these categories. Zoogeographical and topographical factors obscure probable effects of morphological variability on the species richness and artificial propagation of *Salmo trutta fario* L. strongly influences community composition in rivers above 500 m.

In small rivers below 500 m, no distinct relationship existed between morphological variability and species richness (Fig. 3). Smaller water-courses, even in the lowlands, contain naturally fewer species than big streams, and with the present stocking policy, salmonids are favoured on the cost of other species. This could explain the missing relationship between morphological variability and species richness in these river categories. In medium to large rivers of the same altitude, however, the highest number of taxa occurred in rivers with high global variability. In the lowlands of Switzerland, only very few medium and large rivers (30% or less) can be considered as natural or close to natural (Table 1). This is due mainly to channelization and damming during the last 150 years in order to reclaim floodplains for the increasing human population and to the impoundments for hydroelectric powerplants (Vischer, 1986). The present results suggest that in these deteriorated streams the morphological variability of the boundary zones seems to be most important for the species richness of the river.

For conservation purposes, species composition of the fish community is even more important than the number of taxa. Therefore the fraction of threatened species, comprising taxa from categories 1 to 4 in the Red List (Kirchhofer *et al.*, 1990), was related to the parameters describing morphological variability. The percentage of threatened taxa increased with increasing naturalness of the riverbed and banks, with flatter slopes of the banks (providing habitats for larvae and juveniles), with more variable sediments, with more variable sediments, with more irregularities in depths and with higher structural heterogeneity of the riparian zone (Fig. 4). This leads to the assumption that higher morphological variability favours threatened taxa. Six species with different ecological requirements were compared with the "global variability" of the sampling sites in which they occurred. Perch (*Perca fluviatilis* L.), roach (*Rutilus rutilus* L.) and chub (*Leuciscus cephalus* L.) are non specialised, ubiquitous species generally quite common and often abundant in Switzerland. Grayling (*Thymallus thymallus* L.), nase (*Chondrostoma nasus* L.) and riffle minnow (*Alburnoides bipunctatus* BLOCH) are quite rare specialists (category 3 in the Red List) with often declining populations in Switzerland and requiring particular micro-habitats. The grayling and the two threatened cyprinids were more frequent in highly structured habitats and only 20–30% of all occurrences were situated in habitats with a global variability value < 10 (Fig. 5). Perch and the two common cyprinids occurred as well in rivers with high as with low global variability, but more than 50% were found in rivers with low struc-

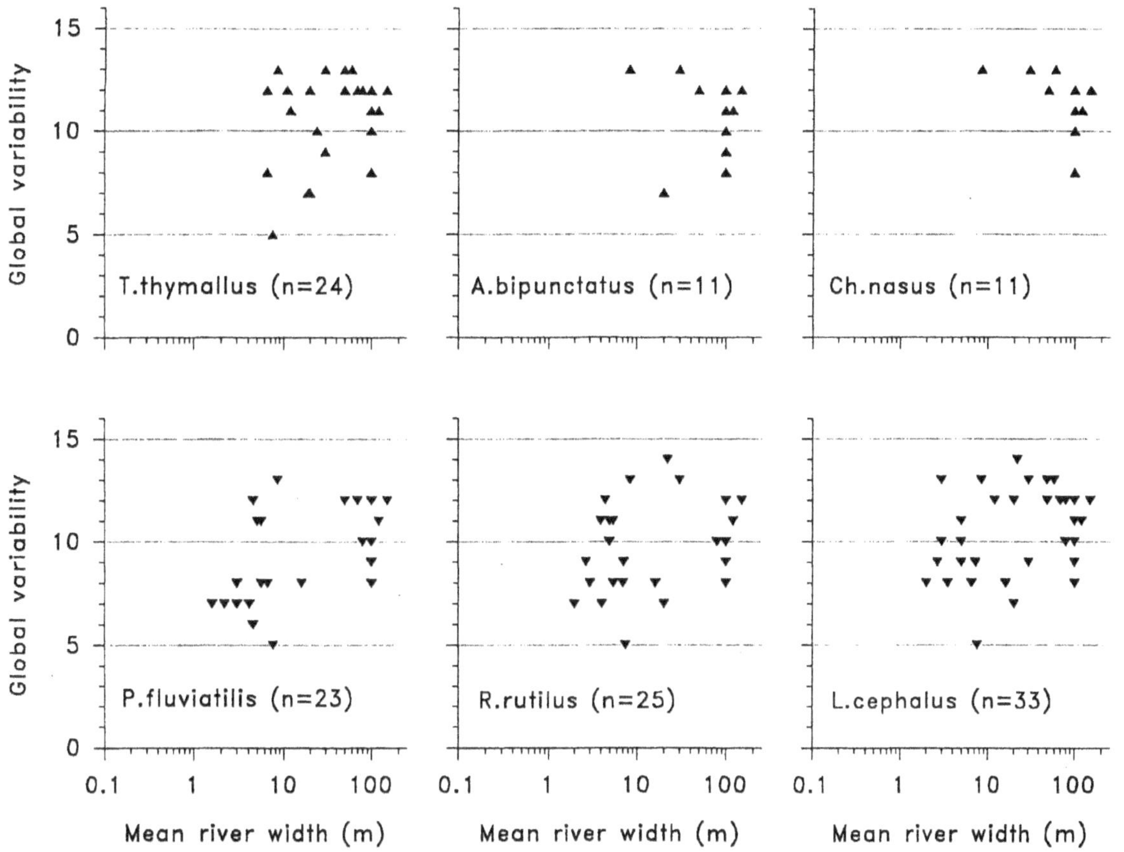

Fig. 5. Global variability versus mean width of sampled river sections for 3 ubiquitous (below) and 3 rare fish species (above); n is the total number of occurrences of the given species out of 222 sampled river sections.

tural diversity. Contrary to the three common species and the grayling, the two threatened cyprinid species were missing in small and medium rivers (less than 10 m width) with low variability of the habitat. The occurrence of grayling in this river categories could be influenced by stocking programmes. This result shows that highly structured riparian zones and river beds are most important for the conservation of specialist species, requiring particular structures such as shallow back-waters with low velocity for larval-stages and juveniles (grayling, nase) or woody-debris and overhanging vegetation as hiding places (riffle minnow).

Discussion and conclusions

The description of morphological variability at a given moment, as used in the present contribution, is quite coarse. Many different and more sophisticated methods to describe river morphology, naturalness or structural heterogeneity have been used in the last few years (e.g. Spiegler *et al.*, 1989; Friedrich & Lacombe, 1992). However, in a nation-wide survey with a high number of sampling sites, a compromise has to be found between the desired high degree of details and the time span and labor available for the study. The present results reveal, that in a representative sample of a larger region, the hypothesis of fish diversity and complexity of the ecotone can be confirmed with this method. Undoubtedly more detailed research will be necessary to get stronger evidence for the suggested hypothesis and to study the reasons for this relation.

Physical variables such as gradient, altitude and temperature are primary factors determining species richness and community composition of the fish fauna in rivers (Sheldon, 1968; Townsend *et al.*, 1983). The present results show that the structurization of the ecotone can be considered as a secondary factor, especially important for the presence of specialized species. The importance of these structures as special habitats for young life-stages of nase (Schiemer & Spindler, 1989) and grayling or hiding places for trout (Peter, 1991) is more or less understood. For other species like riffle-minnow, however, the function of these structures is not yet clear and further research is needed to elaborate concepts for the conservation of typical rheophilic fish communities.

All parameters to describe heterogeneity of the ecotonal zone were related to the degree of modification of the river, and the more natural the studied river section, the higher the morphological variability. In order to conserve not only ubiquitous fish species but the whole fish fauna, including species with high ecological requirements, the relationship between fish diversity and heterogeneity of the ecotone has to be taken into account for river engineering works in the future. To provide suitable habitats for threatened species the banks have to be restored irregularly, creating backwaters with low flowing velocity. The riparian vegetation and alluvial forests which on most river sections are isolated nowadays, have to be reactivated by allowing periodic floodings. Still existing alluvial zones and tributaries have to be reconnected with the main river and irregularities of the river bed and the banks have to be promoted.

Restoration of big and medium rivers with their accompanying floodplains to a natural state seems unrealistic today in Europe (Welcomme, 1992). Conservation of rivers and their typical fauna therefore has to integrate all local authorities interested in the use of this resource. In proceeding step by step and realizing many small improvements in the ecotone, microhabitats of many species and overall species richness could be well improved in Swiss rivers. As the search for a biological indicator of biodiversity in aquatic habitats is still going on (Noss, 1990), these results propose the use of threatened and specialised fish species for monitoring not only the deteriorations, but also the improvements with ecological river-management.

Acknowledgements

The author would like to thank B. Zaugg for making the photographs of the sampling sites in the Suisse Romande available and for the critical review of the manuscript, T. J. Lindt Kirchhofer for the assistance in evaluation and analysis of the data and J. P. Airoldi, J. Thorpe and an anonymous reviewer for valuable remarks on the manuscript.

References

Barila, T. Y., R. D. Williams & J. R. Stauffer, 1981. The influence of stream order and selected stream bed parameters on fish diversity in Raystown branch. Susquehanna river drainage. Pennsylvania. J. appl. Ecol. 18: 125–131.

Friedrich, G., J. Lacombe (eds), 1992. Ökologische Bewertung von Fliessgewässern. G. Fischer, Stuttgart (Limnologie aktuell Bd. 3): 462 pp.

Gorman, O. T. & J. R. Karr, 1978. Habitat Structure and Stream Fish Communities. Ecology 59: 507–515.

Huet, M., 1949. Apercu des relations entre la pente et les populations piscicoles des eaux courantes. Schweiz. Z. Hydrol. 11: 332–351.

Jungwirth, M. & H. Winkler, 1983. Die Bedeutung der Flussbettstruktur für Fischgemeinschaften. Österr. Wasserwirtschaft 35: 229–234.

Jungwirth, M., 1984. Die fischereilichen Verhältnisse in Laufstauen alpiner Flüsse, aufgezeigt am Beispiel der östereichischen Donau. Österr. Wasserwirtschaft 36: 103–111.

Kirchhofer, A., B. Zaugg & J.-C. Pedroli, 1990. Rote Liste der Fische und Rundmäuler der Schweiz. Doc. Faun. Helv. 9: 23 pp.

Naiman, R., H. Décamps & F. Fournier, 1989. The role of land/inland water ecotones in landscape management and restoration: a proposal for collaborative research. Unesco, Paris, MAB Digest 4: 93 pp.

Noss, R. F., 1990. Indicators for monitoring biodiversity: A hierarchical approach. Conservation Biology 4: 355–364.

Pedroli, J.-C., B. Zaugg & A. Kirchhofer, 1991. Verbreitungsatlas der Fische und Rundmäuler der Schweiz. Doc. Faun. Helv. 11: 206 pp.

Peter, A., 1991. Ansprüche von Fischen an die Morphologie und Hydrologie der Bäche. Mitteilungen EAWAG 32: 9–13.

Schiemer, F. & T. Spindler, 1989. Endangered fish species of the Danube river in Austria. Regulated Rivers 4: 397–407.

Schiemer, F. & H. Waidbacher, 1992. Strategies for Conservation of a Danubian Fish Fauna. in P. J. Boon, P. Calow & G. E. Petts (eds), River Conservation and Management. J. Wiley & Sons, Chichester: 363–382.

Sheldon, A. L., 1968. Species diversity and longitudinal succession in stream fishes. Ecology 49: 193–198.

Spiegler, A., G. Imhof, M. Katzmann, B. Pelikan, 1989. Strukturökologische Methode zur Bestandesaufnahme und Bewertung von Fliessgewässern. Bundesministerium für Land- und Forstwirtschaft, Wasserwirtschaftskatatster, Wien: 34 pp.

Townsend, C. R., A. G. Hildrew & J. Francis, 1983. Community structure in some southern English streams: the influence of physicochemical factors. Freshwat. Biol. 13: 521–544.

Vischer, D., 1986. Schweizerische Flusskorrektionen im 18. und 19. Jahrhundert. Mitt. der Versuchsanstalt für Wasserbau. Hydrologie und Glaziologie (ETH) 84: 78 pp.

Welcomme, R. L., 1992. River Conservation – Future Prospects. in P. J. Boon, P. Calow & G. E. Petts (eds), River Conservation and Management. J. Wiley & Sons, Chichester: 453–462.

Hydrobiologia **303**: 111–124, 1995.
F. Schiemer, M. Zalewski & J. E. Thorpe (eds), The Importance of Aquatic-Terrestrial Ecotones for Freshwater Fish.

Ecotones and fluvial regimes in arctic lotic environments

G. Power & M. Power*
Department of Biology, University of Waterloo, Waterloo, Ontario, Canada N2L 3G1
** Department of Agricultural Economics, University of Manitoba, Winnipeg. Manitoba, Canada RT3 2N2*

Key words: hydrology, river ice, fish, benthos

Abstract

The fluvial regime in arctic streams is influenced by size of stream, the presence and storage capacity of lakes in the drainage basin, groundwater, and permafrost. Exchange of material with terrestrial environments depends on the season and is more or less suspended in winter. Permafrost and water storage in ice and snow control run off and erosion. Damage to vegetation by flooding, fire, animal or human activity can affect the permafrost and result in rapid erosion. Living space and conditions in lotic environments are modified by ice and during break up are very labile. Fish and benthic organisms in streams have behavioural and physiological adaptations which allow them to benefit from the wide range of river and stream environments found in the arctic and to tolerate the range of seasonal conditions. Ecotones are weakly developed in arctic streams and are of minor functional significance. Only farther south does riparian vegetation begin to play a role in regulating stream processes and not until south of the tree line can ecotones be considered important modulators of fluvial environments.

Introduction

The decision of the Scientific Advisory Committee of the UNESCO/MAB programme 'Role of Land/Inland Water Ecotones in Landscape Management and Restoration' to stress the relation between fish and riparian ecotones was intended to stimulate a multidisciplinary approach to understanding the connections between land use and water quality. In most parts of the world uncontaminated potable water is already or is quickly becoming a scarce commodity. Canada, containing more freshwater per capita (128 000 m^2) than any other country and ranking third in river flow (Healey & Wallace, 1987), is not excluded from that statement.

Fish accumulate contaminants, are sources of food and recreation, and, because of their position in aquatic food webs, have been proposed as ideal monitors of the state of aquatic systems (Maitland *et al.*, 1981; Marshall *et al.*, 1987; Munkittrick & Dixon, 1989; Munkittrick *et al.*, 1991). Because degradation of freshwaters is inevitably linked to human activities on land, it becomes logical to link land and water through the ecotones that connect them. Amongst ecologists such statements are understood but, when it comes to communication with other specialists, politicians and the public, they may not be.

Confusion arises from use of the term ecotone. Even among ecologists the meaning of the term varies depending on their background. The term was originally proposed by plant community ecologists to describe the special conditions found at the edges of homogeneous plant communities when they merged, sometimes abruptly, into other communities. The example cited in ecology texts (Odum, 1971; Smith, 1990) is the forest edge merging into meadow because this is easily explained in terms of gradients of light, temperature, wind and humidity, and increased biodiversity as organisms from each ecosystem invade the boundary and a few specialist 'edge' species are added. The Fish and Land/Inland Water Ecotones (FLIWE) group have adopted a formal definition (Holland, 1988) which is broad and inclusive, covering both temporal and spacial scales, and emphasizing that an ecotone is a transition between ecological systems. While the temporal scale may not pose a problem (ecotones pass through stages of succession and can be destroyed or created), the spacial scale does. This is apparent in the usage and discussion of the term by various authors in Naiman & Décamps (1990). The definition is 'observer-dependent' (Shugart, 1990); are 'discontinuities' in the groundwater network 'ecotones or not?' (Gilbert *et al.*, 1990); and 'when is an ecotone not an ecotone?' (Petts, 1990). Thorpe (1991) reports Dr R. J. Naiman as favouring a range of scales of eleven orders of magnitude and the 'need to keep a very fluid definition of the concept in mind, and to use the idea in whatever way was most appropriate to understanding the functional problem being studied'. This presents special problems when describing arctic streams in which boundaries between land and water may be indistinct, change seasonally, and move frequently. We prefer to think of riparian ecotones on a scale that is functional in moderating stream environments and is useful for stream restoration.

If ecologists are to successfully communicate with engineers, planners, politicians and the public, they would be wise to do so in terms which have a definite clear meaning which everyone can understand. Trying to convince someone that something you cannot define precisely is important is not likely to succeed. For this reason we are of the opinion that expanding the definition of ecotone to cover all discontinuities in nature is not wise. We feel it is better to refer directly to riparian vegetation, groundwater fluxes, or whatever discontinuity or gradient is under discussion. This avoids any ambiguity of meaning. It is possible that this makes more sense for the Arctic than it would for the Amazon. Arctic ecosystems are moulded more by abiotic than by biotic factors and in geological time are very recent.

This paper describes the hydrological and geomorphological regimes of arctic streams. These habitats are unique because permafrost and river ice exert an overriding influence on all processes. They represent a simpler, physically dominated, early stage in the evolution of more complex ecosystems and, because no distinct riparian ecotones are developed, they provide a useful contrast with running waters to the south.

Lotic arctic environments

Arctic streams and rivers have attracted less attention than lotic environments elsewhere, perhaps because fewer people live in the arctic and study is more difficult. The symposium 'Community, Structure and Function in Temperate and Tropical Streams' organised by the North American Benthological Society (JNABS, 1988), in publishing a series of important papers on stream ecology, including some of the seminal papers contributing ideas to the FLIWE group, omitted the arctic.

Arctic aquatic environments also tend to be ignored or given rather cursory treatment in ecology and limnology texts. There are signs that this is changing, partly because much of the arctic is a cold desert and water supply and waste treatment can be a problem even for small communities. Other impetuses have been provided by resource extraction and hydroelectric development as well as land settlement claims and the interests of indigenous people. An annotated bibliography of arctic stream processes (Scott, 1979) contains a selected summary of papers dealing with the physical dynamics of arctic streams, including Russian and other works not readily available. Arctic hydrology has been dealt with in Ives & Barry (1974), and Prowse & Ommanney (1990) provide a Canadian perspective. Harper (1981) reviewed the ecology of high latitude streams while Craig (1989a, 1989b), Crawford (1989) and Berkes (1990) describe the fish resources and use of them, especially by North American native subsistence fisheries.

The arctic is characterized by low temperatures and low precipitation but, from an ecological perspective, large seasonal variation in light, nutrient supply, low productivity, and the immaturity of ecosystems must be considered (Dunbar, 1968). Arctic environments assumed their present condition as a consequence of glacial events during the pleistocene and many areas have had less than 6000 years to mature. Most organisms are recent colonizers and the biota are selected from those whose dispersal mechanisms allowed their arrival and whose physiology allowed their survival. Among freshwater fish, the number of species in the MacKenzie River basin may be close to 70 because the headwaters are in temperate and sub-arctic areas (Lindsey & McPhail, 1986) but in the high arctic there are less than eight species (Crossman & McAllister, 1986). In many rivers there are no fish or Arctic char (*Salvelinus alpinus*) is the only species. The reasons for this are seasonally impossible conditions and lack of access; most colonization was by euryhaline species via coastal routes (Power *et al.*, 1973).

The landscape in the arctic is still in various stages of post-glacial recovery. Eustatic changes in sea level and isostatic rebound of the land have resulted, in many places, in changes in relative position of sea level of over 300 m during the last 6000 years (Andrews, 1968). An annual cycle of freezing and thawing is believed to shatter rock with entrained water (see Fraser, 1959, for discussion) and sort sediments into patterned ground. Fluvial processes during the brief summer are dominant in sculpting the landscape, especially in the ice-free high arctic (Scott, 1979).

Permafrost affects the hydraulic regime; it is continuous throughout the arctic and can reach 1000 m in thickness, penetrating under the continental shelf. Permafrost is rock or sediment which remains below 0 °C year round. It occurs north of the 0 °C mean annual air temperatures isotherm as discontinuous patches in the south and as continuous frozen ground in the north. Relief, exposure, ground insulation and snow distribution affect the distribution and the depth of the active layer which melts each summer. Lakes and streams transfer heat to the ground and, because of the thermal properties of water, maintain patches or ribbons of unfrozen ground beneath them called taliks. Even in areas of permafrost there is some flow of ground water and pockets of unfrozen water (with higher dissolved solids due to freezing out) may remain. Springs occur where groundwater aquifers break to the surface. The many small lakes and ponds on the tundra are caused by poor drainage and melt water accumulation in depressions which may dry by evaporation or drain as the active layer deepens during the summer. A full description of the effects of permafrost on hydrology is provided by van Everdingen (1987) and Prowse & Ommanney (1990).

During most of the year precipitation is in solid form, is shifted around by wind, and is stored until released by radiant heat during the brief summer. Runoff is rapid over a frozen substrate and more than 50% of the annual discharge may be concentrated in a week or two. Small streams freeze solid in winter; other streams are maintained by springs or lake outflow. Even large rivers may be reduced to a few deep pools. On the Russian and North American mainland, the largest rivers continue to flow, although at very reduced volumes, being sustained by headwaters in subarctic and temperate regions. Figure 1 shows some typical hydrographs. Where springs maintain flows, these often produce downstream icings or aufeis (Fig. 2). When these melt they contribute to maintaining summer discharge (Kane & Slaughter, 1972). Satellite imagery has been used to map icings (Harden *et al.*, 1977) which can attain thicknesses of 56 m and store 25–30% of the annual discharge. Their presence indicates a source of groundwater. The importance of fluvial processes increases in the periglacial environment. In the Mecham River, Cornwallis Island, discharge showed a diurnal pattern relating to snowmelt, and 90% of the annual runoff occurred in early June, initially over ice (McCann *et al.*, 1972).

Movement of sediment is correlated with discharge. In the Colville River, Alaska, 43% of the

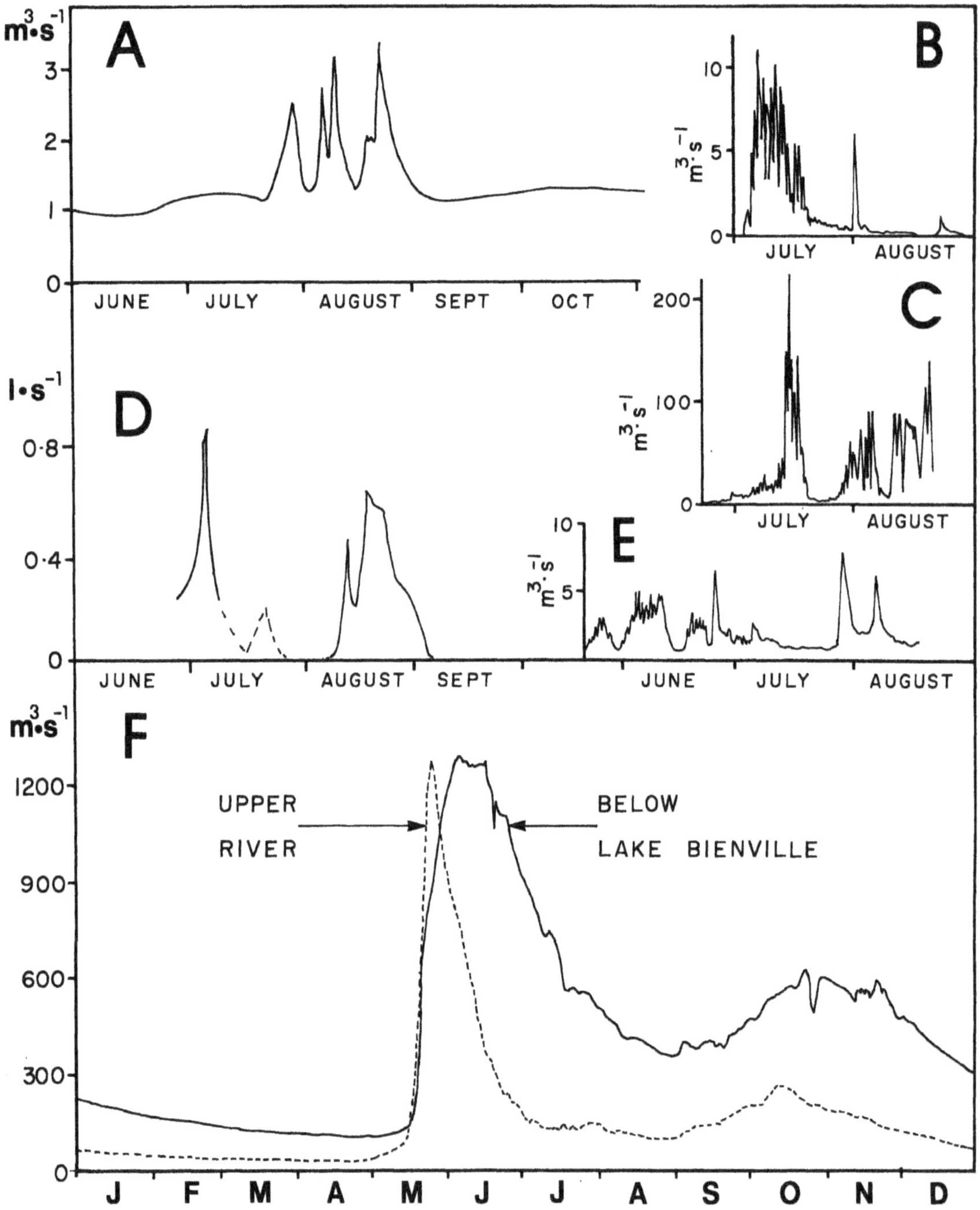

Fig. 1. Hydrographs of typical arctic streams and a subarctic river.

A. Canning Spring, Alaska, 1973, even flow with peaks corresponding to summer storms.
B. McMaster River, Cornwallis Island, N.W.T., 1978, 33 km^2, no winter discharge, snowmelt runoff and peaks due to summer rains.
C. Sverdrup River, Ellesmere Island, N.W.T., 1974, 1630 km^2, snowmelt runoff followed by glacier melt runoff.
D. Stream, Barrow, Alaska, 1973, 1.6 km^2, no winter flow.

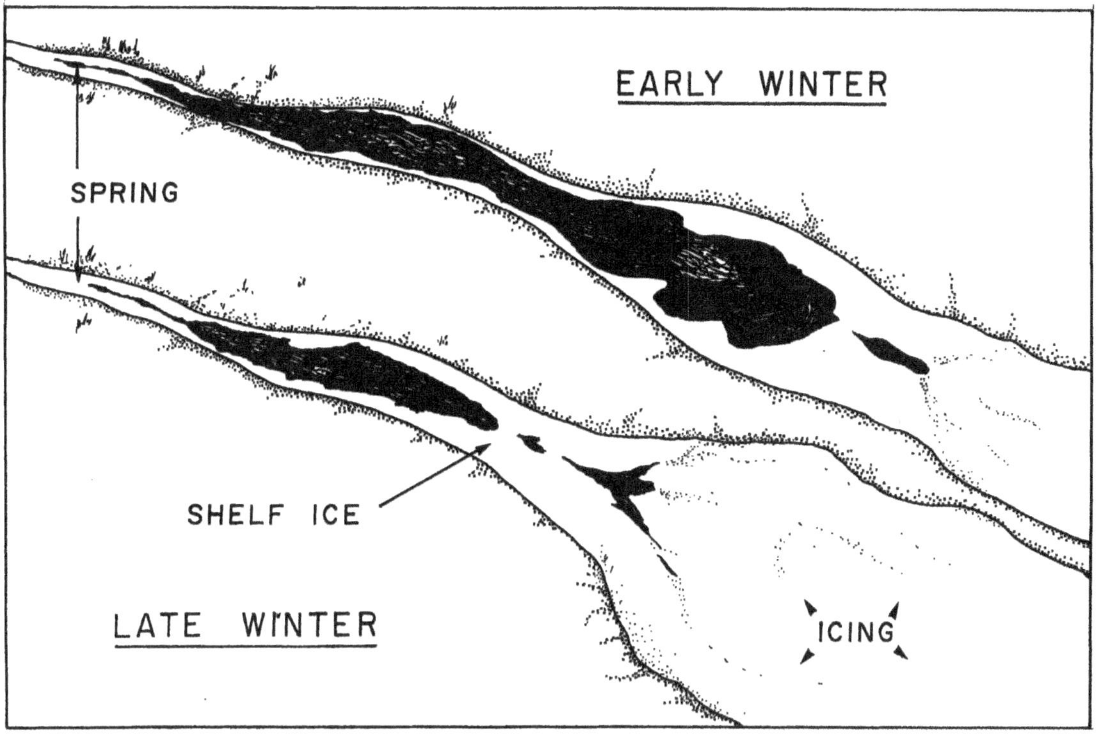

Fig. 2. Diagram of an icing (aufeis) below a spring. In early winter frazil and sheet ice obstruct the flow at a shallow area, water overflows and freezes forming the icing.

discharge and 73% of the suspended sediment load moved to sea in a three week period (Arnborg *et al.*, 1967). Church (1972) reported that sandurs (valley trains) were major post glacial features in the Canadian arctic and snowmelt and summer storm runoff transported most sediment. Storms moved a lot of material because the ground was not protected by vegetation and, below the active layer, the ground was frozen. Heavier than normal rainfall can cause earth flows (Carter & Galloway, 1981). Pulses from disruption of ice ponded water can also be important. Similarly, floods from break-through of snow dammed lakes can produce rapid damage and erosion of unfrozen bed and bank material (Prowse & Ommanney, 1990). Cogley & McCann (1976) describe the effects of an exceptional storm on Ellesmere Island during which local precipitation reached 55 mm. Storm runoff inundated flood plains, caused gully erosion, and massive movement of coarse alluvium.

River channel erosion can be rapid. Using aerial photographs, the maximum erosion on Little Smith Creek, N.W.T., between the 1940s and 1972, was estimated to be about 1 m y^{-1} on the outside curves of meanders (Day & Egginton, 1976). Bank erosion varies according to whether the ground is frozen, thermal exposure, wind

E. Dietrich River, Alaska, 1979, 45 km^2, extended snowmelt through May and June due to cold weather with peaks due to precipitation in summer.
F. Great Whale River, Québec, 1979, 42000 km^2, upper river shows sharp runoff peak which is dampened by Lake Bienville, 2300 km^2, in mid river.

(After Figure 3.1 Prowse and Ommanney (1990) and Hydro Québec.)

direction, and near shore current. Rates of bank loss can reach 30 m y^{-1} (Outhet, 1974, in Scott, 1979). Erosion can also cause substantial changes in permafrost distribution. In the Eagle River, Yukon, as the outer bank of a meander loop eroded into and migrated across the flood plain, the permafrost melted back. Point bars of finer silt material, which formed behind the meander loop, developed permafrost (Crampton, 1979).

Vegetation plays an important role in bank stability by insulating ground in summer and accumulating an insulating blanket of snow in winter. Flooding can transport heat and damage insulating vegetation by siltation (Viereck, 1973). Fire can remove vegetation, although the frequency and extent of fire lessens northwards (Wein, 1976; Payette *et al.*, 1989; Filion *et al.*, 1991). Fire frequency was related to weather and drier, warmer conditions were correlated with larger areas burned. The wildfire rotation period rose from 100 y in the northern boreal forest to 9320 y in the shrub tundra and fires became smaller. The average size of tundra fires in northern Québec in recent years was 81 ± 181 ha and fires tended to avoid the wetter valley bottoms (Payette *et al.*, 1989). The pattern of burns in the Northwest Territories was similar, although exact comparisons are difficult because of the variability of fires on an annual or decadal time scale (Timoney & Wein, 1991). Animals can affect the streamside vegetation by overgrazing or trampling, thus changing the insulating properties of the surface. This has been evident in northern Québec in recent years because of the size of the George River caribou herd (Power & Barton, 1987). Recovery from overgrazing by caribou in the arctic can be rapid if plants maintain viable underground roots. Within a year, willow and grasses had recovered from caribou overgrazing on Rideout Island, N.W.T., but lichen might take 20 y (Henry & Gunn, 1991). Beaver are moving north. They have been observed at the tree line in northern Québec and are increasingly common in subarctic regions of North America. Their effects on hydrology and riparian vegetation are well documented (Naiman *et al.*, 1986, 1988; Johnston & Naiman, 1990; Woo & Waddington, 1990). They can also affect permafrost, bank stability and erosion by their flooding and removal of streamside insulating vegetation. Human activities are increasing the risk of damage to arctic streamside vegetation by use of all-terrain vehicles (ATVs). Especially in summer, these produce track depressions that can become channels for erosion or affect insulation and cause a deepening of the active layer above the permafrost table (Lawson, 1986; Racine & Ahlstrand, 1991). Changes in the thermal regime of the ground can melt permafrost and lead to soil subsidence or increased erosion, especially along stream margins. Retrogressive thaw slumps may follow disturbance and can reach large size with expansion rates of the head walls of about 10 m y^{-1}. Stabilization occurs when the insulating layer of vegetation regenerates and the permafrost has re-established, which can take 40 y (Burn & Friele, 1989).

River ice

Ice is an important environmental component of arctic rivers. Its role in stream ecology has tended to be underestimated. Ice is listed among the factors contributing to patch formation in riparian vegetation by Wissmar & Swanson (1990) but it is not discussed in the Ecology and Management of Aquatic-Terrestrial Ecotones (Naiman & Décamps, 1990). Ice is not restricted to the arctic; it only exerts some of its more extreme effects there. A review of ice effects can contribute ideas about processes which may occur in temperate rivers. As Pinay *et al.* (1990) remarked, there is reason to think that comparative studies across time and space scales will contribute answers to emerging questions in river ecology.

Ice occurs in a number of forms in lotic environments. These are: anchor ice, which forms on the beds of streams, most frequently in shallow riffles; frazil ice, which is in the form of small crystals and is generated in turbulent water; slush, from wind blown snow; and sheet ice, which forms on the surface, spreading out from the edges. As ice expands and thickens, living space for stream biota contracts (Fig. 3). In small

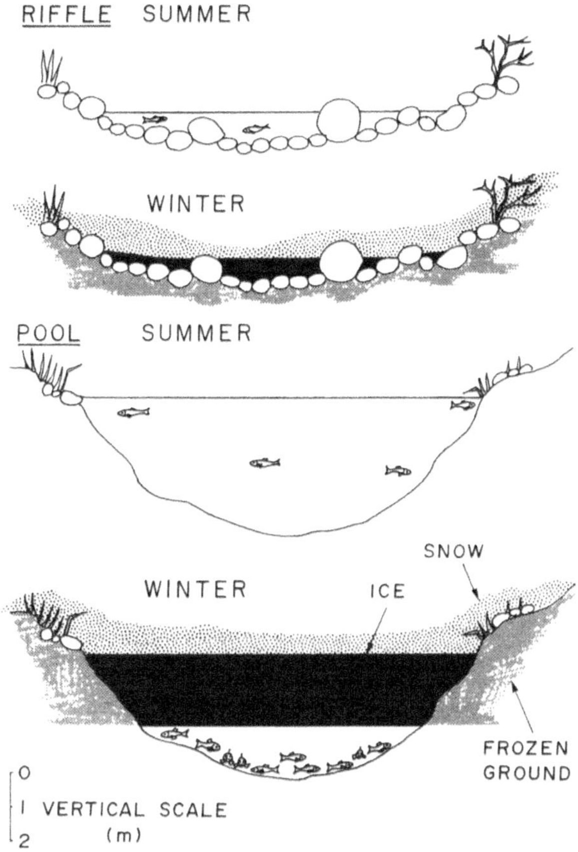

Fig. 3. Living space in an arctic stream riffle disappears in winter and the stream bed may freeze. In a pool a small refugium may remain if it is deeper than the approximately 2 m of ice which forms. Note that only invertebrates capable of withstanding <0 °C temperatures can survive in the frozen ground.

streams, no liquid water may remain and any biota, to survive, must have mechanisms to resist freezing. No fish can tolerate freezing but some arthropods can.

The more primitive orders of Odonata, Plecoptera and Ephemeroptera do not occur in the high arctic nor do Simulidae (Oliver *et al.*, 1964). As the duration and intensity of freezing increases the fauna becomes more specialised (Harper, 1981). Variations in climate may make marginal habitats more or less available. Autumn rainfall had a marked influence on groundwater and the availability of unfrozen refugia in an Alaskan stream (Irons *et al.*, 1989). Arctic stream communities are often dominated by chironomids (Stocker & Hynes, 1976; Miller & Stout, 1989). When habitats freeze temperatures can get very low. Andrews & Rigler (1985) reviewed the literature and believed that many invertebrates could survive temperatures as low as −18 °C. Lotic invertebrates that cannot remain in water have only two options, supercooling or freeze tolerance (Oswood *et al.*, 1991), either of which, within limits, can allow survival. Freshwater fish do not possess these capabilities and must find winter refuge in springs, deeper pools in large rivers, or in lakes.

Fish behaviour associated with overwintering includes migration to the site, suppression of normal aggressive behaviour and tolerating long periods without food. In temperate regions this may mean movement from the preferred riffle habitat to avoid anchor ice, congregating near groundwater inflows, and use of energy stores, or, for small fish, burrowing into the gravel stream bed (Cunjak & Power, 1986, Cunjak, 1988). Arctic fish are well adapted to live for a long period without food (Boivin & Power, 1990).

Catastrophic events for stream fauna are often precipitated by ice. Anchor ice can form overnight and release daily. It can move bed materials and accumulate downstream causing temporary dams. If it builds quickly to any thickness it can restrict flow and divert it to the edges, over top banks and cause erosion. Frazil ice can accumulate (along with slush) in pools where it may form a curtain dam and perhaps play a role in maintaining pool depth (Fig. 4). Too much accumulation reduces the value of the pool as an overwintering site. Ice dams downstream of groundwater maintained flows cause icings. Oxygen depletion can occur beneath an ice cover (Schreier *et al.*, 1980; Whitfield & McNaughton, 1986; Schmidt *et al.*, 1989). The most serious problems relating to ice occur during break-up, which can happen during a mid-winter thaw or in spring. Calkins *et al.* (1989) observed movements of boulders up to 70 kg in a Vermont stream during ice release in spring. Damage to man-made river works usually occurs at this time, especially when ice jams form at constrictions, bridge piers or at the head of shallows. Ice pans raft up, diverting

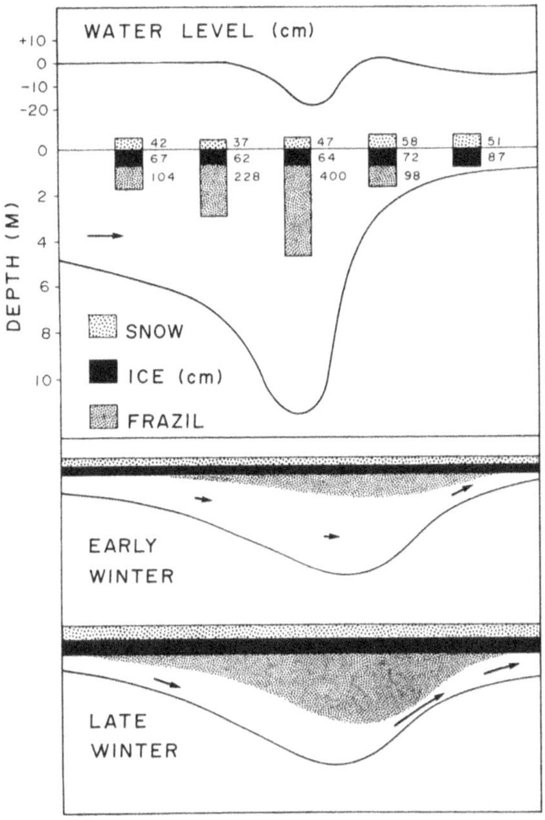

Fig. 4. Upper panel shows ice conditions in Big Hole Pool, Northwest Miramichi River, March 4 1992. Snow, ice and frazil thickness (cm) is recorded every 50 m along the pool. Water levels in the core holes are measured against the ice surface. Lower panel shows a conceptual model of development of ice cover in a pool downstream of a frazil generating riffle(s) during winter. Note the constriction of living space and acceleration of flows which may help scour the bed of the pool as frazil accumulates.

flow into the flood plain. Dewatering occurs downstream. Fish and invertebrates can be stranded on the flood plain or downstream of the jam (Beltaos *et al.*, 1990). Ice scouring can do considerable damage to stream banks, pushing up bed materials, scraping and uprooting stream bank vegetation. Ice push ramparts (Peterson, 1965) are a common feature along the shores of river widenings and are due to wind driven ice rafting up on shore. Similar features can be observed along the banks of large subarctic rivers, such as the Koksoak in northern Québec, where we have observed trees pushed back up to 50 m from the bank and ice produced gouges along the banks up to 500 m long.

Evidence of past water levels can be reconstructed from the effects of ice scour on bank vegetation. Levashov (1966) determined the flood frequency for streams in the Kola Peninsula over the last 300 years using evidence of ice damage to river bank pine trees. Bégin & Payette (1988, 1989, 1991) used this technique to reconstruct changes in water level and climate in subarctic Québec back to the 17th century.

The biological effects of river ice on the population dynamics of the lotic fauna are difficult to evaluate. A conclusion of a Canadian National Research Council Task Force review of the subject, which has not yet been released, was that 'temperate and arctic faunas have evolved to deal with the suite of seasonally predictable conditions' and, only when these are greatly exceeded, is serious damage likely.

Discussion and conclusions

Arctic rivers (Fig. 5) offer a chance to move back in time and view earlier stages of the geomorphological processes and biological successions that have produced the environments we are familiar with farther south. North of the sub-arctic, riparian vegetation is absent or only poorly developed. River processes are very much controlled by climate. For long periods each year, exchange with terrestrial biomes (except for blowing snow and aeolian sediment) is zero for small streams. Springs and exchanges with the groundwater in the talik maintain minimal flows in intermediate rivers. Large mainland rivers are nourished by sub-arctic and temperate headwaters and are not strictly arctic. Runoff is concentrated in a very short period and discharges are strong enough to erode and transport substrate, producing difficult conditions for the biota. Although arctic regions are relatively untouched, they are not immune to human population pressures. Local populations in Canada have a doubling time of about 25 y (Robitaille & Choinier, 1985), and resource needs and the strategic interests of the south are

impacting the region. The possibility of long range transport of atmospheric contaminants (Vermette & Bingham, 1986; Muir *et al.*, 1990) is also of concern. The purity of many arctic waters means that small atmospheric additions can have a disproportionate effect.

Arctic rivers and streams might be viewed as systems without riparian ecotones and not relevant to any discussion of the functional role such ecotones play in modulating terrestrial aquatic exchanges and stabilizing stream channels. We believe this would be a mistake because arctic

Fig. 5. Views of some arctic rivers

A. A small river in S.W. Iceland near Thorsmörk national park showing a braided channel and wide outwash plain.

B. The braided channel of Storelva close to the limit of trees in N. Norway. Riparian ecotones are poorly developed.

C. The Lefroy river at summer discharge of about 20 $m^3 s^{-1}$ almost disappears as it cuts through a boulder moraine in the tundra zone of N. Québec.

D. The Barnoin river at high summer discharge is confined to a definite channel in the bedrock. It is at the tree line in N. Québec. Riparian vegetation is restricted to arctic shrubs and a few small stunted trees in favourable locations away from the coast.

fluvial environments demonstrate conditions in which ecotones are weakly developed and where they are regularly damaged or removed by seasonal weather patterns. Schlosser (1995) makes the point that, in headwater streams, the streamside vegetation is very important in maintaining stream landscape heterogeneity and providing the microhabitats needed by different life history stages of stream fishes. There is clearly less habitat differentiation in arctic streams and few or no fish species. Those fish that do live there must adapt, making temporary use of suitable habitats as they become available during the open water season and retreating to refugia for the rest of the year.

In the arctic ecotonal processes are weighted in favour of pulsed physical movement of inorganic materials from the land to running waters and the organic exchange is rather small. The only exception may be peat eroded from deposits accumulated over long periods of time in a few localities which contributes little to arctic stream food webs. Over a latitudinal or climatic gradient southwards the balance changes, physical transport of materials lessens and large organic debris become much more important in structuring the fluvial environment. Where riparian ecotones are well developed, their removal functionally impairs streams and moves them in the direction of abiotically dominated arctic systems. Biological diversity and productivity is reduced and biomass variability increases. Conversely, restoration projects should allow degraded streams to function normally again.

In conclusion, arctic lotic environments should be included in the UNESCO/MAB FLIWE programme for at least the following reasons:

- they are at one extreme of a continuum, at the end of which abiotic factors predominate;
- they are immature systems showing early phases of the more complex terrestrial-aquatic interconnections that have evolved farther south;
- they are biologically the simplest systems because of their restricted flora and fauna and, therefore, should be easier to understand;
- low temperatures and extreme seasonality in conditions should offer some advantages for study and reveal new adaptations for survival under pulsed periods of high and suspended activity;
- the relative lack of development of nutrient retention systems in arctic rivers implies a greater export of materials to the marine environment and raises questions about the biological inefficiency of immature ecotones and the implications for productivity in bordering environments; and
- because arctic streams are at an extreme, they should be more sensitive to climate change than running waters elsewhere. Changes in the permafrost are likely to have great repercussions in northern fluvial systems, including changes in drainage and sediment transport.

The final point immediately raises questions of considerable interest in an era pre-occupied with the implications of global warming. For example, how quickly will riparian ecotones become established under a warmer climate? Secondly, how effective will such ecotones be in stabilizing the otherwise pliable arctic landscape?

There are, of course, some difficulties and disadvantages associated with arctic research, including a tendency for it to be short term and costly. The list above could probably be lengthened but it is large enough to warrant more discussion.

Acknowledgement

We would like to thank R. A. Cunjak for the data in Fig. 4 and for many ideas and discussions about winter conditions in streams.

References

Andrews, J. T., 1968. Post-glacial rebound in arctic Canada: similarity and prediction of uplift curves. Can. J. Earth Sci. 5: 39–47.

Andrews, D. & F. Rigler, 1985. The effects of an Arctic winter on benthic invertebrates in the littoral zone of Char Lake, Northwest Territories. Can. J. Zool. 63: 2825–2834.

Arnborg, L., H. J. Walker & J. Peippo, 1967. Suspended load

in the Colville River Alaska, 1962. Georgrafiska Annaler, Series A. 49: 131–144.
Bégin, Y., & S. Payette, 1988. Dendroecological evidence of lake-level changes during the last three centuries in subarctic Québec. Quaternary Research 30: 210–220.
Bégin, Y. & S. Payette, 1989. La végétation riveraine du lac à l'Eau Claire, Québec subarctique. Géographie physique et Quaternaire 43: 39–50.
Bégin, Y. & S. Payette, 1991. Population structure of lakeshore willows and ice-push events in subarctic Québec, Canada. Holarct. Ecol. 14: 9–17.
Beltaos, S., R. Gerard, S. Petryk & T. D. Prowse, 1990. Working group on river ice jams, Field studies and research needs. National Hydrology Research Institute Science Report 2, Environment Canada, Saskatoon, Saskatchewan, Canada, 121 pp.
Berkes, F., 1990. Native subsistence fisheries: a synthesis of harvest studies in Canada. Arctic 43: 35–42.
Boivin, T. G. & G. Power, 1990. Winter condition and proximate composition of anadromous arctic charr (*Salvelinus alpinus*) in eastern Ungava Bay, Québec. Can. J. Zool. 68: 2284–2289.
Burn, C. R. & P. A. Friele, 1989. Geomorphology, vegetation succession, soil characteristics and permafrost in retrogressive thaw slumps near Mayo, Yukon Territory. Arctic 42: 31–40.
Calkins, D. J., L. W. Gatto & B. E. Brockett, 1989. Field assessment of fisheries habitat enhancement structures in Bingo Brook, Vermont, after the spring, 1989, ice run. XXIII International Association for Hydraulic Research Congress, Ottawa. August, 1989.
Carter, L. D. & J. P. Galloway, 1981. Earth flows along Henry Creek, Northern Alaska. Arctic 34: 325–328.
Church, M. A., 1972. Baffin Island sandurs: a study of arctic fluvial processes. Geological Survey of Canada. Bulletin 216, 208 pp.
Cogley, J. G. & S. B. McCann, 1976. An exceptional storm and its effects in the Canadian High Arctic. Arctic Alpine Research 8: 105–110.
Craig, P. C., 1989a. An introduction to anadromous fishes in the Alaskan Arctic. In Norton, D. W. (ed.), Research Advances on Anadromous Fish in Arctic Alaska and Canada, Nine papers contributing to an ecological synthesis. Biological Papers of the University of Alaska No. 24: 27–54.
Craig, P. C., 1989b. Subsistence fisheries at coastal villages in the Alaskan arctic, 1970–1986. In Norton, D. W. (ed.), Research Advances on Anadromous Fish in Arctic Alaska and Canada. Nine papers contributing to an ecological synthesis. Biological Papers of the University of Alaska No. 24: 131–152.
Crampton, C. B., 1979. Changes in permafrost distribution produced by a migrating river meander in northern Yukon, Canada. Arctic 32: 148–151.
Crawford, R., 1989. Exploitation of Arctic fishes. Canadian Manuscript Report of Fisheries and Aquatic Sciences 2002, 43 pp.
Crossman, E. J. & D. E. McAllister, 1986. Zoogeography of freshwater fishes of the Hudson Bay drainage, Ungava Bay and the Arctic Archipelao. In Hocutt, C. H. & B. O. Wiley (eds), The Zoogeography of North American Freshwater Fishes. Wiley, New York: 53–104.
Cunjak, R. A. & G. Power, 1986. Winter habitat utilization by stream resident brook trout (*Salvelinus fontinalis*) and brown trout (*Salmo trutta*). Can. J. Fish. aquat. Sci. 43: 1970–1981.
Cunjak, R. A., 1988. Physiological consequences of overwintering in streams: The cost of acclimatization? Can. J. Fish. aquat. Sci. 45: 443–452.
Day, T. J. & P. A. Egginton, 1976. River channel instability studies, District of Mackenzie. Geological Survey of Canada. Paper 76-1C: 207–217.
Dunbar, M. J., 1968. Ecological development in polar regions. Prentice-Hall, New Jersey, 119 pp.
Filion, L., D. Saint-Laurent, M. Desponts & S. Payette, 1991. The late Holocene record of aeolian and fire activity in northern Québec, Canada. Holocene 1: 201–208.
Fraser, J. K., 1959. Freeze-thaw frequencies and mechanical weathering in Canada. Arctic 12: 40–53.
Gilbert, J., M-J. Dole-Olivier, P. Marmonier & P. Vervier, 1990. Surface water-groundwater ecotones. In Naiman, R. J. & H. Décamps (eds), The Ecology and Management of Aquatic-Terrestrial Ecotones. Man and the Biosphere, Vol. 4. Parthenon Publishing Group, UNESCO, Paris: 199–226.
Harden, D., P. Barnes & E. Reimnitz, 1977. Distribution and character of naleds in northeastern Alaska. Arctic 30: 28–40.
Harper, P. P., 1981. Ecology of streams at high latitudes. In Lock, M. A. & D. D. Williams (eds), Perspectives in Running Water Ecology. Plenum Press, New York: 313–337.
Healey, M. C. & R. R. Wallace, 1987. Canadian aquatic resources: an introduction. In Healey, M. C. & R. R. Wallace (eds), Canadian Aquatic Resources. Can. Bull Fish. aquat Sci. 215: 1–11.
Henry, G. H. R. & A. Gunn, 1991. Recovery of tundra vegetation to overgrazing by caribou in arctic Canada. Arctic 44: 38–42.
Holland, M. M. (compiler), 1988. SCOPE/MAB technical consultations on landscape boundaries: report of a SCOPE/MAB workshop on ecotones. Biology International, Special Issue 17: 1–163.
Irons III, J. G., S. R. Ray, L. K. Miller & M. W. Oswood, 1989. Spatial and seasonal patterns of streambed water temperatures in an Alaskan subarctic stream. Headwaters Hydrology American Water Research Association June 1989: 381–390.
Ives, J. D. & R. G. Barry (eds), 1974. Arctic and Alpine Environments. Methuen, London, 997 pp.
J.N.A.B.S., 1988. Community Structure and Functions in Temperate and Tropical Streams. J. North Am. Benthol. Soc. 7: 261–524.

Johnston, C. A. & R. J. Naiman, 1990. Aquatic patch creation in relation to beaver population trends. Ecology 71: 1617–1621.

Kane, D. L. & C. W. Slaughter, 1972. Seasonal regime and hydrological significance of stream icings in central Alaska. In The Role of Snow and Ice in Hydrology: Proceedings of Banff Symposia, Sept. 1972, International Association of Scientific Hydrology 1: 528–540.

Lawson, D. E., 1986. Response of permafrost terrain to disturbance: A synthesis of observations from northern Alaska, U.S.A. Arctic Alpine Research 18: 1–17.

Levashov, A. A., 1966. Approximate determination of high flood frequency in rivers without hydrological observations. Soviet Hydrology: Selected Papers 547–548 (Published in English American Geophysical Union).

Lindsey, C. C. & J. D. McPhail, 1986. Zoogeography of fishes of the Yukon and Mackenzie basins. In Hocutt, C. H. & E. O. Wiley (eds), The Zoogeography of North American Freshwater Fishes. Wiley, New York: 53–104.

Maitland, P. S., H. A. Regier, G. Power & N. A. Nilsson, 1981. A wild salmon, trout and char watch: an international strategy for salmonid conservation. Can. J. Fish. aquat. Sci. 38: 1882–1888.

Marshall, T. R., R. A. Ryder, C. J. Edwards & G. R. Spangler, 1987. Using the lake trout as an indicator of ecosystem health – applications of the dichotomous key. Great Lakes Fisheries Commission Technical Report 49.

McCann, S. B., P. J. Howarth & J. G. Cogley, 1972. Fluvial processes in a periglacial environment, Queen Elizabeth Islands, N.W.T., Canada. Trans. Inst. br. Geographers 55: 69–82.

Miller, M. C. & J. R. Stout, 1989. Variability of macroinvertebrate community composition in an arctic and subarctic stream. Hydrobiologia 172: 111–127.

Muir, D. C. G., C. A. Frod, N. P. Grift, D. A. Metner & W. L. Lockhart, 1990. Geographic variation of chlorinated hydrocarbons in burbot (*Lota lota*) from remote lakes and rivers in Canada. Arch. envir. Contam. Toxic. 19: 530–542.

Munkittrick, K. R. & D. G. Dixon, 1989. Use of white sucker (*Catostomus commersoni*) populations to assess the health of aquatic ecosystems exposed to low-level contaminant stress. Can. J. Fish. aquat. Sci. 46: 1455–1462.

Munkittrick, K. R., C. B. Portt, G. J. Van Der Kraak, I. R-. Smith & D. A. Rokosh, 1991. Impact of bleached kraft mill effluent on population characteristics, liver MFO activity, and serum steroid levels of a Laker Superior white sucker (*Catostomus commersoni*) population. Can. J. Fish. aquat. Sci. 48: 1371–1380.

Naiman, R. J., J. M. Melillo & J. E. Hobbie, 1986. Ecosystem alteration of boreal forest streams by beaver (*Castor canadensis*). Ecology 67: 1254–1269.

Naiman, R. J., C. A. Johnson & J. C. Kelly, 1988. Alteration of north American streams by beaver. BioScience 38: 753–762.

Naiman, R. J. & H. Décamps (eds), 1990. The Ecology and Management of Aquatic-Terrestrial Ecotones. Man and the Biosphere, vol. 4. Parthenon Publishing Group, UNESCO, Paris, 316 pp.

Odum, E. P., 1971. Fundamentals of Ecology, 3rd edn. W. B. Saunders Company, Philadelphia.

Oliver, D. R., P. S. Corbet & J. A. Downes, 1964. Studies on arctic insects: the Lake Hazen project. The Canadian Entomologist 96: 138–139.

Oswood, M. W., L. K. Miller & J. G. Irons, III, 1991. Overwintering of freshwater benthic macroinvertebrates. In R. E. Lee, Jr. & D. L. Denlinger (eds), Insects at Low Temperature, Chapman and Hall, New York and London: 360–375.

Payette, S., C. Morneau, L. Sirois & M. Desponts, 1989. Recent fire history of the northern Québec biomes. Ecology 70: 656–673.

Peterson, J. A., 1965. Ice-push ramparts in the George River Basin, Labrador-Ungava. Arctic 18: 189–193.

Petts, G. E., 1990. The role of ecotones in aquatic landscape management. In R. J. Naiman & H. Décamps (eds), The Ecology and Management of Aquatic-Terrestrial Ecotones. Man and the Biosphere, Vol. 4. Parthenon Publishing Group, UNESCO, Paris: 227–262.

Pinay, G., H. Décamps, E. Chauvet & E. Fustec, 1990. Functions of ecotones in fluvial systems. In R. J. Naiman & H. Décamps (eds), The Ecology and Management of Aquatic-Terrestrial Ecotones. Man and the Biosphere, Vol. 4. Parthenon Publishing Group, UNESCO, Paris: 141–170.

Power, G. & D. R. Barton, 1987. Some effects of physiographic and biotic factors on the distribution of anadromous Arctic char (*Salvelinus alpinus*) in Ungava Bay, Canada. Arctic 40: 198–203.

Power, G., G. F. Pope & B. W. Coad, 1973. Postglacial colonization of the Matamek River by fishes. J. Fish Res. Bd Can. 30: 1586–1589.

Prowse, T. D. & C. S. L. Ommanney (eds), 1990. Northern hydrology: Canadian perspectives. National Hydrology Research Institute Science Report 1. Environment Canada, Saskatoon, Saskatchewan, 308 pp.

Racine, C. H. & G. H. Ahlstrand, 1991. Thaw response of tussock-shrub tundra to experimental All-Terrain Vehicle disturbances in south-central Alaska. Arctic 44: 31–37.

Robitaille, M. & R. Choinier, 1985. An overview of demographic and socio-economic conditions of the Inuit in Canada. Ottawa: Indian and Northern Affairs, Canada, Research Branch Corporate Policy, 112 pp.

Schlosser, I. J., 1995. Critical landscape attributes that influence fish population dynamics in headwater streams. Hydrobiologia 303 (Dev. Hydrobiol. 105): 71–81.

Schmidt. D. R., W. B. Griffiths & L. R. Martin, 1989. Overwintering biology of anadromous fish in the Sagavanirktok River delta, Alaska. In D. W. Norton (ed.), Research Advances on Anadromous Fish in Arctic Alaska and Canada, Nine papers contributing to an ecological synthesis. Biological Papers of the University of Alaska 24: 55–74.

Schreier, H., W. Erlebach & L. Albright, 1980. Variations in water quality during winter in two Yukon rivers with emphasis on dissolved oxygen concentration. Wat. Res. 14: 1345–1351.

Scott, K. M., 1979. Arctic stream processes – an annotated bibliography. Geological survey water-supply paper 2065. U.S. Government Printing Office, Washington, 78 pp.

Shugart, H. H., 1990. Ecological models and the ecotone. In R. J. Naiman & H. Décamps (eds), The Ecology and Management of Aquatic-Terrestrial Ecotones. Man and the Biosphere, Vol. 4. Parthenon Publishing Group, UNESCO, Paris: 23–36.

Smith, R. L., 1990. Ecology and Field Biology, 4th edn. Harper and Row Publishing, New York, 922 pp.

Stocker, S. J. & H. B. N. Hynes, 1976. Studies on the tributaries of Char Lake, Cornwallis Island, Canada. Hydrobiologia 49: 97–102.

Thorpe, J. E., 1991. Review of the introductory session: general considerations. In M. Zalewski, J. E. Thorpe & P. Gaudin (eds), Fish and Land/Inland-Water Ecotones. UNESCO MAB: 11–15.

Timoney, K. P. & R. W. Wein, 1991. The areal pattern of burned tree vegetation in the subarctic region of Northwestern Canada. Arctic 44: 223–230.

van Everdingen, R. O., 1987. The importance of permafrost in the hydrological regime. In M. C. Healey & R. R. Wallace (eds), Canadian Aquatic Resources, Can. Bull. Fish. aquat. Sci. 215: 243–276.

Vermette, S. J. & V. G. Bingham, 1986. Trace elements in Frobisher Bay rainwater. Arctic 39: 109–118.

Viereck, L. A., 1973. Ecological effects of river flooding and forest fires on permafrost in the taiga of Alaska. Proceedings of the 2nd International Permafrost Conference, Yakutsk. National Academy of Science: 60–67.

Wein, R. W., 1976. Frequency and characteristics of arctic tundra fires. Arctic 29: 213–222.

Whitfield, P. H. & B. McNaughton, 1986. Dissolved oxygen depressions under ice cover in two Yukon rivers. Wat. Resour. Res. 22: 1675–1679.

Wissmar, R. C. & F. J. Swanson, 1990. Landscape disturbances and lotic ecotones. In: The ecology and management of aquatic-terrestrial ecotones. In R. J. Naiman & H. Décamps (eds), The Ecology and Management of Aquatic-Terrestrial Ecotones. Man and the Biosphere, Vol. 4. Parthenon Publishing Group, UNESCO, Paris: 65–90.

Woo, M.-K. & J. M. Waddington, 1990. Effect of beaver dams on subarctic wetland hydrology. Arctic 43: 223–230.

Hydrobiologia **303**: 125–137, 1995.
F. Schiemer, M. Zalewski & J. E. Thorpe (eds), The Importance of Aquatic-Terrestrial Ecotones for Freshwater Fish.

Microhabitat use by two small benthic stream fish in a 2nd order stream

Irene Zweimüller
Inst. f. Zoology, Dept. Limnology, A-1090 Vienna, Austria

Key words: microhabitat use, stone loach, gudgeon, discharge fluctuations, hydrology

Abstract

Direct observation of two benthic fish species – the stone loach (*Barbatula barbatula*) and the gudgeon (*Gobio gobio*) – in the field revealed a spatial segregation between the species in a study area with shallow riffle and moderately deeper pool sections. Stone loach generally inhabited shallow, more current-exposed locations and gudgeon preferred deeper, mostly sandy areas. The small individuals of both species were confined to shallow muddy locations and the larger individuals were found in deeper and more current exposed areas.

The main factor affecting microhabitat choice was the flow regime:

– it was a limiting factor for the stone loach, where the discharge rates controlled the presence of fish in the study area.

– distribution patterns of both species were influenced by discharge and by fluctuations in discharge.

The following mechanisms regulating the distribution of stone loach and gudgeon are hypothesized:

(A) Gudgeon: They prefer high water depth, low current velocities and sandy substrate, which strongly limits their spatial niche. Mainly relatively small individuals (size class 2; 6–9 cm) changed microhabitat in relation to environmental parameters. Size class 3 (approx. 9–12 cm) may be interpreted as a rather unpredictable transitory period between juvenile and adult stage. Large gudgeon entered the observation area mainly when discharge rates were high and variable. Increasing discharge rates increased the spatial niche of the large gudgeon.

(B) Stone loach: At low flow rates, the observation area seemed to be an optimal place for the stone loach. Changes in environmental conditions are reflected in the distribution patterns. The transition between juvenile and adult microhabitat use takes place in size class 2. Under high and/or variable flow regime the species was not found in the observation area.

Introduction

The aim of this study was to demonstrate the microhabitat use of two benthic fish species and to find the factors mainly governing it.

The microhabitat approach implies that the habitats small scale variability is meaningful to the organisms studied. However, fish differ in their ability to use microhabitat diversity: smaller or less mobile species or individuals may be more adapted to use the small-scale variability than larger or more mobile ones. With regard to the species examined in the present study, gudgeon are known to exhibit migrations (Stott, 1966), even when they are still young of the year (Copp, 1987), whereas stone loach showed low migration activities and low recolonization capacities in regulated rivulets in the FRG (Brunken, 1988). I hypothesize, that the gudgeon is less efficient in using small-scale variability than stone loach.

In running water ecosystems, abiotic factors (current velocity and therefore substrate composition, food supply, oxygen supply) and the fish association change along three directions:

(a) longitudinally, riffles and pools usually have separate fish communities (e.g. Schlosser, 1985; Grossman *et al.*, 1987a + b; Greenberg, 1991);

(b) laterally, shore microhabitats can be distinguished from main river channel areas (Lindroth, 1955; Bain *et al.*, 1988; Lobb & Orth, 1991);

(c) vertically, surface-orientated, water column and benthic fish species can be distinguished (Grossman *et al.*, 1987a + b; Moyle & Baltz, 1985).

The microhabitat use of different species or even different size classes of a species is a compromise between the physiological needs of the animal and biotic interactions within the ecosystem. Current velocity and water depth set physiological and physical limits to fish distribution. Physiologically important factors such as oxygen supply or temperature affect the distribution of fish species along running waters on a larger spatial scale (e.g. Moyle & Baltz, 1985). On a smaller spatial scale current velocity and water depth proved to be good predictors for fish distribution in many studies (e.g. Bain *et al.*, 1988; Harvey & Stewart, 1991). Substrate is usually regarded as a covariate of the former two parameters. When small-scale gradients in oxygen or temperature exist, these factors may also affect microhabitat choice (Zalewski *et al.*, 1990). Favoured microhabitats may also be sites of optimal food supply (Fausch, 1984; Greenberg, 1991).

The variables mentioned are more or less tightly controlled by the hydrological regime, which is rather variable in low-order steams. Therefore the hydrological regime was characterized by discharge rates and fluctuations in discharge on a short and a long time scale in this study. Hydrological changes like floods change the distribution of organic matter stochastically and reduce the predictability of resources. Under such conditions specialization should be less profitable. This is in agreement with the abiotic-biotic continuum regulation concept, predicting that abiotic factors govern the fish community of temperate streams, biotic factors playing a minor role (Zalewski & Naiman, 1985). However, in many temperate running water ecosystems, flow rates are low, yet stable during summer. Under such conditions (low habitat availability and high energetic costs) biotic interactions may gain importance in structuring the community (e.g. Feminella & Resh, 1990; see Hart, 1983 for a review for benthic invertebrates; Kalleberg, 1958; Fausch, 1984 for benthic fish). In this study the density of conspecifics as well as that of the other species was included as an indication for intra- or interspecific competition for space. Several studies revealed at least some effect of fish predators on their fish prey (Schlosser & Ebel, 1989; Harvey, & Stewart, 1991).

Table 1. Fish assemblage in the rivulet Mauerbach

	A	B	C
SALMONIDAE:			
Brown trout (*Salmo trutta fario*)	+		
Rainbow trout (*Oncorhynchus mykiss*)	+	+	+
CYPRINIDAE:			
Gudgeon (*Gobio gobio*)	+	+	+
Chub (*Leuciscus cephalus*)	+	+	+
Minnow (*Phoxinus phoxinus*)	+	+	+
Tench (*Tinca tinca*)			+
COBITIDAE:			
Stone loach (*Barbatula barbatula*)	+	+	+
GOBIIDAE:			
Miller's thumb (*Cottus gobio*)	+	+	
PERCIDAE:			
Perch (*Perca fluviatilis*)	+	+	

A The whole stretch of the Mauerbach (Hacker & Meisriemler, 1972)
B The observation stretch (Hacker & Meisriemler op. cit.)
C Own observations

Biology of fish species

The two fish species studied have flattened bellies and are relatively small, which enables them to use current protected spaces. The adult stone loach is a crepuscular species (e.g. Burdeyron & Buisson 1982a + b), whereas juvenile stone loach (Zweimüller, 1990) and gudgeon of all size classes are active mainly during daytime. Both species feed predominantly on benthic invertebrates – preferably midge larvae – (Bernet, 1960; Sauvonsaari, 1971; Perrin, 1980; Burdeyron & Buisson, 1982b; Welton *et al.*, 1983), to which the stone loach is directed by chemical stimuli (Smyly, 1955; Filek-Wittinghausen, 1960), the gudgeon by visual stimuli.

Study area

The field study was carried out in a small, mesotrophic 2nd-order stream (Mauerbach) within the Wienerwald area, a forest belt surrounding Vienna. The rivulet, which flows through two small villages, is regulated throughout its course. The gradient is 0.19% for the whole course of the Mauerbach (Pomeisl, 1953) and 0.067% for the study area (Water Authority Vienna). The base flow is approximately 0.2 $m^3\ s^{-1}$. Discharge rates quickly increase to 1–2 $m^3\ s^{-1}$ after rainfall in the catchment area (approx. 40 km^2; Pomeisl, 1953).

The chemical and hydrological conditions of the rivulet are quite typical for low order streams: oxygen saturation is high (>90% saturation) and water temperatures follow air temperature. The conductivity values indicate slight anthropogenic eutrophication (250–668 μS; Dokulil *et al.*, 1977).

The rivulet runs along the foot of the hill covered by a beech community (*Fagus silvatica*, the dominant Wienerwald forest species), which forms the southern shore. The northern shore is lined by gardens. It consists mainly of a steep, several-metres-high slope covered by grass or *Polygonum* sp.. A characteristic shore vegetation is missing due to the steep gradient of the beech covered hill and the narrow transitional zone. The whole observation stretch was at least 3/4 shaded throughout the year.

A former study documented 6 cyprinid, 1 gobiid and 1 percid species in the Mauerbach (Hacker & Meisriemler, 1972; unpublished report), all except the perch (*Perca fluviatilis*) and the miller's thumb (*Cottus gobio*) were recorded for the observation area. A similar species assemblage was found during the observation period (Table 1).

Material and methods

Within a 200 m stretch of the stream, five observation transects (50 cm wide, 3.5–5.5 m long) were established and divided into 16 locations according to substrate characteristics and water depth. These transects included one pool, two riffles and two transects combining erosional and depositional areas. The distribution of 4 size classes (<3, 3–6, 6–9, >9 cm total length) of stone loach and gudgeon was directly observed along these experimental transects 44 times during night checking for fish using a torch. Size class estimates were controlled by photographs from fish caught by dip net. Gudgeon grow to larger sizes than loach (>15 cm total length). Most observed fishes showed no sign of being disturbed, but fled when touched. This is in agreement with the findings of Grossman *et al.* (1987a).

For each of the 16 locations water depth, current velocity and number of fish of each size class and species were noted on a tape recorder. For each observed fish the water depth near the fish was recorded.

Current velocities were measured 1 cm above the substrate by a modified ink injection method at each location. Water depth was measured twice every 20 cm along the transects. Discharge rates were calculated from data of a continuous water level recorder.

Multivariate statistics and electivity index

The spatial distribution of fish was examined by cluster analysis: Squared Euclidian distances and Ward's agglomeration methods were used.

Principal Component Analysis (PCA): The variables were z-transformed before analysis to reduce effects of different scaling. Only factors with eigenvalues greater 1 were extracted.

Discriminant Analysis (DA): A series of DA's was carried out to test whether the fish distribution patterns were predictable by the environmental factors and to get an estimate of how important these factors were for the microhabitat selection. For the prediction of presence/absence of fish the hydrological parameters and temperature were directly entered as discriminatory factors, because the PC-axes are influenced by fish numbers. Due to collinearity of factors (e.g. temperature and discharge rates) the discriminatory value of discharge rates may be overestimated. The largest stone loaches were only once absent from the study site, so no DA on their presence was carried out. For the prediction of distribution patterns problems with collinearity of discriminatory factors were avoided by using the 3 PC-axes derived from the PCA for environmental parameters. The distribution patterns of the size class at the observation date (according to the cluster analysis) were the grouping variable. Clusters with at least 3 members were included in the analysis, with the exception of the smallest size class of stone loach, where clusters with 2 members were also included. A stepwise procedure (Mahalanobis) was applied to improve resolution between nearest groups.

Jacobs' Electivity Index was calculated comparing the percentage distribution of the fish with the area-corrected percentage distribution in the environment for water depth, current velocity and substrate.

Results

Abiotic factors

The median discharge levels were similar from 1982 to 1984, but the number of summer days with high rates (>0.5 $m^3 s^{-1}$) increased dramatically in 1984 (Table 2). In 1985, elevated median rates coincided with many high discharge days during summer.

The mean water depth varied with location and time between 4.5 and 30.6 cm.

The maximal water velocity was 42 cm sec^{-1} (at higher discharge levels increased turbidity made observations impossible).

The 16 locations differed significantly with regard to water depth and velocity (Friedman, $p<0.01$). The observation area was divided into an upstream riffle with two rather shallow transects containing mainly coarse substrate, and a downstream pool with three transects. The downstream section (pool) was deeper: 6 downstream locations were significantly deeper than three upstream locations (Wilcox-Wilcoxon-test, $p<0.05$). Differences in water velocities were associated with substrate type (Friedman, $p<0.001$): gravel locations differed from sand and muddy areas. Velocities at gravel areas were higher in the upstream section (riffle).

Water temperature followed air temperature, ranging from 1.6 to 19.6 °C. No vertical or longitudinal gradients were observed. On several observation dates water, temperatures near the shoreline differed approx. 0.1 °C from those of the main course.

Density of benthic fish

Maximal densities were 3.7 m^{-2} for stone loach and 1.6 m^{-2} for gudgeon. Stone loach showed a relationship between hydrological regime and inhabitation of the observation stretch: in 1984 and 1985 no early stages of loach were observed, although temperatures were comparatively high and favourable for fish reproduction (Table 2). In both years elevated discharge rates occurred during summer.

Whether fish were present in the study area was mainly influenced by discharge rates (significant DA, Table 8, but see Material & methods section) in the case of smaller gudgeon (size class 1 + 2) and by discharge rates and variability in discharge in the case of the stone loach (size class 1–3). When stone loach (size class 1–3) and smaller gudgeon were present, discharge rates were lower than the mean of all observations. When larger gudgeon (size class 3) were found, discharge rates were higher than the mean. Mean discharge fluctuations were either equal or considerably lower for observations with fish in all 7 size classes, where a DA was carried out. The occurance of the largest gudgeon was not significantly predictable by the environmental factors included in the DA. Largest stone loach were nearly always present.

Table 2. Hydrological and temperature conditions during study period

	1982	1983	1984	1985
	Maximal daily discharge rates (l s^{-1})			
Q25%	168	102	146	190
median	190	168	190	256
Q75%	300	256	234	344
maximum	1810	990	1526	4920
	Summer days[2] with discharge >500 l s^{-1}			
	0	2	15	16
	Day degrees[3] for >12°C during summer[2]			
	237	198	258	245

[2] 1.V. to 31.VIII.
[3] Mean air temperature from the nearest meteorological station (Tullnerbach).

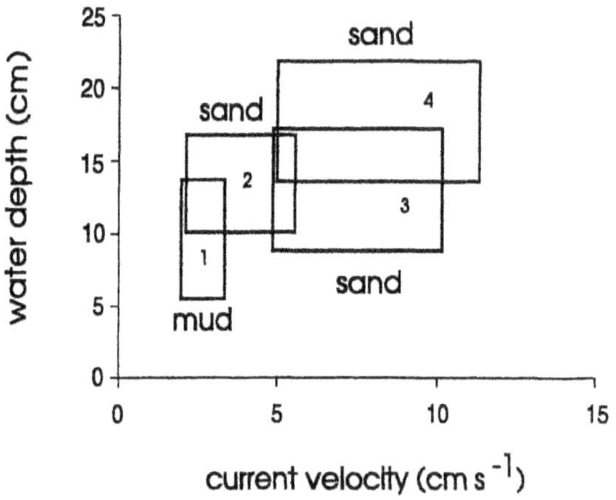

Fig. 1. Spatial niche of the gudgeon. The rectangles include 50% of the observations ($Q_{25\%}$ to $Q_{75\%}$). 1,2,3,4 ..... size class 1, 2, 3, 4. Mud, sand .... substrate used by more then 50% of the size class.

Electivity patterns for water depth, velocity and substrate

Within the comparatively narrow range of observed water depths and velocities, distinct size- and species-specific differences were discovered. Both species showed an ontogenetic shift from shallow areas without water current to deeper, more exposed areas (Tables 3–6, Figs 1 and 2).

The smallest gudgeon (size class 1) were more or less confined to still water. Electivity values for still water therefore were high; areas exhibiting even a slight water current were negatively selected (5–15 cm s^{-1}) or completely avoided (>20 cm s^{-1}). The fish positively selected areas of less than 10 cm water depth, were more or less indifferent to water depths of 10 to

Table 3. Flow rates and mean numbers of fish per observation

	n Obs	Max discharge (l s^{-1})	Gudgeon SC 1	SC 2	SC 3	SC 4	Stone loach SC 1	SC 2	SC 3	SC 4
IV 82	1	189	0.0	1.0	10.0	3.0	0.0	2.0	15.0	2.0
VIII 82	11	91	2.5	4.6	2.3	1.6	1.7	5.1	13.2	10.7
IX 82	3	112	0.3	2.3	1.7	0.7	0.3	3.7	6.0	6.0
II/III 83	3	180	1.0	1.7	1.0	2.0	0.3	1.3	3.3	3.0
IV 83	4	206	0.0	2.3	2.0	3.5	0.5	3.3	4.0	7.5
VII 83	2	129	0.0	1.0	0.0	0.0	0.5	2.0	4.5	11.0
IV 84	1	129	3.0	3.0	1.0	0.0	0.0	0.0	8.0	4.0
X 84	4	150	0.0	2.8	1.0	2.8	0.0	0.3	0.8	4.3
IX 84	2	163	0.5	3.0	0.0	0.0	0.0	0.0	0.5	3.0
IV 85	2	197	0.0	3.5	0.5	0.0	0.0	1.5	0.5	3.0
VII 85	3	123	0.0	1.3	2.3	0.0	0.0	0.0	0.3	3.7
VIII 85	3	174	0.0	1.0	1.3	4.0	0.0	0.0	0.3	5.7
IX 85	2	206	0.0	1.5	4.0	3.5	0.0	0.5	0.0	4.5
X 85	2	146	1.5	0.0	2.5	3.5	0.0	0.0	1.0	3.5
XII 85	1	163	1.0	0.0	0.0	2.0	0.0	0.0	0.0	2.0
SUM	44		39	112	81	81	24	95	230	278

n Obs: number of observations.
Max. discharge: maximal discharge rate at observation date.

Table 4. Electivity for current velocity (a) distribution parameters (m s^{-1}); (b) Jacobs'electivity index

		Gudgeon SC1	SC2	SC3	SC4	Stone loach SC1	SC2	SC3	SC4
(a)	Q25%	0.02	0.02	0.05	0.05	0.03	0.04	0.07	0.07
	median	0.03	0.03	0.07	0.07	0.03	0.11	0.12	0.10
	Q75%	0.03	0.05	0.10	0.11	0.04	0.13	0.14	0.15
(b)	m s^{-1}								
	>= 0.05	**0.98**	**0.90**	0.32	**0.21**	**0.95**	0.38	0.02	0.01
	0.05-0.10	−0.79	−0.17	**0.54**	0.19	−0.69	−0.32	0.00	0.21
	0.10-0.15	−1.00	−1.00	−0.23	0.05	−0.58	**0.61**	**0.57**	**0.32**
	0.15-0.20	−1.00	−0.83	−0.83	−0.04	−1.00	−0.59	−0.59	−0.26
	0.20-0.25	−1.00	−1.00	−0.55	−0.48	−1.00	−1.00	−0.61	−0.64
	0.25-0.30	−1.00	−1.00	−1.00	−1.00	−1.00	−1.00	−0.42	−0.34

15 cm and negatively selected or completely avoided deeper areas.

Gudgeon of size class 2 exhibited a high preference for still water and moderate avoidance of intermediate velocities. They were rare in shallow water and indifferent to deeper water areas.

Size class 3 preferred slight current velocity or still water and moderately avoided water swifter than 15 cm s^{-1}. Compared to size class 2 they expanded the range of velocities preferred (0 to 10 cm s^{-1}) and tolerated (<25 cm s^{-1}). Their distribution with respect to water depth was similar to size class 2.

The largest gudgeon exhibited moderate positive electivity for still water and weak negative electivity for water velocities >15 cm s^{-1}. A pronounced electivity pattern was observed with regard to water

Table 5. Electivity for water depth: a) distribution parameters (cm); b) Jacobs electivity index

		Gudgeon				Stone loach			
		SC1	SC2	SC3	SC4	SC1	SC2	SC3	SC4
(a)	Q25%	5.6	10.0	9.0	13.9	1.7	6.2	6.1	7.7
	median	9.2	12.7	13.0	17.2	4.2	8.3	7.7	11.2
	Q75%	13.6	16.2	16.7	21.7	7.0	14.1	10.5	14.4
(b)	cm								
	>= 5	**0.55**	−0.27	−0.48	−1.00	**0.91**	0.16	0.20	−0.10
	5-10	0.02	0.29	−0.30	−0.73	−0.23	**0.31**	**0.50**	0.04
	10-15	0.19	**0.39**	0.28	0.01	−0.53	−0.29	−0.28	**0.23**
	15-20	−1.00	−0.18	−0.05	0.23	−1.00	0.00	−0.43	−0.22
	20-25	−0.18	0.19	**0.37**	**0.71**	−1.00	−0.60	−0.88	−0.21
	25-30	−1.00	−1.00	−1.00	0.29	−1.00	−1.00	−1.00	−1.00

Table 6. Electivity for substrate: Jacobs'electivity index

		Gudgeon				Stone Loach			
	% O.A.	SC1	SC2	SC3	SC4	SC1	SC2	SC3	SC4
Mud	15	**0.93**	**0.66**	0.21	−0.35	**0.91**	**0.40**	**0.15**	−0.28
Sand	27	−0.18	0.39	**0.61**	**0.66**	−0.30	−0.16	−0.13	**0.16**
Gravel shallow	8	−1.00	0.20	−0.39	−0.75	−1.00	0.09	−0.22	−0.17
Gravel	50	−1.00	−0.95	−0.63	−0.43	−0.83	−0.09	0.14	0.06

O.A.: percentage of observation area.
Bold figures: highest electivtiy index within the size class.

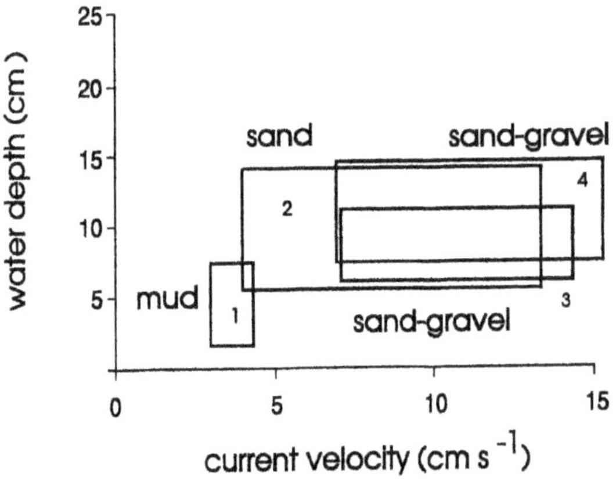

Fig. 2. Spatial niche of the stone loach. The rectangles include 50% of the observations ($Q_{25\%}$ to $Q_{75\%}$). 1,2,3,4 ..... size class 1, 2, 3, 4. Mud, sand, sand-gravel .... substrate used by more then 50% of the size class.

depth: moderate depths were strongly avoided, whereas depths >20 cm were positively selected.

Small stone loach (size class 1) strongly preferred still water situations (<5 cm s^{-1}), and tolerated slight water current (up to 15 cm s^{-1}). No small stone loach were found where water velocities exceeded 20 cm s^{-1}. Most small stone loaches were found in very shallow areas (maximum 5 cm depth) – resulting in highly positive electivity values; water deeper than 10 cm was strongly negatively selected.

The stone loach of size class 2 shifted to much more current-exposed areas. However, the fishes still had high positive electivity values for still water and moderately negative electivity values for swifter water (15–20 cm s^{-1}). This shift towards higher water velocities was associated with a shift to deeper areas (preference for the 5–10 cm water depth).

As the fish grew (size class 3 and 4) their preferred water velocities remained constant (10–15 cm s^{-1}). In size class 3 the preference for intermediate water depth was more pronounced than in size class 2 and accompanied by slightly negative selection for deeper (>20 cm) water. The largest stone loach were nearly indifferent to water depth: only a slight positive selection for 10 to 15 cm depths was registered.

Table 7. Principal Component Analysis for the environmental variables

	Factor 1	Factor 2	Factor 3
Eigenvalue	3.3	1.2	1.1
Pct of Var	41.3	15.3	13.8
Factor Matrix:			
MaxQ	**−0.798**	0.249	−0.081
dQ daily	−0.464	0.224	**0.731**
dQ weekly	0.388	**0.690**	−0.053
Temperature	**0.515**	−0.076	**0.737**
Small sl	**0.845**	0.109	−0.098
Large sl	**0.909**	0.079	0.024
Small g	**0.704**	−0.138	−0.031
Large g	0.010	**0.768**	−0.061

Pct of Var: percentage of variance explained by the factor
MaxQ: maximal discharge at the observation date (m^3 s^{-1})
dQ daily: difference between maximal and minimal discharge at the date of observation
dQ weekly: difference between maximal discharge within the week before and at the date of observation
Temperature: water temperature at the beginning of the observation
Small sl: number of small stone loach (size class 1 and 2)
Large sl: number of large stone loach (size class 3 and 4)
Small g: number of small gudgeon (size class 1 and 2)
Large g: number of large gudgeon (size class 3 and 4)
Bold figures: factor loadings <-0.5 and $>+0.5$

The electivity for substrate was pronounced in gudgeon: more than 2/3 of the gudgeon were observed on fine sediment in all size classes (Table 6, Fig. 1). The smaller fish (size class 1 and 2) occupied sand and mud areas to a similar extent, resulting in high positive electivity for muddy substrates. Size class 2 gudgeon showed a clear positive electivity for shallow gravel areas. The larger gudgeon increasingly preferred sandy areas; size class 4 avoided all other substrates.

In the stone loach substrate electivity was less pronounced, except for the smallest size class, which strongly preferred muddy areas (Table 6, Fig. 2). Size class 4 avoided mud and shallow gravel area.

Distribution patterns

The electivity for water depth, velocity and substrate resulted in species and size-specific distribution patterns along the observation area. A cluster analysis was carried out to obtain a non-subjective grouping. The riffle and pool sections were divided into 4 types of locations (mud, sand, gravel and gravel shallow areas). The percentage distribution of the fish along these 8 variables was used to perform a cluster analysis.

The following types of distribution patterns were distinguished:

1. fish inhabited mainly one substrate type either in the pool or in the riffle section (pool-sand, riffle-coarse...)
2. fish chose their microhabitat according to substrate type throughout the observation area (mud, sand...)
3. fish concentrated in the pool section, where they used all types of substrate
4. water depth was the important factor (pool-deep)

Gudgeon start their life using mainly muddy and sandy locations in the pool section, and these distribution patterns were dominant throughout their development. However, the riffle section and coarse substrate gained importance as the gudgeon grew larger. In size class 2 shallow gravel areas were the dominante microhabitat during 1/5 of the observations. The diversity of distribution patterns was greatest in size class 2 and 3 and decreased in size class 4 (Table 9).

The small stone loaches were found in the fine sediment locations of the riffle section. Size class 2 and 3 shifted their main microhabitat to the pool section, where they inhabited all substrate types. The largest stone loach were found all over the observation area. In contrast to the overall electivity pattern, shallow gravel was an important microhabitat for them during 1/5 of the observations. The diversity of distribution patterns increased with fish size (Table 9).

Effect of environmental factors on distribution patterns

The environmental variables were highly interdependent. To estimate their effect on microhabitat choice they were reduced to independent factors by a PCA including

- maximal discharge rate at sampling date in m^3 s^{-1},
- difference between maximal and minimal discharge rate at the sampling date,
- difference between maximal discharge rate during the week before sampling and the maximal discharge rate at the sampling date,
- temperature at the beginning of the observation,
- number of small (size class 1 and 2) and large (size class 3 and 4) stone loach and gudgeon.

Table 8. Separation of distribution patterns according to environmental conditions - Significance of discrimination and discriminatory value of the environmental factors

	Among distribution patterns				Between observations + and - fish					
	Sign. DA	Sign. of F-test for			Sign. DA	Sign. of F-test for				
		PC 1	PC 2	PC 3		MaxQ	dQ daily	dQ weekly	Temperature
Gudgeon									
Size class 1	0.16	0.16	0.97	0.68	**0.00**	**0.02**	0.20	0.10	0.61
Size class 2	**0.00**	**0.01**	0.11	0.40	**0.00**	**0.00**	0.07	0.32	0.74
Size class 3	0.16	0.57	0.16	0.51	0.09	0.44	0.14	0.23	0.11
Size class 4	**0.02**	**0.04**	0.17	0.33	0.22	0.40	0.28	0.67	0.46
Stone loach									
Size class 1	**0.05**	**0.04**	0.11	0.08	**0.00**	**0.00**	**0.05**	**0.04**	0.1
Size class 2	0.25	0.20	0.38	0.11	**0.00**	**0.00**	**0.00**	**0.02**	0.8
Size class 3	**0.03**	**0.02**	**0.04**	0.55	**0.00**	**0.00**	**0.00**	0.08	0.67
Size class 4	**0.01**	**0.00**	0.72	0.76	–	–	–	–	–

Sign. DA: p-value for significance of discrimination among groups
Sign. of F-test for PC 1 (PC 2, PC 3): *p*-value for univariate F-tests of PC 1 (PC 2, PC 3)
Bold figures: significant at the 5% level
Abbravitations of environmental factors as in table 7

The following 3 axes were extracted by the analysis (Table 7): The first axis (PC 1) had highly negative coefficients for maximal discharge rates and highly positive ones for number of all stone loach and small gudgeon. This axis can be viewed as a seasonal gradient with typical hydrologically stable summer situations (low discharge rates, high stone loach densities) at the positive side of the axis. The second axis (PC 2) is mostly influenced by long-term variability in the flow regime and the number of large gudgeon. The third PCA-axis (PC 3) is explained by short-term fluctuations in discharge and temperature. High values of PC 3 represent summer situations with an unstable flow regime.

Discrimination among distribution patterns

For 5 of the 8 size classes the distribution patterns in the field were predictable by environmental factors (DA, Table 8, see Material & methods section) – mainly the seasonal gradient expressed in PC 1: Stone loach (size classes 1, 3 + 4) were influenced by the actual discharge rates, the number of conspecifics and smaller potential competitors, gudgeon (size classes 2 + 4) by flow rates, number of potential competitors and number of smaller conspecifics. Although the environmental factors failed to predict distribution patterns of the smallest gudgeon size class, seasonality also seemed to have the strongest effect on them (PC 1), whereas small stone loach were strongly affected by seasonality and may have been influenced by short-term variability in discharge (PC 3 for size class 1 + 2). Long-term fluctuations in discharge along with number of large gudgeon (PC 2) seemed to have the strongest effect on middle-sized gudgeon (size class 3). However, a significant separation of distribution patterns according to environmental factors could not be achieved either in this size class or in the stone loach of size class 2.

Microhabitat shifts

Gudgeon

Size class 1: The distribution patterns were not clearly associated with environmental conditions (Table 9).

Size class 2: During stable summer conditions muddy areas in the pool section were the most important microhabitat. When discharge rates increased, the fish invaded coarse substrate. When discharge was high, the fish chose their microhabitat according to discharge stability: the riffle-gravel shallow pattern occurred under more stable discharge rates (short and long-term) as compared to the riffle-gravel pattern (not shown in Table 9).

The larger gudgeon (size class 3) reacted to variability of the flow regime: when long-term fluctuations in discharge were great, a variety of distributions – mainly in the riffle section – occurred. The two

Table 9. Ontogenetical changes in reaction towards environmental factors: Occurrance of distributoin patterns in relation to environmental factors (PC-axes)

GUDGEON

	<====		====>
Size class 1	+ Low discharge, high fish densities	PC 1 indifferent	- high discharge, low fish densities
	MUD (3) **POOL-FINE SEDIMENT (3)**	**POOL-MUD (7)** **RIFFLE-MUD (3)**	POOL-SAND (1)
Size class 2	+ Low discharge, high fish densities	PC 1 indifferent	- high discharge, low fish densities
	POOL-MUD (10) **MUD (3)**	**POOL-FINE SEDIMENT (10)** RIFFLE-SHALLOW (2) FINE SEDIMENT (1)	POOL-GRAVEL SHALLOW (2) RIFFLE-MUD (1) **RIFFLE-GRAVEL SHALLOW (5)** **RIFFLE-GRAVEL (3)**
Size class 3	- stable discharge (weekly), few big gudgeon	PC 2 indifferent	+ fluctuating discharge, many big gudgeon
		POOL-MUD (8) **POOL-SAND (10)**	POOL-GRAVEL (1) **POOL-SAND;RIFFLE-GRAVEL (5)** RIFFLE-MUD (1) **RIFFLE-SAND (6)** RIFFLE-FINE SEDIMENT (2) RIFFLE-GRAVEL SHALLOW (1)
Size class 4	+ Low discharge, high fish densities	PC 1 indifferent	- high discharge, low fish densities
		POOL-SAND (11) POOL-FINE SEDIMENT (2) **POOL-GRAVEL (3)** **POOL-SAND;RIFFLE-GRAVEL (5)**	POOL-GRAVEL SHALLOW (1) **RIFFLE-SAND (7)**

Table 9-2. For clarity only the most important factors according to the DA's (Table 8) are shown. Bold letters indicate distribution patterns included in the DA's. Numbers in brackets indicate number of observations. Further explanations see text.

STONE LOACH

	+ <=== Low discharge, high fish densities	PC 1 indifferent	===> - high discharge, low fish densities
Size class 1	POOL-FINE SEDIMENT (1) **RIFFLE-MUD (6)** **RIFFLE-SAND (2)**	**MUD (2)**	RIFFLE-GRAVEL (1)

	- <=== stable discharge (daily), cool	PC 3 indifferent	===> + fluctuating discharge, warm
Size class 2	**POOL-MUD (6)** RIFFLE-MUD (2)	**POOL-SAND (3)** **POOL-GRAVEL (5)** **POOL-SHALLOW (3)** GRAVEL SHALLOW (1) RIFFLE-SAND (2) RIFFLE-GRAVEL (1)	**POOL (4)**

	+ <=== Low discharge, high fish densities	PC 1 indifferent	===> - high discharge, low fish densities
Size class 3	**POOL (10)** **GRAVEL (5)** RIFFLE-SAND (1)	POOL-MUD (2) POOL-SAND (2) **POOL-GRAVEL SHALLOW (5)**	**POOL-GRAVEL (3)** RIFFLE-MUD (1) RIFFLE-GRAVEL SHALLOW (2) RIFFLE-GRAVEL (1)

	+ <=== Low discharge, high fish densities	PC 1 indifferent	===> - high discharge, low fish densities
Size class 4	**POOL 1 (5)** SAND (2)	**POOL 2 (3)** **POOL-SAND; RIFFLE-GRAVEL (5)** **RIFFLE-SAND (6)** RIFFLE-COARSE (2)	POOL-GRAVEL SHALLOW (1) **POOL-DEEP (4)** **GRAVEL (4)** RIFFLE-MUD (2) **RIFFLE-GRAVEL SHALLOW (8)** RIFFLE-GRAVEL (1)

dominant distribution patterns pool-mud and pool-sand were indifferent to long-term-variability in discharge, but they differed with regard to short-term fluctuations: the muddy areas were mainly inhabited when flow rates were stable (not shown in Table 9).

In the largest gudgeon the distribution pattern riffle-sand was characteristic for high discharge/low fish density-situations. The distribution pattern pool-sand occurred under more stable (both short-term and long-term) discharge conditions than most other patterns (not shown in Table 9).

Stone loach

Size class 1: The smallest stone loach shifted from fine sediment locations in the riffle section (riffle-sand or riffle-mud) under stable summer situations to muddy and sandy areas all over the transect as discharge rates increased. The fish were inhabiting the riffle-section, when discharge rates were fluctuating (PC 2; not shown in Table 9).

Size class 2: Seasonal changes were not detected in this size class, but the fish seemed to react towards variability in flow regime: when flow rates were stable over short periods (PC 3) the fish inhabited mainly shore regions. As variability increased, sandy locations and even coarse substrate became important. The deeper locations (e.g. pool-sand) were used when flow rates were fluctuating over a longer period (PC 2; not shown in Table 9) (versus the gravel-shallow patterns for stable discharge).

Size class 3: The variety in distribution patterns in this size class corresponds with subtle reactions to environmental changes. During stable summer situations the fish mainly occupied deep locations (pool, gravel, riffle-sand). As discharge increased, the fish used mainly shallow locations. The use of shallow gravel areas in both sections was linked to relatively low short-term fluctuations in discharge (PC 3; not shown in Table 9).

Size class 4: Under stable summer conditions, mainly sandy areas (pool, sand, riffle-sand) were inhabited. The response to increasing discharge was a shift to coarser substrate in the riffle section. The use of shallow gravel areas was linked to discharge variability, these areas (riffle-gravel shallow) were frequently used under high short-term-fluctuations (PC 3; not shown in Table 9).

Discussion

The study shows a distinct spatial segregation between species and size classes concerning water depth and current velocity. This is striking, considering the narrow range of water depth and velocity under investigation. The species have similar critical current velocities (approx. 50 to 60 cm s^{-1} for adults of both species; Pavlov *et al.*, 1972; Stahlberg, 1986) and their body size is similar. As hypothesized the stone loach utilizes small-scale variability throughout its ontogeny, whereas gudgeon is restricted to places of suitable depth, current velocity and substrate. The gudgeon is much less efficient in using the velocity-reduced boundary layer and the smaller individuals may therefore tend to avoid even slight water currents in the field. On the other hand the ratio of gill surface to body weight is quite low in stone loach (Jakubowski, 1977; Robotham, 1987). Their preference for relatively high velocities may reflect both a need for well-oxygenated water and their ability to cope with higher water velocities.

The differentiation with regard to water depth can hardly be explained by physiological constraints. A number of authors explain size-specific differences in water depth preferences through differential predation pressure by terrestrial (avoidance of shallow water) or aquatic (avoidance of deeper water) predators. Snakes facultatively feeding on fish (*Natrix natrix*) and trout (*Oncorhychus mykiss, Salmo trutta fario*) were frequently observed in the study area. The two fish species may well be differently susceptable to predation. However, this does not explain the intraspecific differentiation. Another factor governing depth preferences may be differences in feeding strategies: stone loach stay in close contact with the substrate during most of the searching and feeding process, whereas gudgeon swim actively in the water column, most probably searching for prey visually before digging activities start (Zweimüller, 1990). Such a behavior can be better performed in 'deeper' water.

Feeding strategies of strictly benthic fish are also known to affect their microhabitat according to substrate type (Greenberg, 1991). Grain size does affect the feeding success of the large stone loach: on muddy substrate they are unable to fully utilize the food offered in laboratory experiments (oligochaetes), which may partly explain their negative selection for this substrate in the field. The gudgeon showed no substrate-specific differences in feeding efficiency, but spent the least time for intensive searching behavior on sandy substrate (Zweimüller, 1990) – their preferred sediment in the field. Thus, sand is likely to provide the least energy-demanding substrate for their feeding activities. Water depth preferences in the gudgeon may even be a result of species-specific substrate preferences.

At first glance the results of this study support the strong influence of abiotic factors for the microhabitat choice of fish. A seasonal gradient separating the stable summer aspect from situations with higher discharge has the strongest effect on adult fish. Unfortunately competitive interactions within and between the species cannot be ruled out. Under harsh hydrological conditions competitive interactions seem quite unlikely, as both species are not territorial. Under stable summer conditions competitive interactions cannot be ruled out. However, there is little evidence supporting the theory of competitive niche segregation, as both species tend to use the pool section – especially sand areas – under such conditions. Dgebuadze *et al.* (1991) found distinct differences in microhabitat use of a subspecies of the stone loach in two mongolian rivers of similar hydrology. In this multi-species community the stone loach used mainly riffle situations, whereas they were predominantly found in pools, when only two species were present.

Most microhabitat shifts are similar for all size classes: the fish avoid shallow gravel areas when discharge is fluctuating on the short-time-scale (exception: large stone loach). Such a behavior is highly adaptive, as otherwise fish risk getting trapped in micropools during discharge fluctuations (Bain *et al.*, 1988 for larger rivers). Shore locations were frequently used under high discharge conditions in both species. Especially the largest stone loach deviated from their overall electivity patterns (avoidance of shallow gravel and mud areas) when discharge was high and use shore habitats. Both species tended to use the riffle section mainly at elevated discharge rates. For the larger gudgeon this section may have been available only during these periods (depth preferences). Very high flow rates have strong effects on pools, where higher scoring forces are exerted on the substrate than in riffles. Although the high discharge rates during observations are far from exerting scoring forces, microhabitat shifts may be preadaptive for higher discharge rates.

Many aspects of microhabitat choice in the two species remain unclear: why do large stone loach react differently to discharge fluctuations? Why do long-term fluctuations in discharge have other effects than short-term fluctuations? Is there any adaptive value in the diversity of distribution patterns, which did not seem to be effected by the environmental parameters?

The results also document the limitations of the microhabitat approach: for the gudgeon the study area was too small to contain representative reproduction areas. In both species it is doubtful whether winter habitats were fully included. However, the study shows the importance of small-scale diversity for at least one of the species (stone loach) and indirectly stresses the necessity of migration possibilities for the other (gudgeon).

Acknowledgments

The Water Authority Vienna (Dr Peckarek) provided data from a continuous water level recorder. I want to thank Drs B. Higler, H. Keckeis, M. Stachowitsch and two reviewers for constructive criticism on an earlier version of the manuscript and for improving my English. Dr Dgebuadze helped in providing his paper on distribution patterns of the stone loach. Finally I want to thank all participants of the Ecotone meeting in Lunz for the stimulating atmosphere and the fruitful discussions.

References

Bain, M. B., J. T. Finn & H. E. Booke, 1988. Stream regulation and fish community structure. Ecology 69: 382–392.

Bernet, B., 1960. Recherches biologiques sur les populations de Gobio gobio (Linne 1758) de la Nivelle (fleuve cotier du Pays Basque). Ann. Stat. c. Hydrobiol. appl. 8: 243–252.

Brunken, H., 1988. Movement patterns of the stone loach *Noemacheilus barbatulus* (L.). Abstr. VI Congr. Europ. Ichthyologists.

Burdeyron, H. & B. Buisson, 1982a. On a Circadian Endogenous Locomotor Rhythm of Loaches (*Noemacheilus barbatulus* L., Pisces, Cobitidae). Zool. Jb. Physiol. 86: 82–89.

Burdeyron, H. & B. Buisson, 1982b. Etude du rythme alimentaire circadien d'un poisson benthique dulciquicole carnivore, la loche (*Noemacheilus barbatulus*) dans son milieu naturel. Annales de la Station Biologique Besse-en-Chandresse 16: 171–181.

Copp, G. H., 1987. Le rôle et le fonçtionnement des milieux aquatiques du Haut-Rhône français comme sites de reproduction et de nurserie pour les poissons du fleuve. pH-Thesis Univ. Lyon I, 97 pp.

Dgebuadze, J. J, W. A. Kjasminas, A. Dulmaa & W. J. Jermochin, 1991. Chislennost i biomassa ryb dvukh malykh rek Mongolii [Number and biomass of fish in two small Mongolian rivers, in Russian]. Voprosy Ichtiologii 31: 674–677, in Russian.

Dokulil, M., 1977. Chemisch-bakteriologischer Langzeitvergleich zweier Wienerwaldbäche (Mauer- und Liesingbach). Gewäss. Abwäss. 62/63: 97–110.

Fausch, K. D., 1984. Profitable stream positions for salmonids relating specific growth rates to net energy gain. Can. J. Zool. 62: 441–451.

Feminella, J. W. & V. H. Resh, 1990. Hydrologic influences, disturbance, and intraspecific competition in a stream caddisfly population. Ecology 71: 2083–2094.

Filek-Wittinghausen, W., 1960. Beiträge zum Verhalten von *Nemachilus barbatula* (L.) mit vergleichenden Beobachtungen an anderen Vertretern der Fam. Cobitidae. ph-Thesis, Univ. Vienna, 120 pp.

Greenberg, L. A., 1991. Habitat use and feeding behavior of thirteen species of benthic stream fishes. Envir. Biol. Fishes 31: 389–401.

Grossman, G. D., A. D. Sostoa, M. C. Freeman & J. Lobon-Cervia, 1987a. Microhabitat use in a mediterranean riverine fish assemblage – Fishes of the upper Matarraña. Oecologia 73: 501–512.

Grossman, G. D., A. D. Sostoa, M. C. Freeman & J. Lobon-Cervia, 1987b. Microhabitat use in a mediterranean riverine fish assemblage – Fishes of the lower Matarraña. Oecologia 73: 491–500.

Hart, D. D., 1983. The importance of competitive interactions within stream populations and communities. In Barnes, J. R. & Minshall G. W. (eds), Stream Ecology. Plenum Press, New York and London, 399 pp.

Harvey, B. C. & A. J. Stewart, 1991. Fish size and habitat depth relationships in headwater streams. Oecologia 87: 336–342.

Jacobs, J., 1974. Quantitative measurements of food selection: a modification of the forage ratio and Ivlev's electivity index. Oecologia 14: 413–417.

Jakubowski, M., 1977. Size and vascularization of respiratory surfaces of gills and skin in some Cobitids (Cobitidae, Pisces). Bull. Acad. Pol. Sci Ser. Sci. Biol. 25: 307–316.

Lindroth, A., 1955. Distribution, territorial behaviour and movements of sea trout fry in the River Indalsïven. Rep. Inst. Fw. Res. Drottningholm 36: 104–119.

Lobb, M. D. & D. J. Orth, 1991. Habitat Use by an Assemblage of Fish in a Large Warmwater Stream. Trans. am. Fish. Soc. 120: 65–78.

Kalleberg, H., 1958. Observations in a stream tank of territoriality and competition in juvenile salmon and trout (*Salmo salar* L. and *Salmo trutta* L.). Rep. Inst. Fw. Res. Drottningholm 39: 55–98.

Moyle, P. B. & D. M. Baltz, 1985. Microhabitat Use by an Assemblage of California Stream Fishes: Developing Criteria for Instream Flow Determinations. Trans. am. Fish. Soc. 114: 695–704.

Pavlov, D. S., Yu. N. Sbikin, A. Ye. Vashchinnikov & A. D. Mochek, 1972. The effect of light intensity and water temperature on the current velocities critical to fish. J. Ichthyol. 12: 703–711.

Perrin, J. F., 1980. Structure et fonçtionnement des ećosystèmes du Haut-Rhône français. 14 – Etude des préférences alimentaires de la loche franche (*Noemacheilus barbatulus* L.) par une méthode des points modifiée. Hydrobiologia 71: 217–224.

Pomeisl, E., 1953. Der Mauerbach. In G. Pleskot (ed.), Beiträge zur Limnologie der Wienerwaldbäche. Wetter und Leben, Sonderheft 2: 103–121.

Robotham, P. W. J., 1987. The dimensions of the gills of two species of loach, *Noemacheilus barbatulus* and *Cobitis taenia*. J. exp. Biol. 76: 181–184.

Sauvonsaari, J., 1971. Biology of the stone loach (*Noemacheilus barbatulus* L.) in the lakes Päijänne and Pälkkänevesi, southern Finland. Acta zool. fenn. 8: 187–193.

Schlosser, I. J., 1985. Flow regime, juvenile abundance and the assemblage structure of stream fishes. Ecology 66: 1484–1490.

Schlosser, I. J. & K. K. Ebel, 1989. Effects of flow regime and cyprinid predation on a headwater stream. Ecological Monographs 59: 41–57.

Smyly, M., 1955. On the biology of the stone loach *Nemachilus barbatula* (L.). J. anim. Ecol. 24: 167–186.

Stahlberg, S., 1986. Das Schwimmverhalten von Bachschmerle (*Noemacheilus barbatulus*) und Gründling (*Gobio gobio*) in einem Strömungskanal. Mitteilungen des Instituts für Wasserwirtschaft, Hydrologie und Landwirtschaftlichen Wasserbau der Universität Hannover 61: 221–296.

Stott, B., 1966. The movements and population densities of roach (*Rutilus rutilus* (L.)) and gudgeon (*Gobio gobio* (L.)) in the river Mole. J. Fish Biol. 27: 785–792.

Welton, J. S., C. A. Miller & E. L. Rich, 1983. Food and habitat partitioning in two small benthic fishes, *Noemacheilus barbatula* (L.) and *Cottus gobio* L. Arch Hydrobiol. 97: 434–454.

Zalewski, M. & R. I. Naiman 1985. The regulation of riverine fish community by a continuum of abiotic-biotic factors. In Alabaster, J. S. (ed.), Habitat modification and freshwater fisheries. Butterworth Scientific Ltd, London: 3–9.

Zalewski, M., P. Frankiewicz, M. Przybylski, J. Banbura & M. Nowak, 1990. Structure and dynamics of fish communities in temperate rivers in relation to the abiotic-biotic regulatory continuum concept. Pol. Arch. Hydrobiol. 37: 151–176.

Zweimüller, I., 1990. Mikrohabitatwahl bei der Schmerle (*Noemacheilus barbatulus* (L.)) und dem Gründling (*Gobio gobio* (L.)) im Mauerbach. pH-Thesis Univ. Vienna, 145 pp.

Large rivers and floodplains

Hydrobiologia **303**: 141–148, 1995.
F. Schiemer, M. Zalewski & J. E. Thorpe (eds). The Importance of Aquatic-Terrestrial Ecotones for Freshwater Fish

The dependence of fish community structure and dynamics on floodplain and riparian ecotone zone in Parana River, Brazil

Angelo A. Agostinho & Maciej Zalewski
Univ. Est. Maringa, NUPELIA, Maringa-PR, Parana, Brazil, Department of Applied Ecology, University of Łódź, Poland

Abstract

Flood intensity was a main factor determining access to the riparian/floodplain ecotonal resources of the upper Parana River, Brazil, and in consequence determining yield of the dominant trophic groups, which are fish feeding on flooded grasslands and on riparian fruits and leaves. Prey and predator density both declined in years of low floods, and predators did not recover until the next year of high flood, more slowly than in the case prey fish, most probably due to differences in life cycle length. The life cycle of one of the most important species, Curimba (*Prochilodus lineatus* – Characidae) depends on access to the floodplain lagoons and return to the reservoir after maturation for intensive growth. The riparian/floodplain habitat complexity and connectivity has great significance for fish community dynamics and fisheries yield, and may serve a reason to create a national park in the last floodplain section of the upper Parana River.

Introduction

The dependence of fish yield in large river systems on floodplain habitats is a well known phenomenon (Welcomme, 1979), but there have been few in-depth analyses of the ecological affinity of fish species for different types of floodplain and ecotonal habitats (Schiemer & Spindler, 1989).

The floodplain and riparian zone of the tropical river Parana has been an integral part of the ecosystem, determining to a great extent the dynamics of its fish communities (Agostinho *et al.*, 1993).

The construction of dams started early this century, reduced much of the fluvial habitat to a less diversified lentic environment, which has been reflected in the fish community (Agostinho *et al.*, 1994). The dams have obstructed migration of some commercially valuable fish species (*Pseudoplatystoma corruscans* – Pimelodidae, *Salminus maxillosus* – Characidae) in upper part of this river, seriously reducing its fisheries yield (Lowe-McConnell, 1987).

Another, even more important factor, which reduces biodiversity and productivity of fish communities is the decline of diversity of land/water ecotones due to the deforestation of the floodplain, and much of the reservoir shoreline. It influences fish communities directly by reducing the terrestrial food supply and the spawning and rearing substrate. Recently, some attempts have been made by the management of the Itaipu Binacional to reforest the Brazilian shore of the reservoir, to reduce land erosion and increase biodiversity of terrestrial and aquatic habitats (Muller & Zelazowski, 1989).

This paper presents preliminary results of studies on the dependence of fish communities in the Parana River on riparian/floodplain ecotone resources, for the maintenance of their diversity. The data form the background for the possible creation of a National Park Reserve in the last floodplain part of the upper Parana River.

Study area

The Parana River extends over 809 km in Brazilian territory, of which only 500 km is running water. In 1995 a new dam will be closed at Porto Primavera, further restricting the flowing stretch to only 200 km. Finally, in about 2010, the planned Ilha Grande dam project starting just above the Itaipu reservoir, will destroy the last lotic segment of this river in Brazil.

Sites were studied in the lower part of this stretch, including its free course, its floodplain and Itaipu reservoir. Fourteen stations, situated in different environments, including rivers (4 stations), floodplain channels (3), floodplain pools (4) and the reservoir (3), were investigated through two years (Fig. 1). The mesotrophic Itaipu Reservoir (Surehma, 1989) closed in 1982, has an area of 1460 km^2, is 151 km long and has a maximum depth of 170 m (mean 21.5 m). Its turnover time is 40 days. Its upper border covered the Sete Quedas Falls that was a barrier for fish migration upstream and delimited two ichthyological provinces (Bonneto, 1986).

Materials and methods

Standardized catches (CPUE = number or biomass / 1000 m^2 gillnets) were made from October 1986 to September 1988 by 10 gillnets (20.0 m × 2.0 m) of stretched mesh sizes ranging from 3.0 to 16.0 cm, operated during 24 hours a month/station.

All fishes were measured immediately after transport to the field laboratory. body, gonads and gut content were weighed in the field and stomach contents were preserved in formalin and investigated in the laboratory.

The fishes were classified in three trophic groups on the basis of the predominant food: aquatic (planktivores, benthivores, piscivores, part of insectivores); iliophagous (*aufwuchs* and mud eating); and ecotonal (foraging in flooded areas on fruits, leaves and insects falling into the water – including detritivores, herbivores, part of omnivores, and some insectivores). Stomach contents were analyzed by the methods described by Hyslop (1980), and detailed in Hahn (1991).

Results

Shift in community structure as an effect of river impoundment

Itaipu has been one of the biggest reservoirs in the world, situated below that section of the diversified floodplain of the Parana River where it is an almost natural tropical river. This has had a positive effect on the fish community biodiversity and especially on maintenance of the migratory species of great commercial value, such as Curimba (*Prochilodus lineatus* – Prochilodontidae), whose different life stages utilize floodplain habitats, but when mature became an important fisheries resource in the reservoir. Probably due to the permanent connection of the reservoir with the floodplain, several other migratory species which were expected to disappear at impoundment, still occur in the reservoir and are an important component of the fisheries yield (Agostinho *et al.*, 1994).

Despite this connection, during the first six years after the dam was closed, the species richness fell from 113 (ITAIPU BINACIONAL, 1979) to only 83. The two commercially valuable species of Characidae in the riverine fishery, both fructivorous – *Piaractus mesopotamicus* and *Brycon orbignyanus*, were completely eliminated. What is more important, only one of the ten most frequently collected species in the river remains among the main species in the reservoir (Agostinho *et al.*, 1994), namely *Plagioscion squamosissimus* – Sciaenidae, which was not native, but introduced in to the Parana in 1967 (Cruz *et al.*, 1990).

The density of the planktivorous *Hypophthalmus edentatus* – Hypophthalmidae increased sharply after conversion of the lotic habitat into the lentic (Fig. 2). The second dominant was the insectivorous *Auchenipterus nuchalis* – Auchenipteridae a species of short life span and internal fertilization.

The mud feeders were reduced severely in proportion from 57% to 8–20% (Hahn, 1991) and replaced in the reservoir by insectivores. Predators still remain the most important group as before impoundment, but their species composition (19) has changed. The main predators of commercial value are: jau – *Paulicea luetkeni* (Pimelodidae), barbado – *Pinirampus pirinampu* (Pimelodidae) and curvina - *Plagioscion squamosissimus*. Dourado-cachorro - *Rhaphiodon vulpinus* (Cynodontidae) is also frequent but less important as are piranha – *Serrasalmus marginatus* (Serrasalmidae) which fishermen consider as 'pest' species. The ichthyocoenosis is still changing under the strong influence of floods above the reservoir.

Much more drastic changes in fish community composition are expected after 2010, when the Reservoir Ilha Grande floods the last big section of natural floodplain of the Parana.

Effect of flood intensity on fish community dynamics

Access to the various floodplain habitats has been very variable from year to year depending on flood inten-

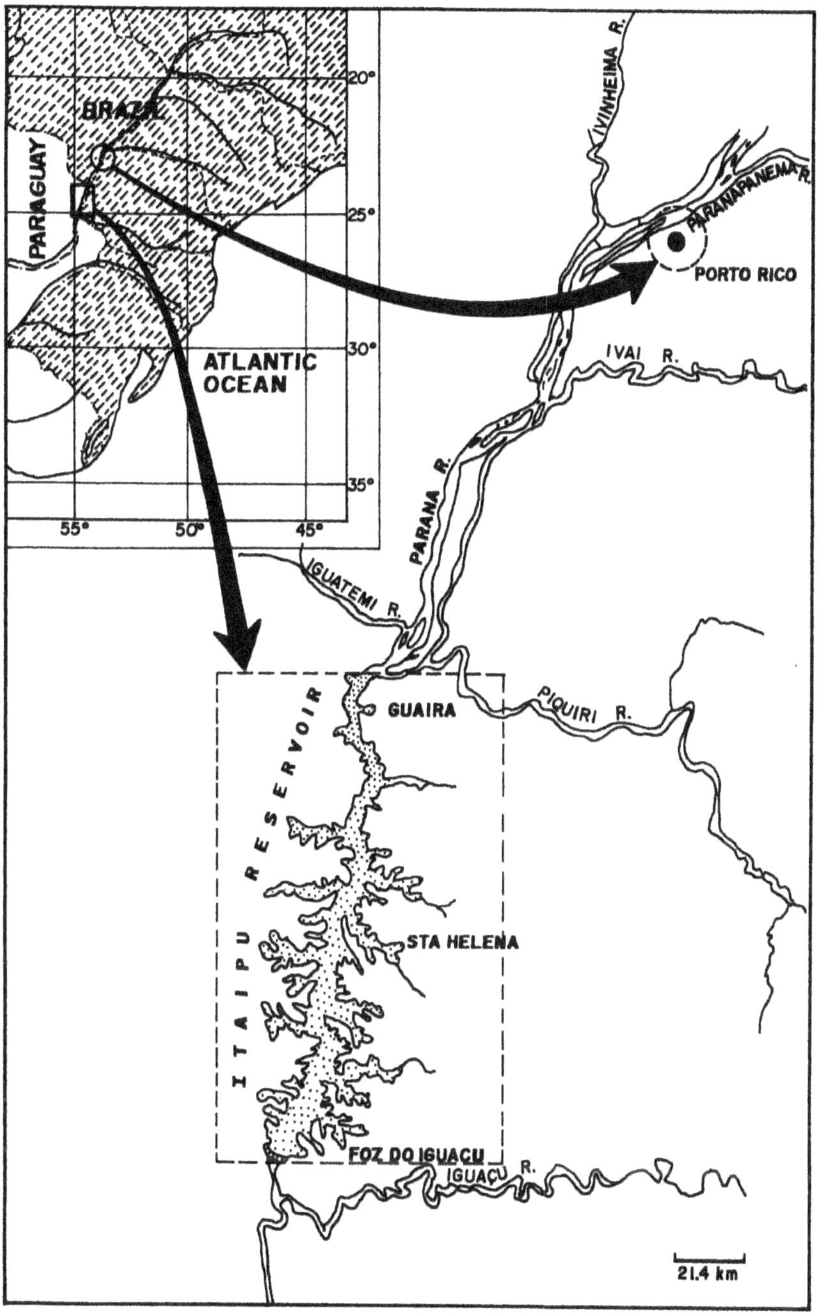

Fig. 1. The map of the upper Parana River and Itaipu Reservoir. The areas of most intensive research are the floodplain in the Porto Rico region and Itaipu Reservoir.

sity. The fish statistics data from years of differing flood intensity suggest a long term effect of preventing access to the floodplain ecotone on the fish community (Fig. 3). During autumn 1987, when the flood was very low, the density of prey fish declined much more than in a year of normal flood (1988), probably due to intense predator pressure at their migration in the main channel and to higher efficiency of the fisheries. In the year of normal flow the main factors increasing mortality in autumn were increasing hydrological

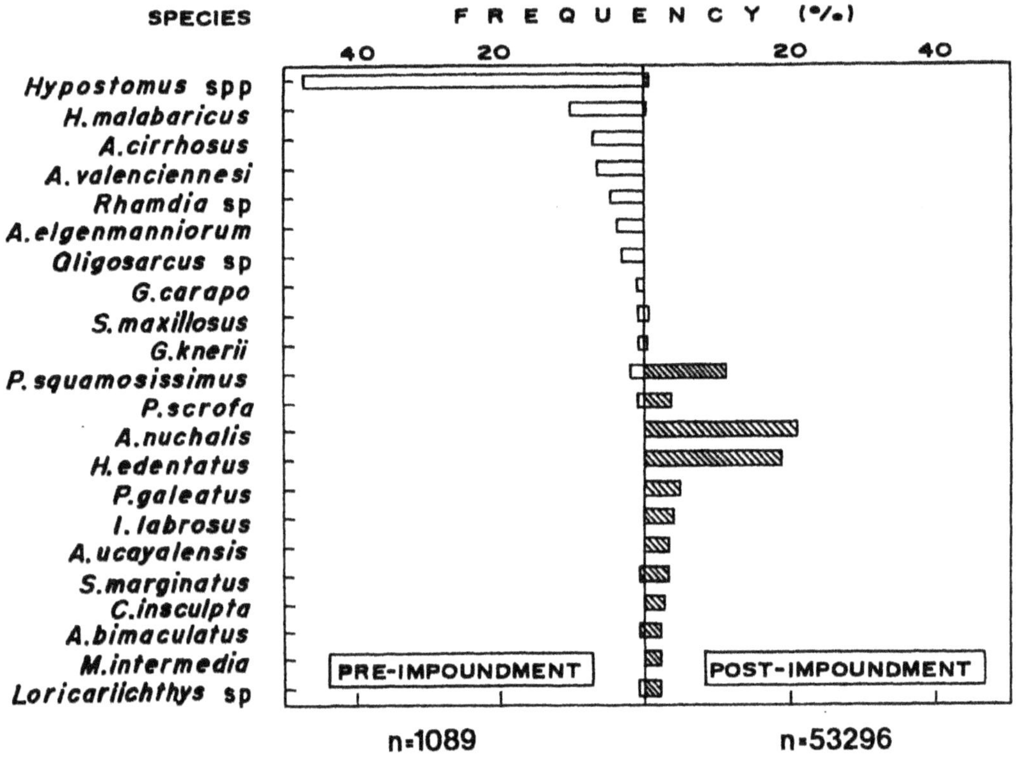

Fig. 2. The shift in fish community structure (main species) as an effect of impoundment of Itaipu Reservoir (after Agostinho *et al.*, 1994).

stress – declining water level, decreasing temperature and declining oxygen level.

The biomass of prey fish started to rise immediately after the high flood in 1988, possibly due to high prey survival in the diversified floodplain habitats. The situation was different for predatory fish. In the year of low flood their density also declined sharply, probably due to high efficiency of the fishing at the mouth of the floodplain channels where predators are especially concentrated when preying on fish descending from the drying floodplain. In such a situation of high concentration, cannibalism may increase significantly. In 1988, when the flood occurred, predator biomass achieved only half of that of 1987. This difference in the pattern of predator and prey density fluctuation might be a result of differences in the life cycle length. Predators are usually larger and mature later, than prey fish. However, the different pattern of dispersion during and after low and high floods might influence these patterns. Those preliminary observations need further profound studies.

The floodplain connection and fish life cycle

Riverine fishes in all types of climate are adapted to use optimal habitats at different stages of ontogeny (Northcote, 1978; Thorpe, 1988). That is why the floodplain connection is an important factor in the maintenance of at least some species in the river system as for example curimba in the Parana River (Fig. 4). Juvenile fish, up to the time of their first maturation at age 2 (L2) live in floodplain lagoons. Then they migrate on the high flood via the anastomosing floodplain channels to the main river and finally downstream to the reservoir. The flood's intensity determining access to the marginal habitats, is critical for a successful life cycle and thus for high fish yield. During low flood years most fertile floodplain lagoons are isolated from the river system and curimba yield may be reduced by the lack of the access to their optimal habitat and declining water level in the lagoons, which expose trapped fish to piscivorous birds and other predators. High mortality of trapped fish also occurs through desiccation. The

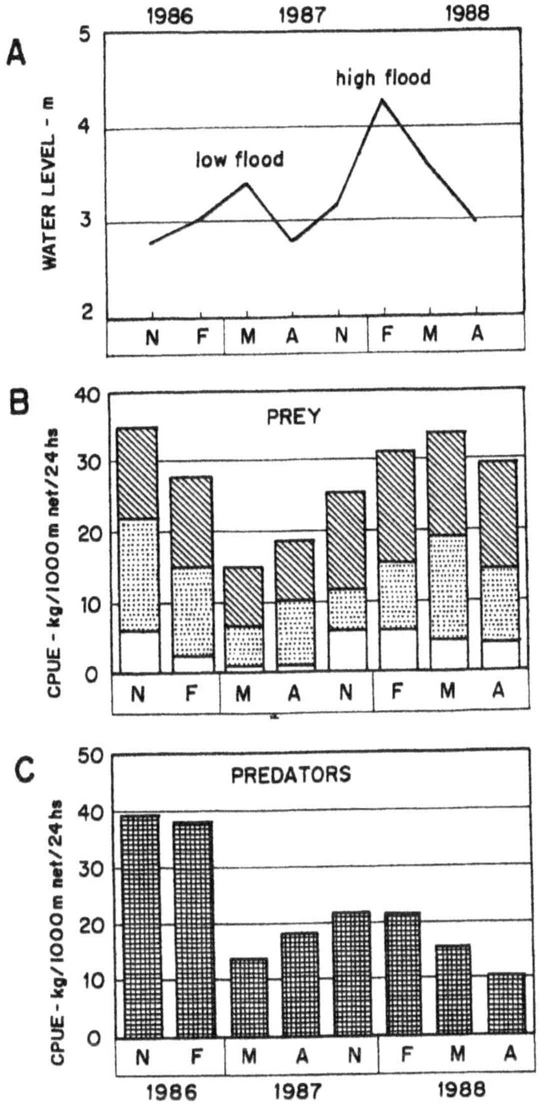

Fig. 3. Changes in density of fish of different trophic groups from floodplain habitats at upper Parana River, coincident with different flood levels: A. Changes in annual average water level in 1987 and 1988; B. Catchability of the nonpredatory trophic groups (prey) on the low flood in 1987 and the high in 1988 (see text for explanation) □ -aquatic; ▦ -iliophagous (floodedgrass + *aufwuchs*); ▧ - ecotonal (terrestrial); C. Decline in predatory fish biomass in the year after the low flood 1987. CPUE = catch per unit effort expressed as fish biomass by 1000 m^2 of gillnets at 24 hours. (n_{86-87} = 15,737; n_{87-88} = 17,079).

loss of fish biomass has been evaluated to be approximately four times that of the of professional fisheries (Bonetto *et al.*, 1969).

The importance of ecotonal food resources in the different types of habitats

To evaluate the importance of the land/water ecotones as habitats and sources of food for fish in the Parana River, its floodplain and Itaipu Reservoir, the occurrence of the main trophic groups have been analyzed by CPUE (Fig. 5). The combined data from the years of low and high flood demonstrate that highest CPUE occurred in lagoons and channels. The fish biomass was highest in the main river channel, but the density of all four groups increased proportionally from reservoir to river, to channels, and to the lagoons, with the biggest increase among fish feeding directly on ecotonal resources (fruits and leaves) and on flooded grasslands (iliophagous). This finding is consistent with the assumption that marginal ecotonal habitats are important rearing places where, under good trophic conditions, juvenile fish of all feeding groups may avoid predators, which are the main component of the fish assemblage in the river. In consequence the reduction of floodplain connection and/or the degradation of the riparian ecotone structure may lead to drastic reduction of the diversity of the fish community and its productivity within the whole Brazilian segment of the Parana River system.

Discussion

In tropical floodplain rivers, where the fish fauna is most diverse, the largest fraction of fish communities relies on floodplain ecotone resources (Lowe-McConnell, 1987). Recent progress in understanding ecological processes in floodplain rivers (Junk *et al.*, 1989) and acceleration of damming, canalization and deforestation in floodplains create the necessity for more insight into the dynamics of the top trophic levels-fish in these old aquatic ecosystems. These human impacts modify the hydrological regime and habitat complexity for fish along the whole river continuum. Schiemer and Zalewski (1992) distinguished a hierarchy of factors dependent on riparian/floodplain zone complexities for fish. In the diverse natural river environment, the most common compensatory response to both abiotic and biotic stresses, particularly severe on the early development stages, has been migration (Zalewski & Naiman, 1985). That component of the riverine system which compensates most for environmental harshness and provides suitable habitat is the floodplain and riparian zone. The steady reduction

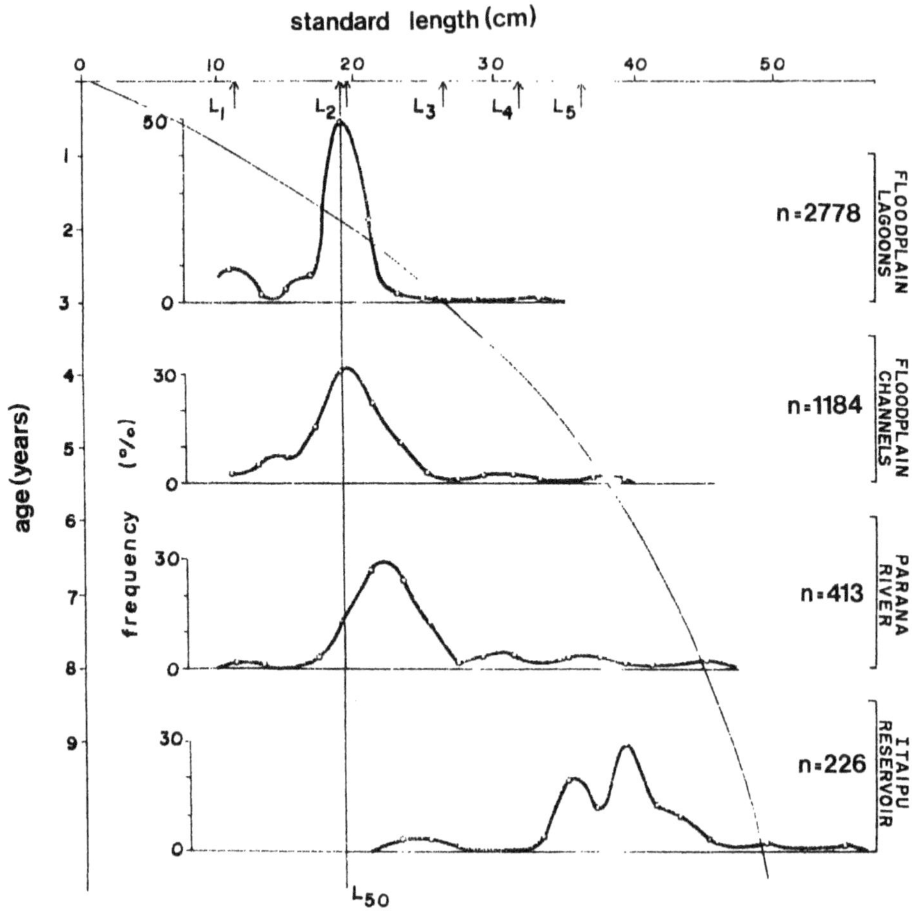

Fig. 4. The life cycle of curimba (*Prochilodus scrofa*) in different types of habitats in the middle Parana River, presented as a change in frequency distribution in different habitats. The first maturation (L50) and growth curve are superimposed on the frequency diagram (after Agostinho *et al.*, 1993).

of this zone reduces seriously biodiversity and quality/quantity of fish yield. This is because, in tropical rivers, fish dependent on the riparian zone for food are a significant percentage of fish communities, and because those fish are themselves important in the predators' diet. In consequence reduction in their occurrence will decrease the commercially important fisheries yield in the upper Parana River of dourado (Characiformes) and pintado (Siluriformes).

In the Plata River basin, Quiros & Baigun (1985) demonstrated that abundance of fish community, composed similarly as in the upper Parana, mainly by iliophagous and detritivorous species, the fish abundance is correlated with amount of organic matter in water column. In turn the complexity of the land/water ecotonal zone (floodplain) is the main source of the organic matter for the river (Junk *et al.*, 1989).

In north-eastern Brazilian reservoirs, the highest fish yield occurs if the number of predators is between 2 and 4 (Pinto Paiva *et al.*, 1994). Probably fish yield in Itaipu could be significantly increased by reducing the 19 predatory species here.

At this region some preimpoundment studies were done (Merona, 1985) and some evaluation of the early effect of closing river by dam on downstream fish communities (Merona, Carvalho & Bittencourt, 1987).

The high vulnerability of some riverine fish species to habitat modifications is mostly due to their complicated life cycle and this may be exaggerated by intensive exploitation. The life cycle of curimba is strictly dependent on the existence of a diverse floodplain above the reservoir. Hence future development plans threaten to eliminate this species and to reduce biodiversity and productivity of fish communities seriously. Further conversion of floodplain wetlands into

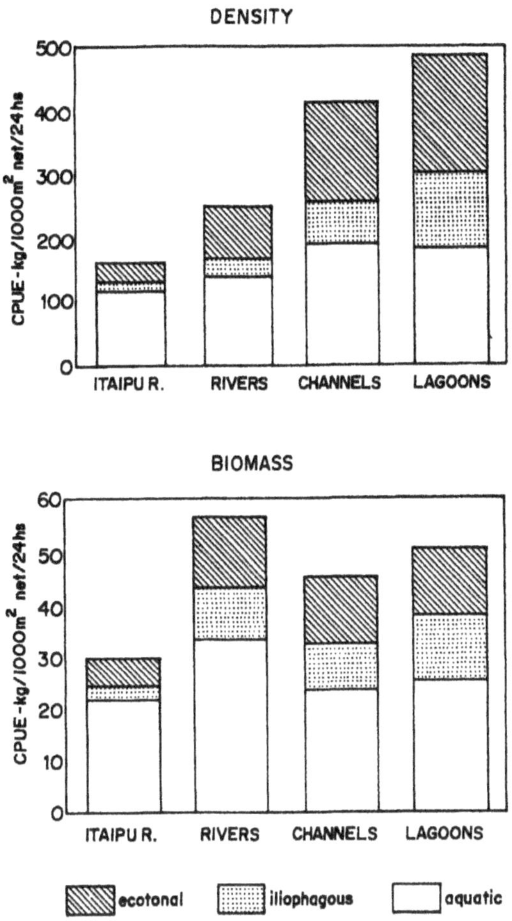

Fig. 5. The gradual increase of fish density along the habitat gradient, with increasing heterogeneity of ecotones. CPUE is catch per unit effort expressed as fish number or biomass by 1000 m^2 of gillnets at 24 hours($n_{Itaipu\ R}$ = 18 069; n_{rivers} = 4,107; $n_{channels}$ = 10,363; $n_{lagoons}$ = 18,346).

pasture land, and then dam construction in 2010 above Itaipu reservoir will destroy the last floodplain section of the Parana River. Global changes of the climate may reduce further the flood intensities in this region.

For efficient protective measures to maintain the diverse fish fauna of the middle Parana River, more quantitative data on dependence of fish on ecotonal floodplain and riparian habitat is necessary. Such data will be background for the long term strategy of restoration and management of land/inland water ecotones in the last floodplain section of the Parana River.

Acknowledgments

We would like to express our thanks to John Thorpe, Rosemary Lowe-McConnell, Kirk Winemiller and anonymous reviewer, for helpful suggestions, and English improvement.

References

Agostinho, A. A., H. F. Julio Jr, M. Petrere 1994. Itaipu Reservoir (Brazil): Impacts of the impoundment on the fish fauna and fisheries. In: I. G. Cowx (ed.), Rehabilitation of Freshwater Fisheries. Fishing News Books, Bodman, UK: 120–129.

Agostinho, A. A., A. E. Vazzoler, L. C. Gomes, E. K. Okada, 1993. Estratifacacion especial y comportamiento de *Prochilodus scrofa* en distintas fases del ciclo de vida, en la planicie de inundacion del alto rio Parana y embalses de Itaipu, Parana, Brasil, Revue Hydrobiol. trop. 26(1): 79–90.

Bonetto, A. A., 1986. The Parana River System. In: Davies, B. R. and Walker, K. F. (eds), The ecology of river systems. Dr W. Junk Publishers, Dordrecht: 541–556.

Bonetto, A. A., E. Cordiviola de Yuan, C. Pignalberi, O. Oliveros, 1969. Ciclos hidrologicos del rio Parana y las poblaciones de peces contenidas en lagunas temporarias de su valle de inundacion. Buenos Aires. Physis. XXIX 78: 213–223.

Cruz, J. A., J. R. Verani, L. Girardi, C. E. C. Torloni, 1990. Lewantamento da ictiofauna e aspectos da dinamica de populacao de algumas especies do reservatorio de Promissao, SP (la.Etapa). Sao Paulo. Companhia Energetica de Sao Paulo/Universidade Federal de Sao Carlos. Pesquisa e Desenvolvimento 52: 1–78.

Hahn, N. S., 1991. Alimentacao e dinamica da nutricao da curvina *Plagioscion squamosissimus* (Heckel, 1840) (Pisces, Perciformes) e aspectos da estrutura trofica da ictiofauna acompanhante no rio Parana. (Ph. D. Thesis, Univ. Est. Paulista, Sao Paulo, Brasil).

Hyslop, E. J., 1980. Stomach contents analysis, a review of methods and their application. J. Fish Biol. 17: 411–429.

ITAIPU BINACIONAL, 1979. Ictiofauna: inventario ambiental. (Report). Sao Paulo, Companhia de Tecnologia e Saneamento Ambiental/CETESB, 89 pp.

Junk, W. J., P. Bayley, R. E. Sparks, 1989. The flood pulse concept in river floodplain system. Can. Spec. Publ. Fish. aquat. Sci. V 106: 110–127.

Lowe-McConnell, R. H., 1987. Ecological studies in tropical fish communities. New York, Cambridge University press, 382 pp.

Merona, B., 1985. Les peuplements de poissons et la peche dans le bas Tocantins (Amazonie Bresilienne) avant la fermeture du barrage de Tucurui. Verh. int. Ver. Limnol. 22: 2698–2703.

Merona, B, J. L. Carvalho, M. M. Bittencourt, 1987. Les effets immédiats de la fermeture du barrage de Tucurui (Bresil) sur l'ichtyofaune en aval. Revue Hydrobiol. Trop. 20: 73–84.

Muller, A. C., V. H. Zelazowski, 1989. Reflorestamento ecologico da faixa de protecao do reservatorio de Itaipu – ME. Sao Paulo.

Northcote, T. G., 1978. Migratory strategies and production in freshwater fishes. In: Ecology of Freshwater Fish Production (S. Gerking ed.) Oxford, Blackwell: 326–359.

Pinto Paiva, M., M. Petrere Jr., A. J. Petenate, F. H. Nepomuceno, E. A. Vasconcelos 1994. Number of predatory fish species and de fish yield of large North-Eastern Brazilian Reservoirs. In: I.

G. Cowx (ed.), Rehabilitation of Freshwater Fisheries. Fishing News Books, Bodman, UK: 120–129.

Quiros, R., C. Baigun, 1985. Fish abundance related to organic matter in the Plata river basin, South America. Trans. am. Fish. Soc. 114: 377–387.

Schiemer, F., T. Spindler (1989). Endangered Fish species of the Danube River in Austria. Regulated Rivers: Research and Management. 4: 397–407.

Schiemer, F., M. Zalewski, 1992. The importance of land/water ecotones for fish communities dynamics in rivers. Neth. J. Zool. 42: 323–335.

SUREHMA, 1989. Estudos Limnologicos do reservatorio de Itaipu, Parana, Brasil, no periodo de julho/87 a junho/88 (Report). Curitiba/Parana, Superintendencia de Recursos Hidricos e Meio Ambiente, 187 pp.

Thorpe, J. E., 1988. Salmon migration. Sci. Progr. Oxf. 72: 345–370.

Welcomme, R. L., 1979. Fisheries Ecology of Floodplain Rivers. London, Longman, 317 pp.

Zalewski M., R. J. Naiman, 1985. The regulation of riverine fish communities by a continuum of abiotic-biotic factors. In: Alabaster, J. S. (ed.), Habitat modification and freshwater fisheries, 3–9, london, Butterworths Scientific Ltd.

Zalewski M., J. Thorpe, P. Gaudin, 1991. Fish and Land/inland Water Ecotones, Univ. Lodz., Univ. Sterling, Univ. Claude Bernard, Lyon, 101 pp.

Hydrobiologia **303**: 149–157, 1995.
F. Schiemer, M. Zalewski & J. E. Thorpe (eds), The Importance of Aquatic-Terrestrial Ecotones for Freshwater Fish.

An outline on lateral fish migrations within the Central Delta of the Niger River, Mali

Vincent Bénech[1] & Milan Peňáz[2]
[1] *O.R.S.T.O.M. Institut Français de Recherche Scientifique pour le Développement en Coopération, 213, rue La Fayette, 75480 Paris cedex 10, France;* [2] *Institute of Systematic and Ecological Biology, Academy of Sciences of the Czech Republic, 603 655 Brno, The Czech Republic*

Key words: floodplain, hydrological cycle, aquatic-terrestrial transition zone

Abstract

The lateral migration of fishes was studied during one entire flood cycle (1991) in a typical floodplain sector of the Central Delta of Niger at Batamani, Mali. The intensity and temporal course of migrations, as well as the species and size composition of migrants, were found to be correlated with the hydrological regime (river and floodplain water levels, direction and velocity of water flow in a temporary canal). Four different types of migratory patterns were described and their typical representatives indicated. The significant role of the river as a spawning site and that of the flood plain as a nursery for most locally important fishes was also accentuated.

Introduction

The Central Delta of the Niger in Mali represents a large-scale 'aquatic/terrestrial transition zone' (ATTZ) as defined by Junk *et al.* (1989), with remarkable temporal and spatial dynamics and with extreme interchange between aquatic and terrestrial ecosystems. Its water area fluctuates from less than 1000 km^2 at the end of dry season up to a mean of 12400 km^2 (but up to nearly 36000 km^2 before '70) at the culmination of flood (Poncet, pers. comm.), with the water level fluctuating within 5–6 m, and with floods normally lasting almost 6 months. Regularly flooded biotopes play an important role in the spawning and nursery of fishes, strongly influencing their production and exploitation and having a strategic importance for the nutrition of local inhabitants. This floodplain function is associated with massive lateral migrations of fishes exerted in both directions between the main channel of the Niger and the inundated flood plain, mainly via temporary streams (locally called 'marigots'), changing its current direction in the course of the flood cycle.

The aim of the present investigation was to determine the types of migration undertaken by fish and to identify the environmental mechanism associated with the migrations. The results presented here are only an outline of our research activities on lateral fish migrations, which are a part of a multidisciplinary study of fisheries in the Central Delta of the Niger River (Bénech *et al.*, 1992).

Study area

Our observations were carried out in the course of one entire flood season, from August to

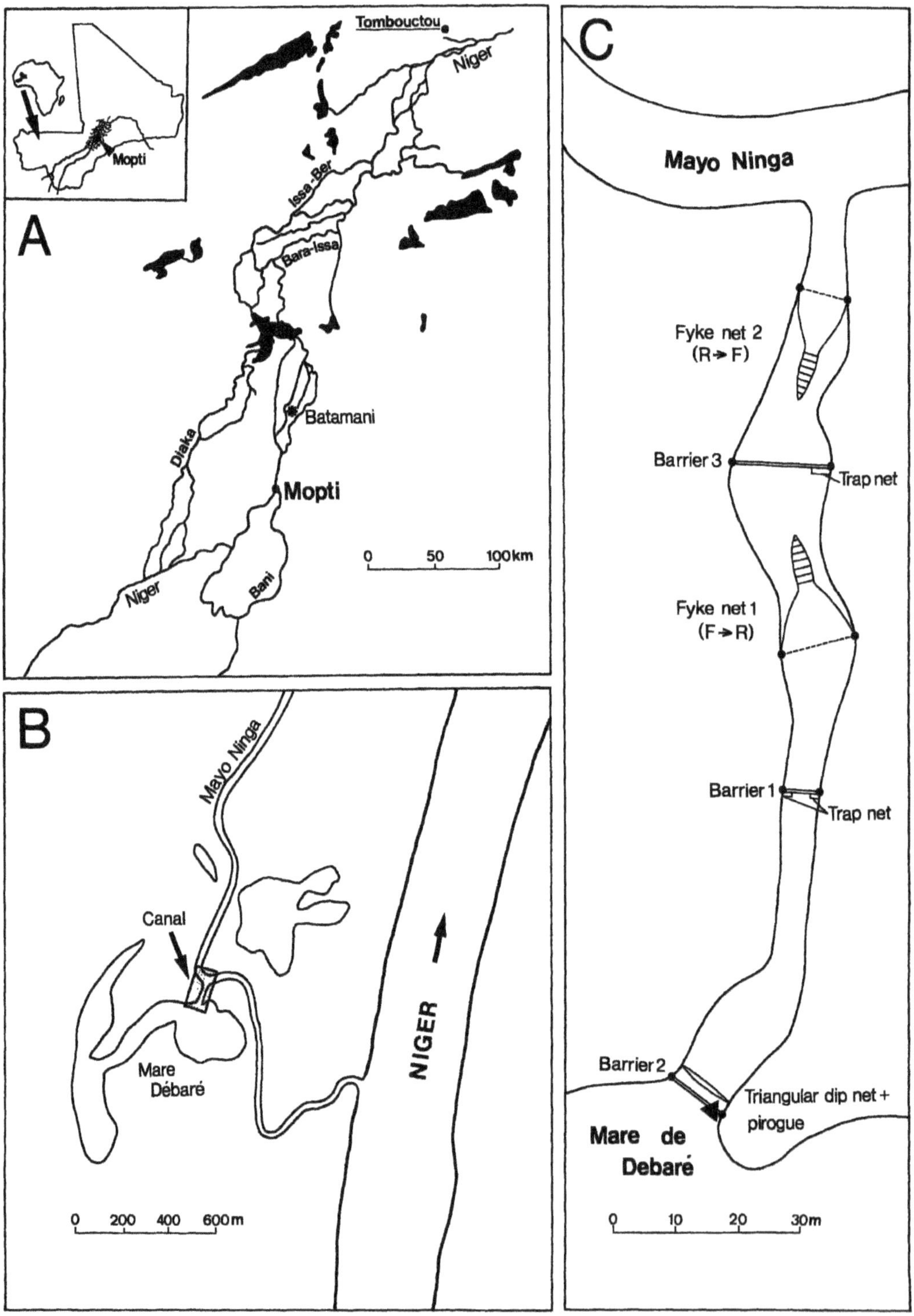

Fig. 1. A. Location of the Batamani study area in the Central Delta of the River Niger (Mali). B. Temporary canal connecting the 'Mayo Ninga' side arm of the Niger with the floodplain sector Mare de Debaré. C. The position of experimental and local fishermens' fishing devices in the canal (see text).

Table 1. Results of experimental fyke net fishings in the canal connecting the river and the flood plain. (Summarized catches during entire survey, 15 Aug.–13 Dec. 1991).

Period (direction of current in the canal	Direction of migration							
	R→F (Fyke net 2)				F→R (Fyke net 1)			
	N		W, kg		N		W, kg	
	a	b	a	b	a	b	a	b
Increasing flood 15 aug.–29 Oct. (R→F)	18077	1506	40.98	3.42	9870	987	23.28	2.33
Decreasing flood 30 Oct.–13 Dec. (F→R)	1398	155	8.53	0.95	33118	3680	402.76	44.75
All	19475	927	49.51	2.36	42988	2263	426.04	22.42

a. Overall catch during all fishings.
b. Catch per one 24-h cycle (C.P.U.E.).

December 1991, at a site located on the marigot connecting the Niger side arm Mayo Ninga with the floodplain sector 'Mare de Débaré' near the village of Batamani (Fig. 1). The study site is a relatively large and isolated floodplain sector about 70 km downstream of Mopti, with a 70 hectares of permanent lake. The lake is connected to the Niger by a single, short canal (150 m long, 3–4 m wide), which could be controlled for fish migrations with relative easy. The water in this marigot flowed from the river towards the flood plain (R→F) during the period 5 August to 29 October and reversely (F→R) during the remaining period of the flood season, 30 October 1991 to 13 December 1991.

Table 2. The species predominating in catches by fyke nets.

R→F			F→R		
Species	Relative abundance (%)	Relative biomass (%)	Species	Relative abundance (%)	Relative biomass (%)
Brycinus leuciscus	33.45	24.01	*Brycinus leuciscus*	19.50	3.88
Pellonula leonensis	26.21	9.85	*Siluranodon auritus*	15.53	11.07
Barbus spp.	10.55	22.00	*Pellonula leonensis*	8.73	0.78
Tilapia sensu lato	4.81	5.51	*Tilapia* sensu lato	7.85	6.84
Hydrocynus sp.	2.89	5.69	*Schilbe intermedius*	7.16	12.73
Micralestes elongatus	2.50	1.00	*Barbus* spp.	6.15	1.23
Schilbe intermedius	2.32	5.15	*Pollimyrus isidori*	6.06	1.71
Pollimyrus isidori	2.14	2.49	*Leptocypris niloticus*	6.02	0.79
Bagrus (bayad)	2.01	4.50	*Hemisynodontis membranaceus*	4.66	23.62
Siluranodon auritus	1.80	4.59	*Physalia pellucida*	2.56	0.36
Other species	11.32	15.21	Other species	15.78	36.99

Remark: the species are ordered according to the decreasing frequency.

Material and methods

A system of two fyke nets (locally called 'gangui'; mesh size 8 mm), installed facing opposite directions across the canal (R→F and F→R) and closing its entire cross sections, served as the basic fishing device to record simultaneously the two-way lateral migrations of fishes (Fig. 1). These

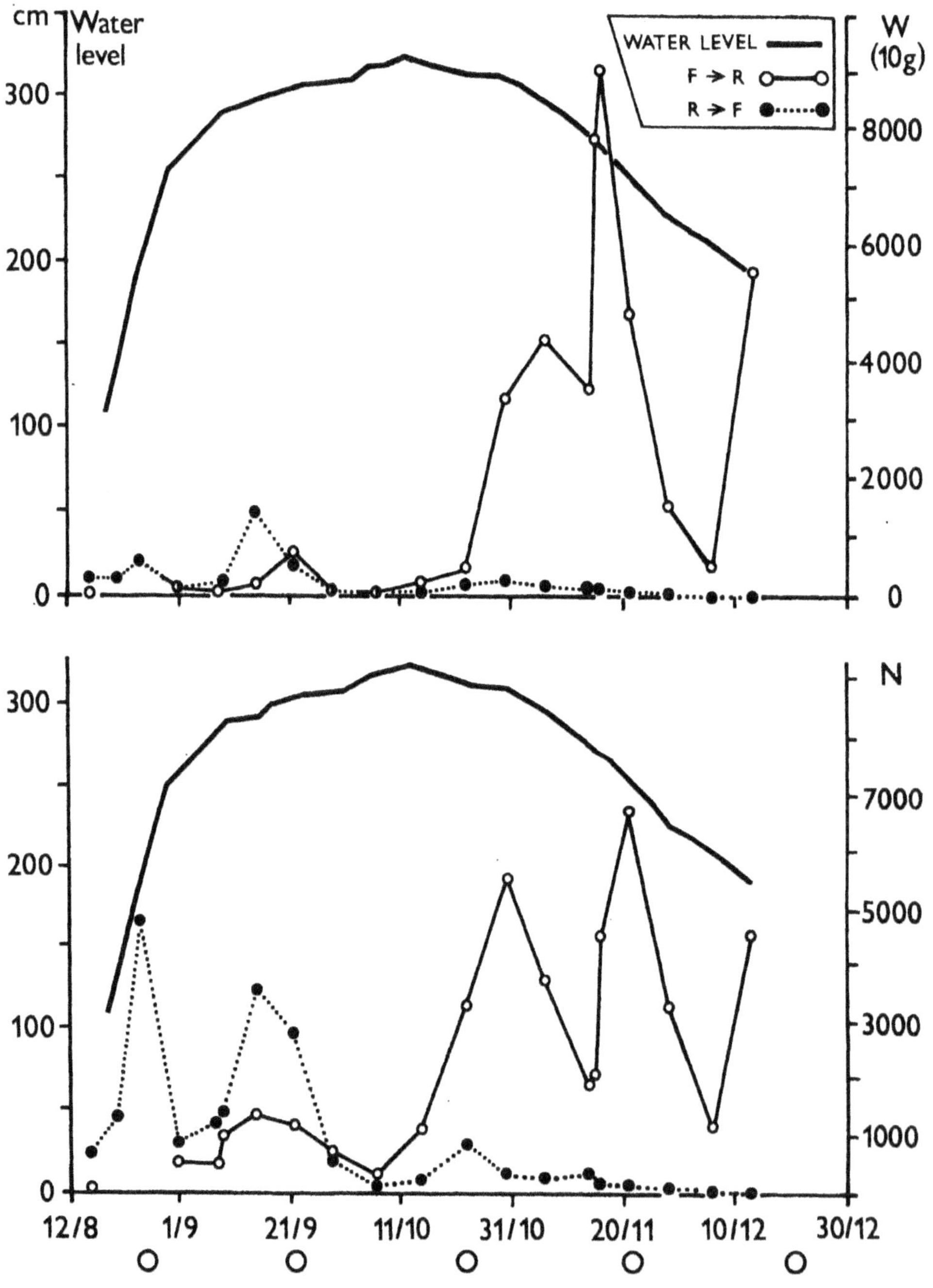

Fig. 2. Total catches obtained from fyke nets in the canal connecting the Niger and the flood plain. Ordinate – weight and number; abscisse – date. White circles indicate the days within the full moon phase.

were installed 21-times for 24-h cycles between 16 August and 13 December 1991, at a comparable lunar phase and emptied every 3 hours. One of the fyke nets (R→F) was complemented by a triangular drift net with fine mesh (mosquito net), which was exposed every 2 hours for 20 minutes in order to record the passage of fishes in early stages of development. All fish were identified, measured for SL and weighed; sex and sexual maturity of adults were recorded.

Results

In the fyke nets catches, 55 taxa of fish were identified. In total, during the whole survey, 19475 specimens (49.51 kg) of fish were captured migrating towards the flood plain and 42988 (426.04 kg) migrating in the opposite direction towards the river (Table 1).

The characid *Brycinus leuciscus* was the most abundant species migrating from the river towards the flood plain (33.45%; Table 2), followed by a small-sized clupeid species *Pellonula leonensis* and *Barbus* spp. (26.21 and 10.55%). Similarly, the biomass share was highest in these three species.

B. leuciscus, *Siluranodon auritus* and *P. leonensis* were the most abundant migrants in the opposite F→R direction, moving from flood plain towards the river (19.50, 15.53 and 8.73%). However, *Hemisynodontis membranaceus*, *Schilbe intermedius* and *S. auritus* dominated the biomass (23.62, 12.73 and 11.07%) (Table 2). The highest ratio between the biomasses of the fish migrating from and towards the flood plain (F→R/R→F) was found in *Mormyrus rume* (3131/3.47 = 902), *Citharinus citharus* (18968/25.94 = 731), *Auchenoglanis occidentalis* (36726/53.23 = 690) and *H. membranaceus* (100647/260.58 = 386).

The temporal course of migrations was correlated in most fishes with the hydrological regime (water level, direction and velocity of water current in the marigot) and, to some extent, also with the lunar cycle (Fig. 2). Sequential migratory waves of different cohorts were observed in some species (e.g. *B. leuciscus* (see Fig. 3) and *Labeo senegalensis*), whereas rather constant migratory

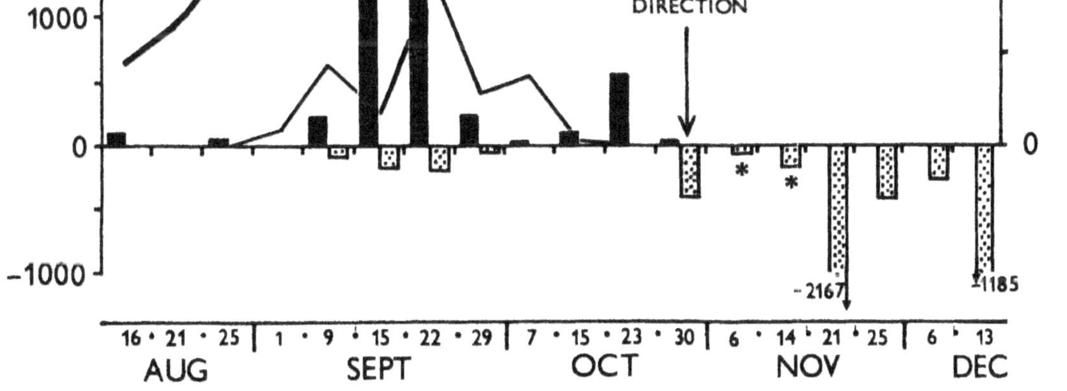

Fig. 3. Temporal course of catches of *Brycinus leuciscus* migrating in both directions through the canal connecting the Niger and the flood plain. The catches by fyke nets 1 & 2 are indicated in histograms, those by mosquito drift net as linegraphs. The days within the full moon phase are indicated in fig. 2.

patterns were observed in others (*H. membranaceus*, *Auchenoglanis* sp.).

In general, four types of migratory patterns could be recognized (Fig. 4):

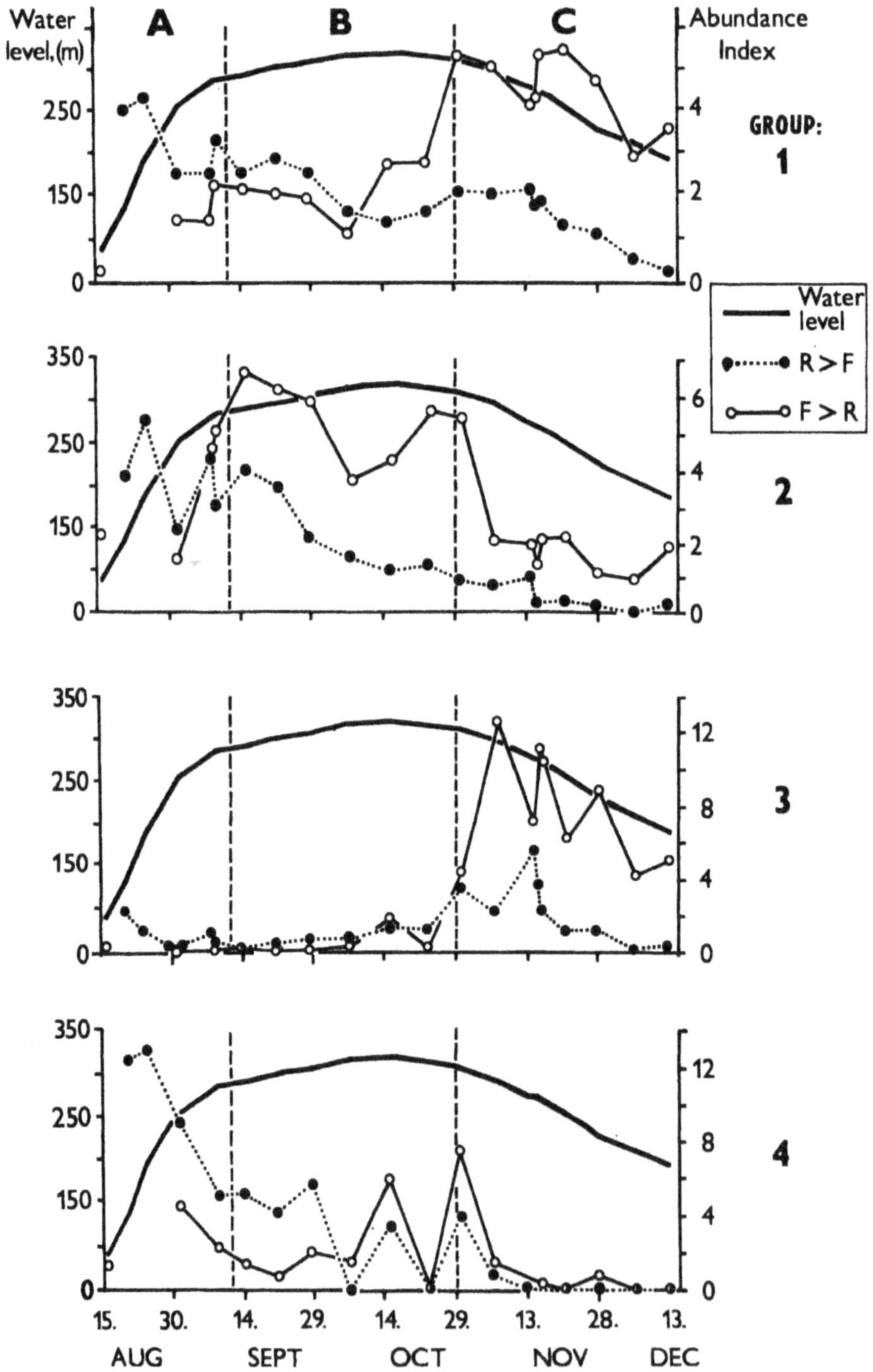

Fig. 4. Four types of migration patterns (based on fish numbers) observed in the canal connecting the Niger and the flood plain. Typical representatives belonging to each group are indicated in the text. A, B, C – phases of hydrological cycle.

Group 1. Fishes of this group were characterized by intensive migration from the river towards the flood plain (R→F) during the rather long period of rising water level. The migratory activity decreased during the flood peak but started again in the reversedirection F→R as the water level fell. Typical representatives were *B. leuciscus*, *S. intermedius*, *Tilapia* sensu lato, *L. senegalensis*, *Lates niloticus*. These migrants included only juveniles of *Labeo* and *Lates*, and both adults and juveniles in three other species.

Group 2. These fish species invaded the flood plain very early during the inundation process up to the beginning of the high water period. The return F→R migration was initiated rather early, immediately with the slackening of rise in water level and it decreases considerably just after the inversion of current. This group consisted mainly of small-sized pelagophilous fishes, such as *P. leonensis*, *Micralestes elongatus*, *Chelaethiops bibie* and also of some commercially important species (only juveniles) such as *Alestes* sp., *Brycinus nurse* and *Hydrocynus* sp. Of the two *Hydrocynus* species, only *H. brevis* entered the floodplain.

Group 3. In these fish species, only a flood plain to river migration was registered; the river to flood plain migration either did not take place or they drift undetected as larvae during the beginning of the flood (e.g. *H. membranaceus*) or they are permanent residents in flooded biotopes all-year-round, (e.g. *Brienomyrus niger*, *Clarias* sp.).

Group 4. This group is composed mainly of small-sized fish species that inhabit running waters, entering the marigot and the transient zone between river and flood plain temporarily, but never occupying the internal parts of the flood plain. They disappear from this transient zone immediately after the water starts to descend (*Nannocharax occidentalis*, *Pollimyrus petricolus*, *Raiamas senegalensis*, *Mochokus niloticus*).

Discussion

Spatially, the Central Delta of the Niger represents a giant ATTZ (Junk *et al.*, 1989) with remarkable temporal dynamics, an extreme interchange between aquatic and terrestrial ecosystems and with substantial allochthonous energetic and mineral inputs. Its complex functioning seems to be well interpreted by the flood pulse concept (Junk *et al.*, 1989, Welcomme *et al.*, 1989). The seasonal pulsing of flood flowing onto the flood plain is the driving force controlling the river-floodplain complex. As water inundates the flood plain, the flood pulse produces a 'moving littoral ecosystem unit', which prevents permanent stagnation and allows the rapid recycling of nutrients and organic matter. This prosperous function is strongly dependent upon the existence and intensity of rains and the strength of floods within the vast territory of the upstream catchment of the Niger. The flood pulse has unfortunately tended to decrease in recent years or was even lacking in some years ('Sahelian Drought' in 1972–1973). This was reflected in the decreased rate of fish production (Dansoko *et al.*, 1976; Welcomme, 1986; Laé, 1992). The lateral migration of fishes represents a fundamental mechanism that ensures the production of fish within the riverine and floodplain ecosystems, this resource being exploitated by traditional local fisheries. Our research corroborates that of a preliminary survey carried out in 1990 (Bénech *et al.*, 1992), but the present study emphasizes the significant competitive and nutritional role of highly dominant, small-sized and commercially insignificant fish species.

Our results are comparable with those from some other flood plains, especially from the Yaéré in North Cameroun (Durand, 1970, 1971; Bénech & Quensiére 1983a, b, 1987). The temporal course of migrations were found to be strongly dependent upon the hydrological cycle and, in some extent, exhibited also the biological rhythmics associated with lunar cycles. We observed a massive colonization at the end of the period characterized by the rapid flood increase during mid-September and the movements in opposite direction after the changed direction of water current during last days of October. The intensity of the return migration, flood plain-to-river, in some species, e.g. *B. leuciscus*, was considerably influenced by lunar cycles (Daget, 1952; Ghazai *et al.*,

1991). Four types of migratory pattern were distinguished, based upon two different ways of floodplain colonization: 1) Species predominantly of little size, belonging to groups 2 and 4; they inhabit the canal and the transient zone between the canal and the floodplain lake; their floodplain-to-river migration was completed immediately after the reversal of current in the marigot had occurred; 2) Species of great size (groups 1 and 3) invaded deeply into the flood plain, where they rapidly grew; their return migration F→R started relatively late during the period of decreasing flood.

Although we started our sampling on 15 August only, *i.e.* 10 days after the onset of the flood at Batamani, it was still a long time prior to the termination of the reproduction season, which in most fish species lasts till September (Bénech, 1990). Also noteworthy was the near absence of large specimens, except for several tilapias migrating into the flood plain to spawn. On the contrary, a considerable colonization of floodplain waters by the earliest juvenile stages was observed already at the beginning of the flood (August). Those fish species, being of greatest interest to fishermen, spawned mainly in the river. However, the inundated flood plain provides these species with juvenile progeny.

The surprising absence (or very low numbers only) of ripe fish migrating towards the flood plain to spawn suggests that a considerable portion of recruitment in fish populations in the flood plain originates from the main river, with larvae and young juveniles entering the flood plain for nursery and growth, as well as from fish resident in the floodplain lakes.

Finally, some imperfections perceived during the present survey should be mentioned: (1) we did not succeed in recording the migration of fish in the first ten days of the flood, a period considered important by local fishermen for some species (Mormyridae, *Clarias*, *Tilapia*); and (2) the fyke net appears to be selective, with a reduced efficiency for some species, namely for *Clarias* and *Tilapia*. This disadvantage was limited partly by the simultaneous capture of fish by other methods.

Despite these imperfections, the sampling methods used permitted us to make a comprehensive description of the lateral fish migrations patterns in relation with hydrological cycle, particularly the movements directing towards the flood plain, *i.e.* the less studied phase of migratory activity so far (Welcomme, 1979). Contrary to most other studies devoted to lateral migrations, the movements of fish were registered simultaneously in both directions during the present survey, including the small-sized species and juveniles that are usually ignored by the local commercial fishery as well by scientists

References

Bénech, V., 1990. Contribution à la connaissance de la reproduction de quelques espèces d'intérêt halieutique dans le Delta Central du Niger. Atelier d'études des pêches artisanales. O.R.S.T.O.M/IER, 20–23 novembre 1990, Bamako, 17 pp.

Bénech, V. & J. Quensière, 1983a. Migrations de poissons vers le lac Tchad à la décrue de la plaine inondée du Nord-Cameroun. II. Comportement et rythmes d'activité des principales espèces. Revue Hydrobiol. trop. 16: 79–101.

Bénech, V. & J. Quensière, 1983b. Migrations de poissons vers le lac Tchad à la décrue de la plaine inondée du Nord-Cameroun. III. Variations annuelles en fonction de l'hydrologie. Rev. Hydrobiol. trop. 16: 287–316.

Bénech, V. & J. Quensière, 1987. Dynamique des peuplements ichtyologiques de la région du lac Tchad (1966–78). Influence de la sécheresse Sahélienne. Thèse de doctorat d'état es-sciences naturelles. Univ. des Sciences et Techniques de Lille-Flandres-Artois, 680 pp.

Bénech, V., M. Peňáz, J. Wuillot, 1992. Juvenile fish assemblages of a tropical floodplain river – the Central Delta of Niger, Mali – A preliminary study. Folia Zool. 41: 161–174.

Bénech, V., M. Peňáz, P. Le Hong Chuong, 1992. Les migrations latérales des poissons dans le delta central du Niger. Etude ponctuelle à Batamani (août–décembre 1991). O.R-.S.T.O.M. Bamako, 27 pp. + annexes.

Daget, J., 1952. Mémoire sur la biologie des poissons du Niger Moyen. I. Biologie et croissance des espèces du genre Alestes. Bull. IFAN, 14: 191–225.

Dansoko, D., H. Breman, J. Daget, 1976. Influence de la sécheresse sur les populations d'Hydrocynus dans le Delta Central du Niger. Cah. O.R.S.T.O.M., ser. Hydrobiol. 10: 71–76.

Durand, J. R., 1970. Les peuplements ichtyologiques de l'El Beid, première note: présentation du milieu et résultats généraux. Cah. O.R.S.T.O.M., sér. Hydrobiol. 4: 3–26.

Durand, J. R., 1971: Les peuplements ichtyologiques de l'El Beid, 2ème note: variations inter et intra-spécifiques. Cah. O.R.S.T.O.M., sér. Hydrobiol., 5: 147–159.

Ghazaï, M. A., V. Bénech, D. Paugy, 1991: L'alimentation des Brycinus leuciscus (Teleostei: Characidae) au Mali: aspects qualitatifs, quantitatifs et comportementaux. Ichthyol. Explor. Freshwaters 2: 47–54.

Junk, W. J., P. B. Bayley, R. E. Sparks, 1989. The flood pulse concept in river-floodplain systems. Dodge, D. P. (ed.): Proceedings of the International Large River Symposium (LARS) Can. Spec. Publ. Fish. aquat. Sci. 106: 110–127.

Laé, R., 1992. Influence de l'hydrobiologie sur l'évolution des pêcheries du Delta Central du Niger de 1966 à 1989. Aquatic Living Ressources 5: 115–126.

Welcomme, R. L., 1979. Fisheries ecology of floodplain rivers. Longman, London, 317 pp.

Welcomme, R. L., 1986: The effects of the Sahelian drought on the fishery of the Central delta of the Niger river. Aquat. Fish. Manag. 17: 147–154.

Welcomme, R. L., R. A. Ryder, J. A. Sedell, 1989. Dynamics of fish assemblages in river systems – a synthesis. In Dodge, D. P. (ed.): Proceeding of the International Large River Symposium (LARS). Can. Spec. Publ. Fish. aquat. Sci. 106: 569–577.

Wuillot, J. & M. Kanta, 1991. Repartition spatiale et évolution saisonnniére des populations d'alevins dans les peuplements végétaux du Delta Central du Niger. Rapport préliminaire, O.R.S.T.O.M./IER, Bamako, 44 pp.

Hydrobiologia **303**: 159–161, 1995.
F. Schiemer, M. Zalewski & J. E. Thorpe (eds). The Importance of Aquatic-Terrestrial Ecotones for Freshwater Fish

The influence of high waters on stream fish populations in regulated rivers

Thomas Spindler
Fisheries and Freshwater Ecology, Unterolberndorf 93, A-2123 Kreuttal, Austria

Key words: population dynamics, Danube, catch curve, silver bream, floods, freshwater

Extended abstract

The hydrological regime of the originally, non-regulated Danube was characterised by a continuous swelling of the river due to snow break in the Alpes in spring time, which resulted in a high-water-period for several weeks. At that time, huge flood areas were inundated along the main channel, which play a major role as spawning areas for phytophilic fish species. The subsequently slow droping of the water levels during the summer months allowed the adults, larvae and juveniles to migrate into the main river or backwaters. Since 1870, this course of natural occurring fluctuations in waterlevels has been considerably altered by man-made regulation and canalization, which lead to a distinct increase of the rapidity of the high water events. For example: during the high water from August 1985, the water level increased within one day from 340 cm up to 780 cm in Vienna, and lasted only for four days. The question was addressed, to which extent the changing hydrological conditions are reflected in year-class strength of the fish populations.

A one-year-study on the abundance and age-structure of fishes was carried out in the alluvial zone of the free flowing part of the River Danube downstreams of Vienna (Austria). Fish were collected using gill nets with mesh sizes 40, 45, and 50 mm as well as three-wall-nets with mesh sizes 180/28/180 mm. Scales were taken from each individual fish for age determinations, using the methods described by Bagenal (1980). Standardisation of data were calculated according to Ricker (1975). The best representative for this investigations seemed to be the silver bream (*Blicca bjoerkna*). This eurytopic, phytophilic species is bound to flooded vegetation in sheltered backwaters for spawning substrates and it is rather common in the backwaters as well as in the main river (Schiemer & Spindler, 1989). Additionally, the silver bream is very catchable by the fishing gears in use, and it is not attractive to the fisherman. It can be assumed, that exclusively natural causes are reflected in the population structure of this species. *Blicca bjoerkna* inhabits the whole waterbody, so that random sampling could be presumed.

The catch curve from the silver bream is shown in Fig. 1. The steeply ascending left limb of the curve represent age classes, which are incompletely captured by the methods: i.e. they are taken less frequently in relation to their abundance (Ricker, 1975). Therefore only the right limb is representative for population structure. This part of the figure shows a rather irregular course, indicating distinct differences in survival rates between the successive age classes. Arrows sign years of apparently higher or lower survival.

Year-class strength were considerably lower in the years 1981 and 1985. In this years, the highest water-levels were exceeded by floods during the summer months. Though it is obvious, that summer high water events have tremendous effects on year class strength of fishes, as a high portion of the juveniles is killed, because of the superior mechanical load (velocity, bed load, wash of the waves etc.). This was sustained by the studies of juvenile populations in the Danube, that were accidentally carried out in the high water year 1985 (Spindler, 1988): after the flood, the density of juveniles was reduced to 50% of the former populations. This was due to the rapidity of discharge and rigorous straightening and damming, which reduced the essential refuge areas to a large extent (Schiemer *et al.*, 1991).

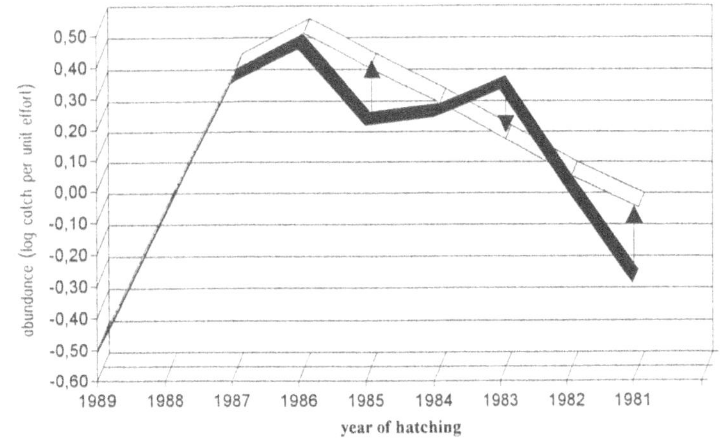

Fig. 1. Catch curve of the danubian silver bream (Blicca bjoerkna) from Austrian backwaters. The log frequency of standardized catches is plotted against year-classes. Arrows symbolize deviations because of environmental influences.

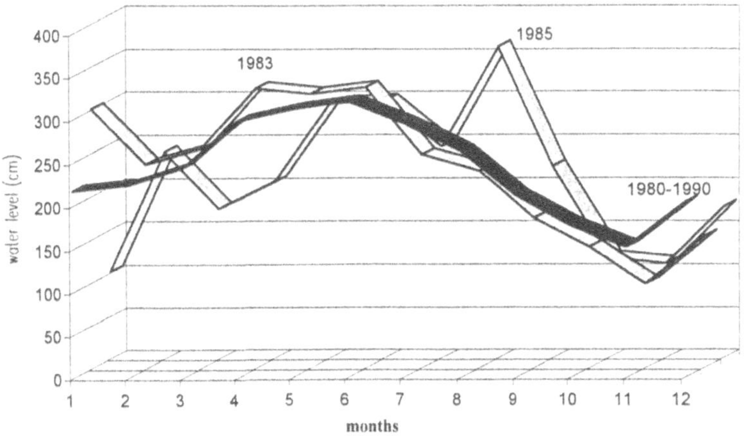

Fig. 2. Mean monthly water levels of the River Danube in Vienna from the years 1983 and 1985 in comparison to the monthly means of the decade from 1980 to 1989 according to datasets from the Austrian water-ways-administration.

On the other hand, year-class strength of the year 1983 is surprisingly high. But the hydrological conditions of that year were remarkable (Fig. 2): during the first six month of the year the water levels were above, and during the second half, waterlevels were below the ten-years average. Additionally, no high water event could be observed. Slightly higher water levels at the spawning time in spring are flooding the shore vegetation and provide excellent spawning conditions for several phytophilic species. A long duration time of this moderate flood will be an optimal basis for the development of eggs and larvae. Further on, significant lower water levels during summer and autumn (without floods) are enhancing the structural diversity, due to the bareing of gravel banks and the formation of soft hydrological gradients, which are essential for survival of the riverine fish fry. Such years with favorable hydrological conditions can compensate the losses due high waters to some extent.

References

Bagenal, T. ,1980. Methods for the Assessment of Fish Production in Fresh Waters. IBP Handbook 3. Blackwell, Scientific Publ., 365 pp.

Ricker, W. E., 1975. Computation and Interpretation of Biological Statistics of Fish Populations. Bull. Fisheries Res. Board Canada 191: 382 pp. Ottawa.

Schiemer, F. & T. Spindler, 1989. Endangered fish species of the Danube River in Austria. Regulated Rivers: Research and Management 4: 397–407.
Schiemer, F., T. Spindler, H. Wintersberger, A. Schneider & A. Chovance, 1991. Fish fry associations: Important indicators for the ecological status of large rivers. Verh. Internat. Verein. Limnol. 24: 2497–2500.
Spindler, T., 1988. Ökologie der Brutfische in der Donau bei Wien. Ph. D. Dissertation, Univ. of Vienna, 129 pp.

Hydrobiologia **303**: 163–179, 1995.
F. Schiemer, M. Zalewski & J. E. Thorpe (eds), The Importance of Aquatic-Terrestrial Ecotones for Freshwater Fish.

Extent and role for fish populations of riverine ecotones along the Sinnamary river (French Guiana)

L. Tito de Morais, M. Lointier & M. Hoff
Centre ORSTOM BP 165, 97323 Cayenne Cedex, French Guiana

Key words: ecotones, fish assemblages, plant communities

Abstract

The Sinnamary is a small tropical river (250 km long, watershed of 6500 km^2) where a dam is to be impounded in 1994.

The extent of naturally periodically flooded areas along the Sinnamary river is estimated by using remote sensing (Landsat TM satellite data) and botanical groupings are described on *in situ* observations, ground truth and collections. The fish assemblages of these areas are described and their density estimated by rotenone sampling. The role of this ecotone to fish populations is discussed.

The extent of the inundated areas appears to be greater than previously estimated by classical mapping.

As expected, the fish fauna is mostly composed of fish juveniles and adults of dwarf (< 3 cm) and small (< 15 cm) species. Large predators where not found. Due to these characteristics, fish are abundant but their biomass is low in naturally inundated areas. However, juveniles of some species found only in these zones. Therefore their adult populations will probably be very sensitive to any modification of these land/water ecotones.

Introduction

Holland's definition of an ecotone (di Castri *et al.*, 1988) as used by the Scientific Advisory Committee of the UNESCO MAB programme is difficult to use 'to determine, for example, the spatial limits of the riparian ecotone' (Thorpe, 1991). In the present paper, an ecotone is the overlap zone between land and water. Along the Sinnamary it includes several types of riparian forest. Their spatial limits extend from the minimum low water mark at peak droughts to the maximum high water mark at peak floods (see Thorpe, 1991 for a detailed discussion on ecotone definitions). These limits were estimated using remote sensing techniques. Each vegetation type appears in different colours depending on its position along the gradient reaching from the permanently flooded areas (swamps) to never flooded forests (upland moist forest).

Riparian zones are known to be of great importance as providers of food resources and habitats for the freshwater fish fauna. According to temperate studies, it has been hypothesised that in response to greater availability of prey, production and abundance of fish may be greater in places that are relatively unshaded by the riparian canopy (Murphy *et al.*, 1981 and Tchaplinsky & Hartman, 1983 in Gregory *et al.*, 1991; Schiemer & Zalewski, 1992). This pattern is probably different in tropical areas where allochthonous food sources (mostly fruits, leaves, ants and other terrestrial insects) are of major importance to the fish fauna (Lowe-McConnell, 1987).

Riparian ecotones are areas of high species richness of plants and are not easily delineated (Gregory *et al.*, 1991). The mosaic structure of vegetation types is difficult to estimate by field sampling, especially in the tropical evergreen forest where access to the upper sections of rivers is often very difficult.

After ground validation and precise sampling of some selected areas along the rivers, the use of remote sensing techniques can be very useful in allowing extrapolations to larger areas of difficult access.

Remote sensing Landsat Thematic Mapper (TM) images have a low resolution of 30 × 30 m per pixel, but they have three infrared bands which are particularly suited for differentiating vegetation.

The purpose of the present study was to help answer the following questions:

- what are the different riparian plant communities along the Sinnamary river?
- do these communities have distinct species compositions and what is their species richness?
- can the surface areas of these communities be estimated by remote sensing techniques? and
- are these communities associated with specific fish assemblages and, if they are, can the importance of such associations to the fish populations be estimated?

Description of sites studied

The Sinnamary river drains the Guiana Shield in French Guiana, a department of France in north eastern South America (Fig. 1). It is a small river by South American standards (250 km in length with a watershed of 6565 km^2). A dam is under construction at Petit Saut (Fig. 2) for hydroelectric purposes. The planned reservoir will have an area of 300 km^2 and will be closed in 1994. Downstream from the dam location, the river flows mostly through the coastal plains in unconstrained reaches (Fig. 2). The section of the river upstream of Petit-Saut is a succession of rapids and small pools in constrained reaches. Upstream from the Takari-Tanté rapids (which are the upper limit of the future reservoir), there is a succession of unconstrained reaches followed by a succession of rapids and pools in constrained reaches. The headwaters are almost unexplored. Complex flood plains are never found along this river. Very few secondary channels of limited extent are found near some rapids and in some unconstrained reaches in the upper river.

The flora of the Sinnamary river basin is poorly known but this zone is considered to have a relatively low plant species richness because of the lack of high hills. Only 809 species have been inventoried among the riparian plant communities (Hoff, 1991) of the area. In contrast, over 2000 species (30% of the total Guianan flora) have been recorded from similar communities in other zones of French Guiana where collecting effort has been greater (Hoff *et al.*, 1990). The flora of the Trinité mountains on the crest line between the Sinnamary and Mana watersheds is not considered in this paper.

The ecology of the French Guiana fish fauna is still poorly known. After the historical fish 'census' of the middle of the century (Puyo, 1949), several fish surveys have been made and field guides to the fish of French Guiana have been published in the last ten years (Le Bail *et al.*, 1984; Planquette *et al.*, 1991). However, biological and ecological studies of fish are scarce. Because of the Petit Saut dam project, the Sinnamary river has received more attention and many aspects of fish ecology are under study. We have focused our studies at three sites on the upper reaches of the Sinnamary River (Figs 1 and 2): near the Takari-Tanté rapids, in the Saut-Dalles area upstream from the rapids, and in a flatter zone at 4°30′ North. In these three areas, all riparian types of vegetation of the upper part of the Sinnamary River are represented.

Geomorphology of the Sinnamary river banks is similar to other French Guiana river banks. Fluvial terraces and inundated margins are mainly made of Demerara clays in the lower reaches. The upstream zones are made of ferralitic soils on a

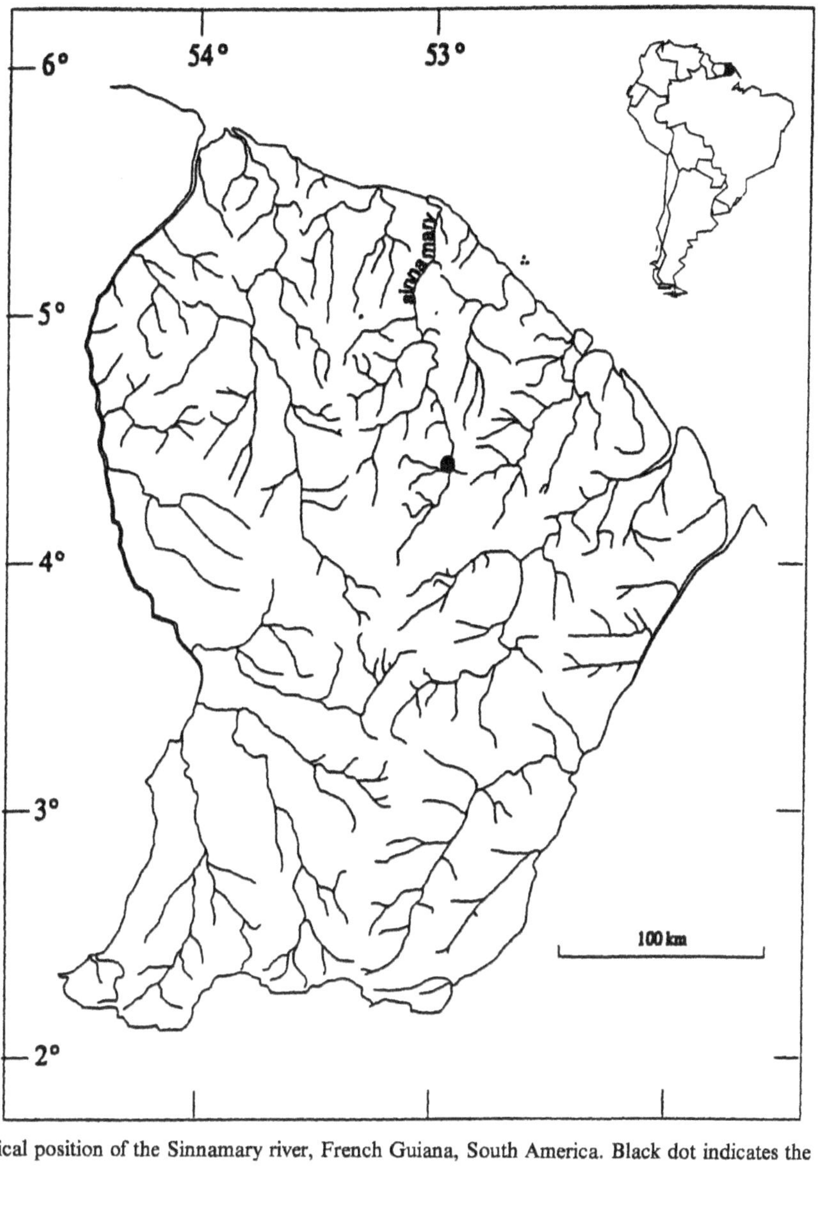

Fig. 1. Geographical position of the Sinnamary river, French Guiana, South America. Black dot indicates the study area in the upper river.

crystalline platform with alluvial terraces of coarse sediments (Blancaneaux, 1981).

Information on hydrology and water chemistry was taken from published data or obtained by other ORSTOM teams and used with permission.

The water level has been recorded continuously at Saut Dalles since November 1990, and rainfall both at Saut Dalles and in the upper basin (Laboratory of hydrology ORSTOM, Cayenne).

Material and methods

The flora

Plant assemblages in the study sites where determined by:

- direct observation and sampling of the riverine plant communities along the river margins,
- sampling along field transects perpendicular to

Table 1. Species richness of the fish fauna of five large rivers of French Guiana. Modified after Hiez & Dubreuil (1964) and Planquette *et al.* (1985), and [a] this paper, [b] Boujard *et al.* (1990a), and [c] Boujard *et al.* (1990b).

	Maroni	Oyapock	Approuague	Sinnamary	Kourou
Total fish	186	133	139	116 (126)[a]	73
Characoids	97	57 (60)[b]	62[b]	48	29
Siluriforms	52	35	–	35	25
Cichlids	17	19	–	9	5
Others	20	22	77[c]	24	14
Watershed area (km^2)	65830	26820	10850	6565	2000
River length	490	370	270	250	112

the river margins, ending, when possible, above the maximum high water mark at peak floods,
- sampling along transects parallel to the river to determine the extent of each plant assemblage.

The length of field transects was reported on maps, plants determined to species level and incorporated into the ORSTOM database.

Floristic communities have been sampled for many years in French Guiana. Over 100 000 specimens have been collected, most of which are conserved in the ORSTOM herbarium in Cayenne. A database (Hoff *et al.*, 1989) includes: species name, sampling locality, geographical coordinates, date, altitude, vegetation type, plant height, and plant habit (Granville *et al.*, 1990).

This database allows the calculation, for each plant community, of the total species richness at regional and local levels (Hoff & Brisse, 1990). The calculations made for each plant community are compared with the overall distribution of each species in French Guiana given by Cremers & Hoff (1990, 1992, in press).

The extent of each vegetation type in each selected area was used to describe the ecotones where fish samplings were performed.

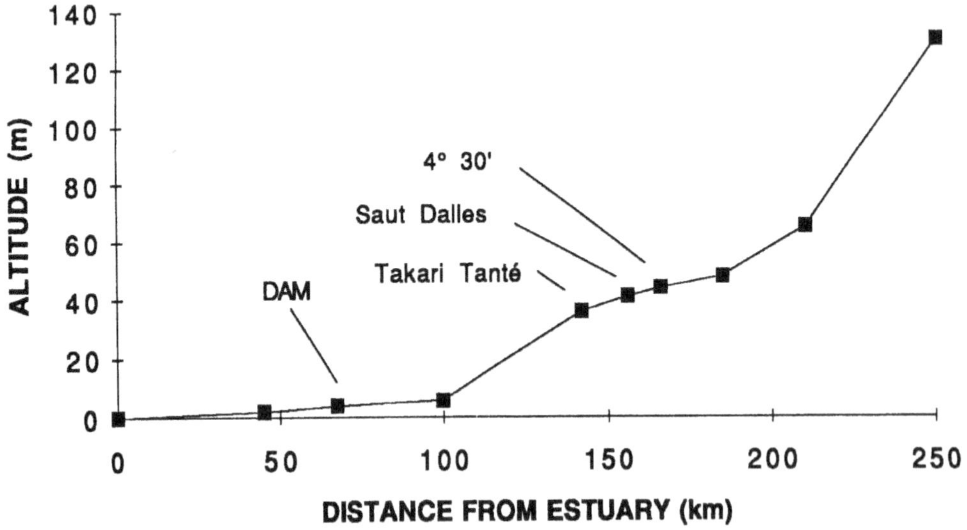

Fig. 2. Profile of the Sinnamary river with indication of the study sites and location of the dam. River length: 250 km. Watershed area: 6565 km^2. Area of the future reservoir: 300 km^2.

Remote sensing

The above floristic samplings and measurements were used as ground truth to allow for the estimation of the areas of each floristic formation at larger scales using Landsat TM images in July 1988. The image was without clouds and of excellent quality for a tropical climate.

Several classifications were tested for the best thematic results. Bands 4, 5 and 7 were then selected by Sheffield's method. A principal component analysis was performed on the three bands selected. Enhancement of colours and trichromic compositions were applied to band 7 and to the 1st principal component of bands 4 and 5 of the images taken in July 1988. Visual analysis then allowed selection of the best colour composition.

In each area (Takari Tanté rapids, Saut Dalles and 4° 30′ zone) a strip of 20 pixels on each side of the river was retained for the study (Plate I). This stripe included most of the inundated and flooded areas along the river.

The percent of each floristic formation (upland moist forest, inundated forest, flooded forest, and forested and unforested swamps) was then calculated for each study area.

The fish fauna

The fish fauna was sampled by using gillnets in open formations and along the stream margin. Gillnets 25 m long and 2,2 to 2,5 m in height with mesh sizes 10, 15, 20, 25, 30, 35, 40, 50, 60 and 70 mm from knot to knot, were employed.

At each sampling, gillnets were fished for 72 h at the same place, set in the morning, and fish removed every 24 h. In inundated forest and in channels within inundated areas fish were sampled using rotenone and a 5 mm mesh net enclosing an area of 300 to 500 m^2. Each sampling session began in the morning and lasted for 4 to 6 hours. The enclosing net was left in place during the night, and the day after the remaining fish were collected.

Nine gillnet samplings were made both in the dry and wet season: twice in November 1989, and once in June, July, and December 1990, June, August and November 1991 and May 1992.

Rotenone samplings in the inundated forest and in the channels within it were performed 4 times in each area during the wet season in 1990 and 1991.

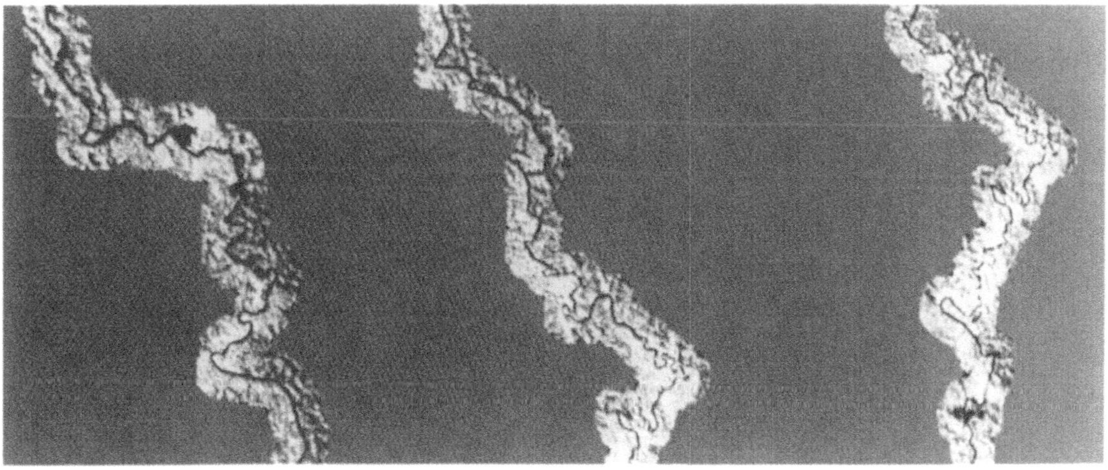

Plate 1. Remote sensing Landsat TM images (INPE/Brazil and C. Charron, remote sensing unit at ORSTOM Cayenne). Right: 4° 30′ zone, center: Saut Dalles, left: Takari Tanté rapids. Vegetation types: upland moist forest (originally green pixels), flooded and swamp forest (originally blue pixels), forested and unforested swamps (originally orange and white pixels).

The fish were identified, weighed to the nearest 0.1 g and their length measured to the nearest mm.

Results

Hydrology

Inundation of the flooded forests started when water level reached 2.40 m at Saut Dalles (Fig. 3). Short periods (less than 3 days) associated to water levels at Saut Dalles lower than 2.40 m still allow the forest flooding in the sites studied.

In 1991, 52 days of flooding of the forest where recorded. Eleven high water peak floods over 2.4 m at Saut Dalles where observed from 1 January to 4 August. Most of the peaks lasted only for very short periods (1 to 8 days). The longest period (31 days) was recorded from 23 May to 21 June.

In 1992, the flooding period of the forest was shorter and started later than in 1991. Forty-four days of flooding were recorded. Only 5 high water peak floods over 2.40 m at Saut Dalles were observed from 17 February to 31 May. Four peaks lasted only for very short periods (3 to 9 days). The longest period of flooding (21 days) lasted from 14 April to 4 May.

Mean annual rainfall (1956–1976) in the Sinnamary basin was 2777 mm (Ecerex, 1983) (Fig. 4).

Day values of percent water saturation in oxygen (at 25 °C) range from 67% to 95% in the main channel, 76% to 93% in creeks, and from 50% to 78% in the inundated forest (Richard, 1992). At night, percent saturation droped to 22% in the inundated forest while it remained at higher levels in the main channel.

Water temperatures at surface ranged from 24.3 °C (in January) to 26.8 °C (in November) during the day, both in the main channel and the inundated forest or in the channels within it (Richard, 1992). Creeks always had lower temperatures (23.3 °C to 24.2 °C). Temperature droped at night and might reach 23.5 °C in the early morning (November).

The Sinnamary river has low iron and silica

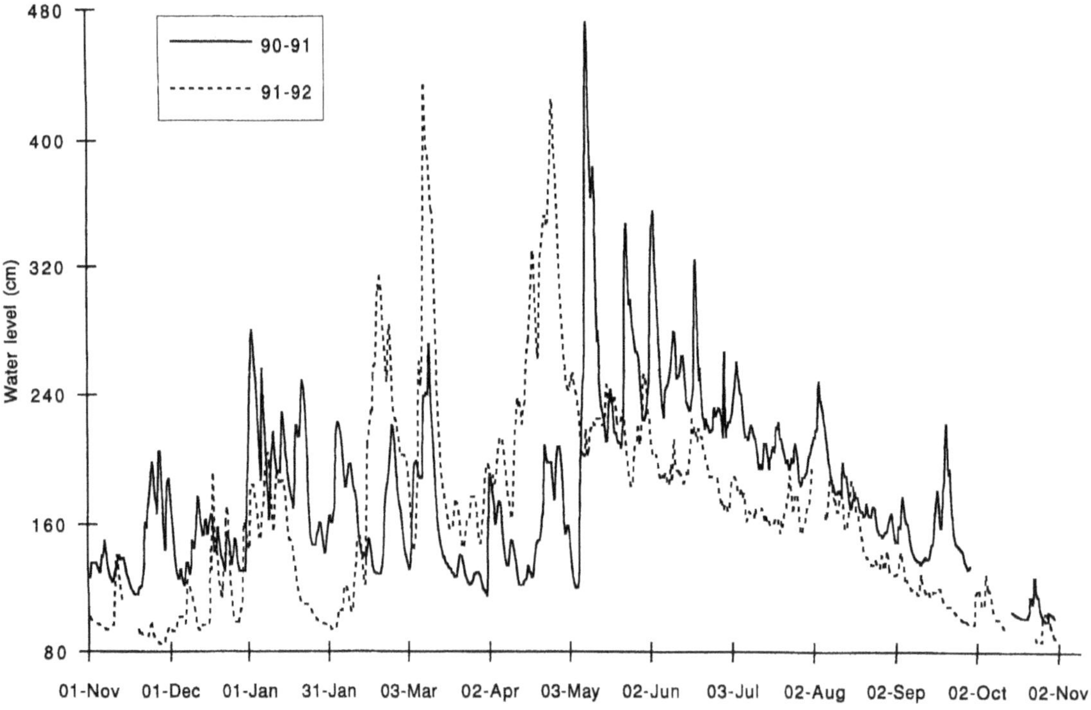

Fig. 3. Water level at Saut Dalles from November 1990 through November 1992.

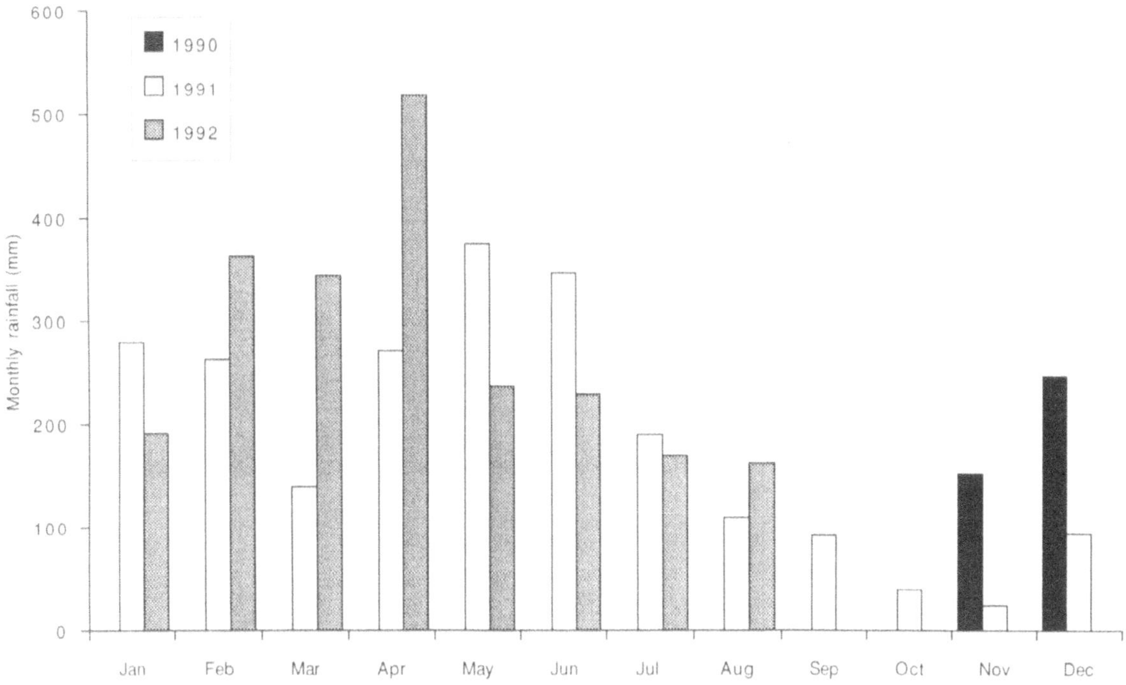

Fig. 4. Monthly rainfall in the upper Sinnamary basin upstream of Saut Dalles from November 1990 through August 1992.

content (Humbel, 1989; Richard, 1992), and the waters were close to 'clear' (Sioli, 1984). The mean hydrochemical characteristics (Humbel, 1989) were: 10 mg l^{-1} suspended matter, 4 mg l^{-1} silica, 0.5 mg l^{-1} iron, 7 mg l^{-1} Cl^{-1}. Richard (1992) found similar values for these parameters both in the main channel and in the inundated forest. She indicated also values for other parameters: pH ranges from 5.3 to 5.7, conductivity from 14.9 μS to 26.6 μS, ammonia from 6.6 μatg l^{-1} to 66.6 μatg l^{-1}, nitrite from 0 to 0.34 μatg l^{-1}, nitrate from 0 to 22.68 μatg l^{-1}, phosphates from 0 to 0.98 μatg l^{-1} and sulfate from 0.37 mg l^{-1} to 2.67 mg l^{-1}. Values recorded in creeks were always lower than in the main channel.

The flora

Forests associated with streams and rivers occupy up to 20% of the French Guianan forest vegetation (Table 2). According to de Granville (1979) (after Schnell, 1965), four forests types are associated with streams and rivers (Fig. 5):

– The low elevation upland moist forests are 'terra firme' groupings well developed on the concave margins of the rivers where they are influenced by erosion. Their border is an arborescent assemblage directly influenced by the river but not flooded by it (Oldeman, 1972). It is a high species richness riverine ecotone with forest trees and heliophilic herbs and vines ordinarily found in the canopy (Table 3).

– The permanent swamp forests vary greatly in surface, from very restricted areas to large extents. They are characterised by a constant wet layer in the underlying sediment. The understory is particularly rich in monocotyledons (Table 4, A).

– The flooded forests grow on the riparian flats, regularly submerged during the rainy season (from January to June). They are well developed on the convex margins of the rivers. Understory species,

Table 2. Species richness of the riverine flora of the upper Sinnamary compared to that of the whole French Guiana. Notes: Open formations include only vascular plants. Forest includes also bryophytes and hepatics. Figure for total number of species for French Guiana is only for the same vegetation types and not the whole flora. Species numbers do not add up because many species are shared by more than one type of vegetation.

	Whole French Guiana (1)	Whole river (2)	Mid and upper river (3)	(3)/(1) %
Open formations	357	–	29	8
Margins	286	58	25	8.7
Sand	37	–	14	38
Mud	11	–	0	0
Rock	34	–	4	12
River bed	30	–	0	0
Rocks in rapids	39	–	4	10
Swamps	31	–	1	0.03
Forest assemblages	1844	–	231	12.5
Swamp forests	203	60	45	22
Flood forests	233	38	20	8.6
Upland forest margins	1723	363	98	5.7
Swampy palm forests	100	–	–	–
Total number of samples	6658	1815	345	5.2
Total number of species	1948	809	251	12.9
Samples per species	3.4	2.24	1.37	

especially monocotyledons, are scarce and adapted to the alternation of wet and dry conditions. Pteridophytes appear to be well adapted to such conditions (Table 4, B).

– The 'pinotières' are palm swamps associated with the low muddy backwaters of most creeks. This community has a very small extent in the study area. Species richness is low (mostly dominated by *Euterpe oleracea* Martius) and is considered here as part of the permanent swamp forest.

Other typical riparian vegetal assemblages are more open formations with smaller plants (shrubs, herbs and moss):

– The backwaters and swamps with small sized vegetation, are composed of shrub and treelets associated with permanent swamp forests. They are dominated by *Triplaris weigeltiana* (Reichenbach) O. Kuntze and, along the river margins, by several species of *Inga* (*I. disticha* Bentham, *I. nobilis* Willdenow, *I. pezizifera* Bentham and mainly *I. sertulifera* De Candolle). Other common species are: *Croton cuneatus* Klotzsch, *Allamanda cathartica* Linnaeus and in the lower strata *Montrichardia arborescens* (Linnaeus) Schott and *Bonafousia siphilitica* (Linnaeus f.) Allorge. Herbaceous species are absent in this assemblage.

– The sandy margins of the active channel are important communities in the study area. The lower story is dominated by *Licania leptostachya* Bentham and *Calliandra surinamensis* Bentham. Herbaceous species are abundant (*Sauvagesia erecta* Linnaeus, *Cyperus miliifolius* Poeppig & Endlicher, *Rhynchospora cephalotes* (Linnaeus) Vahl and *Pariana campestris* Aublet but only one terrestrial fern, *Thelypteris serrata* (Cavanilles) Alston was found.

– The rocky margins have a few rupicolous species. *Turnera rupestris* Aublet is found in the lowest sections of the study area. Characteristic species of this community are: *Hypolytrum longifolium*

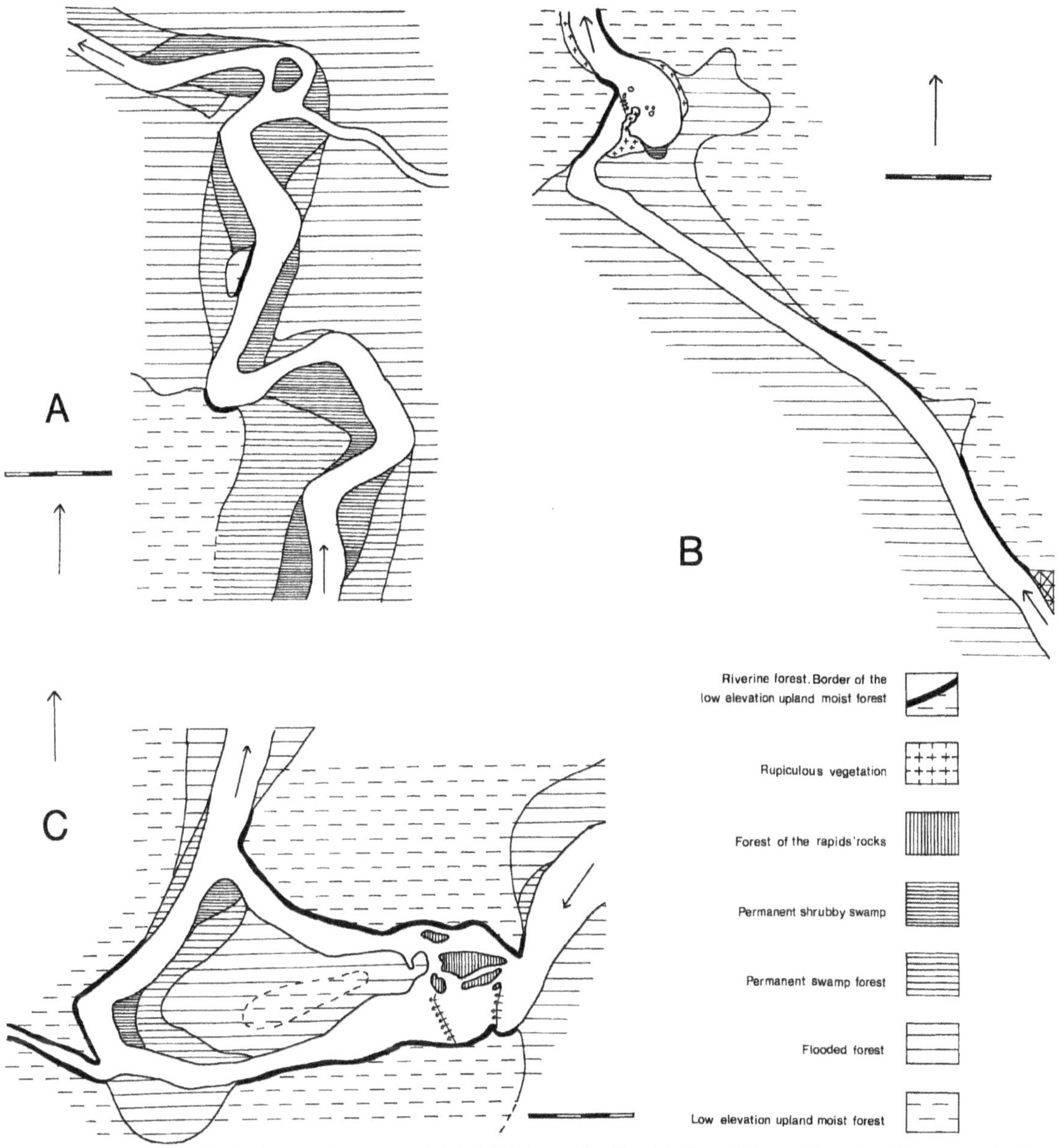

Fig. 5. Vegetation types in the three studied areas. (A) 4° 30′ N zone, (B) Saut Dalles, (C) Takari Tanté Rapids. Arrows indicate North and flow direction. Scale: each division is 40 m.

Riverine forest. Border of the low elevation upland moist forest. Rupiculous vegetation. Forest on rocks in rapids. Permanent shrubby swamp. Permanent swamp forest. Flooded forest. Low elevation upland moist forest.

(L. C. Richard) Nees (Cyperaceae) and a fern, *Trichomanes hostmannianum* (Klotzsch) Kunze.

– The rupicolous vegetation of the rapids is composed mainly of four species: *Henriettea succosa* (Aublet) De Candolle, *Calliandra surinamensis* Bentham and *Oldenlandia lancifolia* (K. Schumann) De Candolle in the sheltered zones, and *Mourera fluviatilis* Aublet along rocks in rapids.

Table 3. Main species of the riverine margins of the upland moist forest in the upper Sinnamary river.

Trees:	
Caesalpinaceae	Epiphytes
Eperua falcata Aublet	Araceae
E. rubiginosa Miquel	*Philodendron grandifolium* (N.J. Jacquin) Schott
Dicorynia guianensis Amshoff	*P. squamiferum* Poeppig
Macrolobium bifolium (Aublet) Persoon	Bromeliaceae
Swartzia benthamiana Miquel	*Pitcairnia leprieurii* Baker
Vouacapoua americana Aublet	*Tillandsia monadelpha* (Brongniart) Urban
Chrysobalanaceae	Orchidaceae
Hirtella bicornis Martius & Zuccarini	*Psygmorchis pusilla* (Linnaeus) Dodson & Dressler
H. hispidula Miquel	Epiphytic pteridophytes
Licania canescens R. Benoist	*Trichomanes crispum* Linnaeus
L. granvillei G.T. Prance	*T. elegans* L.C. Richard
L. leptotaschya Bentham	*Campyloneuron phyllitidis* (Linnaeus) C. Presl
L. licaniiflora (Sagot) Blake	*C. repens* (Aublet) C. Presl
L. majuscula Sagot	*Cochlidium linearifolium* (Desvaux) Maxon & C. Christensen
L. membranacea Sagot ex Lanessan	*Antrophyum guyanense* Hieronymus
L. pallida Spruce ex Sagot	Terrestrial herbs
Paypayrola guianensis Aublet	Cyperaceae
Posoqueria latifolia (Rudge) Roemer & Schultes	*Mapania sylvatica* Aublet
Arecaceae	Poaceae
Geonoma baculifera (Poiteau) Kunth	*Ichnanthus pallens* (Swartz) Munro ex Bentham
G. deversa (Poiteau) Kunth	Zingiberaceae
G. poiteauna Kunth	*Costus congestiflorus* L.C. Richard ex Gagnepain
Lianas	Terrestrial pteridophytes
Fabaceae	*Metaxya rostrata* (Kunth) C. Presl
Dioclea sp.	*Cyathea spectabilis* (Kunze) Domin
Malpighiaceae	*C. surinamensis* (Miquel) Domin
Banisteriopsis lucida (L.C. Richard) Small	*Selaginella falcata* (Palisot de Beauvois) Spring
Heteropterys acutifolia Adr. Jussieu	
H. macradena (De Candolle) W.R. Anderson	
H. nervosa Adr. Jussieu	

Remote sensing

The remote sensing Landsat TM images (Plate I) allowed the discrimination between riparian forest and flooded and swamp areas. The riparian forest was represented by the green pixels. Distinction is not easy within the last group. The permanent swamp forest was sometimes associated with the flooded forest (blue pixels) and sometimes with the unforested swamps (orange and white pixels).

In an area of approximately 15 km^2 at each study zone, it was possible to distinguish between the different riverine associations using the infrared bands (Table 5). The texture of the image in other areas and a finer spectral analysis of the study areas will probably allow a better discrimination in the near future.

The fish fauna

The total number of freshwater and estuarine species of fish in French Guiana is close to 600. The Sinnamary River with *ca* 130 species appears to be high in fish diversity (Table 1).

Rotenone samplings in the flooded areas

At the end of the wet season (June to early August) 49 different species were captured by roten-

Table 4. Main species of the permanent swamp forests (A) and the flooded forests (B) in the upper Sinnamary river basin.

A Permanent swamp forests
Trees
Symphonia globulifera Linnaeus f.
Caryocar microcarpum Ducke
Eschweilera pedicellata (Richard) S. Mori
Lecythis corrugata Poiteau
Couratari gloriosa Sandwith
Bertiera guianensis Aublet
Cordia nodosa Lamark
Monocotyledons of the understory
Rapatea paludosa Aublet
R. ulei Pilger
Marantaceae
Calathea cyclophora Baker
C. elliptica (Roscoe) K. Schumann
C. propinqua (Poeppig & Endlicher) Koernicke
Musaceae
Heliconia lourteigiae Mello & E. Santos
H. spathocircinata Aristeguieta
Zingiberaceae
Costus congestiflorus L.C. Richard ex Gagnepain
Arecaceae
Geonoma baculifera (Poiteau) Kunth
B Flooded forests
Trees
Quararibea lasiocalyx (K. Schumann) Vischer
Heisteria acuminata (Humboldt & Bonpland) Engler
Treelets
Psychotria racemosa (Aublet) Raeusch
Leandra agrestis (Aublet) Raddi
Monocotyledons of the understory
Arecaceae
Geonoma deversa (Poiteau) Kunth
Poaceae
Hypolytrum longifolium (L.C. Richard) Nees
Pteridophytes
Triplophyllum funestum (Kunze) Holltum
Lindsaea lancea (Linnaeus) Beddome
Lomagramma guianensis (Aublet) Ching
Lomariopsis japurensis (Martius) J. Smith
Lygodium volubile Swartz
Adiantum latifolium Lamarck
A. obliquum Willdenow
Selaginella parkeri (Hooker & Greville) Spring

one sampling in the flooded forest (20 to 24 at each sampling). In the channels within the flooded forest up to 54 species (28 to 35 at each sampling) were caught (Table 6). Nevertheless, the species composition was similar in the two zones (40 species in common).

The fish were similar in size at both locations. Over 40% were juveniles of larger fish found in the main stream as adults. Others were dwarf (<30 mm in standard length, *i.e.* the total length without the caudal fin) and small (<150 mm in SL) species.

Conversely, the catches were higher both in numbers and in weight in the channels than in the forest itself.

Gillnet samplings

Gillnet captures were highly variable (32 to 309 fish per capture – 11 to 39 species). Most fishes were adults.

Although the number of individuals was low, the species richness was similar to that of the rotenone samplings (Table 7). Forty-one species were found in the margins of the upland forest and 52 species in the margins of the swamp zone. Because of the greater selectivity of gill-nets, larger fish were captured with this sampling method.

At the species level (Table 8), the faunal composition was different at each sampling site. A total of 92 different taxa were collected but only 15 species were common to the four zones.

Discussion

Water temperature was particularly stable in the study area. The river flood regime was intermediate between equatorial and austral regimes and directly correlated with rainfall and runoff water (Hiez & Dubreuil, 1964). Flow predictability and flood frequency were intermediate according to Poff & Ward's (1989) classification.

The input into the aquatic system of allochtonous food sources (especially insects) caught by the rising flood in the areas adjacent to the river may be important (Welcomme, 1985). But in the Sinnamary river, flood plains are scarce and of small extent in the upper reaches. In these zones, the continuous input of insects and plant litter

Slide 1. Upland moist forest on slope (4)30′ zone).

Slide 9. Swamps with Triplaris weigeltiana and Inga sertulifera at the base (4° 30′ zone).

Slide 3. Contrast between the left and right banks. Upland moist forest (left) and flooded forest (right) (4° 30′ zone).

Side 15: Rupicoluos vegetation of the rapids.

Slide 6. Swamps and permanent swamp forest (4° 30′ zone).

Slide 17. Open formations and sandy floristic assemblages below the Parasol rapids.

Table 5. Surfaces and relative proportions of the three main forest types according to remote sensing calculations (C. Charron, remote sensing unit at ORSTOM Cayenne).

	Takari Tanté rapids	Saut Dalles rapids	4° 30′ section
D.E.	142	156	166
Surface	14.5	14.5	14.5
Upland	7.50	5.60	4.64
U%	70	50	50
Flooded/Swamp	0.75	2.17	0.87
F/S%	7	20	10
Swamps	2.60	3.05	3.63
S%	23	30	40
Others/water	3.85	3.63	5.36

D.E.: Distance to estuary (km).
Surface: Surface of the study section (km^2).
Upland Surface of the upland forest in the studied section.
U%: Percent surface of the upland forest.
Flooded/Swamp Surface of the flooded and swamp forest in the studied section (km^2).
F/S%: Percent surface of the flooded/swamp forest.
Swamps: Surface of the forested and unforested swamps in the studied section (km^2).
S%: Percent surface of the forested and unforested swamps.
Others/Water Surface of the other vegetation types and the river channel.

from the river banks may be of primary importance and the river continuum concept (Vannote *et al.*, 1980) provides a more appropriate template for description than the flood pulse concept of Junk *et al.* (1989).

However, the original continuum concept needs to be adapted to the differences in the riparian vegetation (Barmuta & Lake, 1982; Wiley *et al.*, 1990; Bayley & Li, 1992).

The riparian flora of the Sinnamary is still insufficiently investigated (Table 2). The number of individuals collected for each species (1.37) is much lower than that of the whole country (3.4) which demonstrates a lower sampling effort in this area. Nevertheless, close to 13% of the total number of species of plants known from French Guiana have been collected here. Indeed, three communities appear to be particularly rich: the sandy margins (38% of the total Guianan species), the flooded forests (22%), and to a less extent the rocky edges (12 %).

By contrast, the low backwater communities both forested (the 'pinotières') and unforested ('pri pris') were scarce. The low rainfall in the upper Sinnamary and its geomorphology may ex-

Table 6. Rotenone samplings in the inundated forest (swamp and flooded forest) at 4° 30′ and in the channels within the inundated forest.

	Date	N	S	W (g)	SL (mm) ± STD	N m^{-2}	W m^{-2}
Inundated forest	June 90	92	23	180	40.61 ± 27.95	0.3	0.6
	July 90	284	20	534	35.90 ± 16.52	0.7	1.3
	June 91	122	24	203	46.11 ± 38.43	0.4	0.7
	June 91	628	22	3576	37.98 ± 36.12	1.6	8.9
Channels within the inundated forest	June 90	623	28	6230	63.81 ± 34.80	3.1	31.2
	June 91	896	35	2028	48.27 ± 53.68	2.1	4.8
	Aug 91	1742	32	2959	34.06 ± 33.36	8.7	14.8
	Aug 91	1605	30	5944	25.73 + 38.09	8.0	26.5

Date: sampling date.
N: total number of fishes.
S: species richness.
W: total weight of fishes (g).
SL ± STD: mean standard length ± standard deviation.
N m^{-2}: number of fishes per m^{-2}.
W m^{-2}: weight of fishes in g per m^{-2}.

Table 7. Gillnet samplings at the river margins. (A) Riverine margins of the low elevation upland moist forest. (B) Riverine margins of the swamp zones (both forests and shrubs).

Sampling date	Effort	N	S	W	SL ± STD	CPUE N 100 m^{-2}	CPUE W 100 m^{-2}
(A)							
Nov 89	2475	238	24	19456.00	134.86 ± 58.47	10	778.24
July 90	990	309	39	7025.00	97.10 ± 33.77	31	702.50
May 92	625	213	12	40611.50	145.00 ± 75.97	34	6497.84
(B)							
Nov 89	990	109	29	37105.00	184.60 ± 127.92	11	3710.50
June 90	1980	80	26	1803.00	106.05 ± 43.94	4	90.15
Dec 90	1980	296	28	31449.00	152.52 ± 93.99	15	1572.45
June 91	1485	35	11	7310.60	197.00 ± 88.57	2	489.81
Aug 91	1980	32	13	3535.00	146.00 ± 70.50	2	176.75
Nov 91	1875	215	30	50702.20	155.00 ± 93.63	11	2535.11

Date: sampling date.
Effort: fishing effort, *i.e.* total gillnet area (m^2).
N: total number of fishes captured.
S: species richness.
W: total weight of fishes (g).
SL ± STD: mean standard length of fishes ± standard deviation (mm).
CPUE: catch per unit effort:
N 100 m^{-2}: in number of fishes per 100 m^{-2}.
W 100 m^{-2}: in g per 100 m^{-2}.

plain their scarcity. These groupings are indeed well developed in the coastal areas where they are highly diversified (Champeau *et al.*, 1991).

In the permanent swamp forest, no adults of large fishes were found. In these areas, fish abundance varied from 0.31 fish m^2 to 1.6 fish m^2 and from 6 kg ha^{-1} to 89 kg ha^{-1}. The main role of these ecotones is probably to provide shelter to young and small fishes. The reproductive success of these populations will probably be very sensitive to any modification of these land/water ecotones (Schiemer & Zalewski, 1992).

The inundated forest and the channels within it shared 40 species. Besides the juveniles and the dwarf species under 30 mm, the fish fauna characteristic of the inundated forest and the channels was composed mainly of cichlids (*Aequidens guianensis*, *A. maronii*, *Nannacara anomala*), the small traira (*Hoplias malabaricus*) which inhabits river stretches with slow currents and low oxygen levels, *Moenkhausia oligolepis* which is very abundant in small creeks and *Hypopomus beebei*, a bottom living gymnotiform. An intermediate fauna of species from the main channel: *Leporinus friderici*, *L. granti*, *Hemiodopsis quadrimaculatus*, *Chilodus zunevei*, *Poptella orbicularis* and *Bryconops sp. aff. caudomaculatus* was also found in the channels within the inundated forest. All these species were rather ubiquitous and widespread throughout the river basin.

The fish fauna associated with the riparian ecotones of the river margins was rich and diversified. The stomach contents of the fish contained large amounts of terrestrial insects, mostly ants (Horeau, comm. pers.). These ecotones were probably of major importance in providing a diversified food source for fish.

The upland forest and the flooded forest margins had 33 fish species in common. This fauna was composed of large fish (such as *Myleus ternetzi* and the giant traira *Hoplias aimara*) and some fish associated with rocky substrates (*Deuterodon* sp.) and fast currents (*Acestrorhynchus falcatus*, *A. microlepis*, *Triportheus rotundatus*, *Pimelodus ornatus* and *Doras carinatus*). The river margin of the upland moist forest appeared to be richer in large granivorous species such as *Myleus ternetzi*. The flooded zone's margins were richer in small

Table 8. Total number of fish captured in the river margins of the upland forest (Up) in the margins of the swamp zones (Sz) and in the channels within the inundated forest (Ch) and in the inundated forest itself (If).

	Up	Sz	Ch	If		Up	Sz	Ch	If
Acestrorhynchus falcatus	18	23	3	2	*Hypostomus plecostomus*	4	23	0	0
Acestrorhynchus microlepis	75	184	0	0	*Lasiancistrus sp.*	0	2	0	0
Acestrorhynchus sp.	1	14	2	5	*Leporinus despaxi*	5	17	8	1
Aequidens maronii	0	1	16	5	*Leporinus fasciatus*	1	1	0	0
Aequidens sp.	0	1	0	0	*Leporinus friderici*	13	10	23	0
Aequidens sp. *aff. guianensis*	2	4	184	100	*Leporinus granti*	3	13	48	2
Ancistrus hoplogenys	0	0	18	0	*Leporinus* sp.	5	2	3	0
Ancistrus sp.	0	2	0	0	*Lithoxus* sp.	2	0	1	0
Anostomidae sp.	0	0	1	16	*Loricariidae* spp.	3	2	0	0
Anostomus brevior	0	6	22	2	*Moenkhausia barbouri*	0	1	0	0
Apistogramma sp.	0	0	0	1	*Moenkhausia collettii*	0	0	217	64
Astyanax sp. 1	2	9	20	8	*Moenkhausia comma*	3	0	10	1
Astyanax sp. 2 *aff. polylepis*	0	0	27	5	*Moenkhausia georgiae*	0	1	0	0
Auchenipterus nuchalis	18	23	0	0	*Moenkhausia* n. sp. *aff. barbouri*	0	0	13	0
Bivibranchia bimaculata	21	17	0	0	*Moenkhausia oligolepis*	1	9	107	126
Bryconops sp. 1 *aff. melanurus*	3	53	0	0	*Moenkhausia* sp.	0	2	0	0
Bryconops sp. 2 *aff. caudomaculatus*	153	28	28	2	*Moenkhausia* sp. *aff. surinamensis*	0	24	8	1
Bryconops sp. 3 *aff. affinis*	0	3	0	0	*Myleus rhomboidalis*	3	2	0	0
Characidae nain sp. 1	0	0	10	0	*Myleus ternetzi*	78	10	0	0
Characidae nain sp. 6	0	0	6	0	*Nannacara anomala*	0	1	38	1
Characidae nain spp.	0	0	33	0	*Parupygus savannensis*	2	2	14	0
Characidae sp.	0	0	3	0	*Phenacogaster megalostictus*	0	0	1257	26
Characidae sp. 4	0	0	0	254	*Pimelodella cristata*	11	26	16	11
Characidium sp. *gr. fasciatum*	0	0	64	11	*Pimelodella gracilis*	0	4	0	0
Chilodus zunevei	15	14	10	4	*Pimelodus ornatus*	13	4	0	0
Corydoras octocirrus	3	13	0	2	*Poptella orbicularis*	9	45	11	5
Crenicichla saxatilis	5	5	21	6	*Pristella maxillaris*	0	0	91	3
Ctenobrycon spilurus	0	0	0	42	*Pseudopimelodus albomarginatus*	0	0	1	0
Cyphocharax cyprinoides	1	0	0	0	*Pseudopimelodus raninus*	0	2	21	3
Cyphocharax helleri	0	63	32	7	*Pseudopimelodus* sp.	0	0	1	0
Cyphocharax sp.	0	0	305	132	*Pseudopimelodus* sp. 1	0	0	0	1
Cyphocharax spilurus	38	58	106	2	*Pseudopimelodus zungaro*	0	0	0	1
Deuterodon sp.	12	15	0	0	*Pseudopristella simulata*	0	0	1593	107
Doras carinatus	31	18	0	0	*Pyrrhulina filamentosa*	0	0	1	3
Eigenmannia virescens	0	4	0	0	*Rhamdia quelen*	0	0	1	1
Geophagus surinamensis	2	2	0	0	*Rivulus xiphidius*	0	0	8	2
Gymnotus carapo	0	0	56	2	*Satanoperca* sp. *aff. leucosticta*	0	1	0	0
Harttia surinamensis	6	0	0	0	*Siluriforme*	0	0	4	0
Hemiancistrus sp.	13	0	0	0	*Sternopygus macrurus*	0	0	5	1
Hemicetopsis sp.	0	0	0	1	*Synbranchus marmoratus*	0	0	3	0
Hemigrammus ocellifer	0	0	85	1	*Tatia intermedia*	0	8	136	0
Hemiodidae sp.	0	0	0	1	*Tetragonopterus chalceus*	1	0	0	0
Hemiodopsis quadrimaculatus	10	27	12	6	*Trichomycterus guianense*	0	0	2	1
Hemiodus unimaculatus	1	0	0	0	*Triportheus rotundatus*	44	10	0	2
Hoplias aimara	31	40	57	64	*Utiaritichthys* sp.	0	1	0	0
Hoplias malabaricus	1	0	25	18	Unidentified	5	0	57	44
Hypopomus beebei	0	0	23	20					

sized omnivorous species such as *Moenkhausia sp. aff. surinamensis*.

Conclusion

(1) The flora of the Sinnamary basin comprised 809 known species, 251 of which were associated with the river margins. The total plant species richness of the area, based on similar areas of other parts of French Guiana, is expected to reach 2000 species. Along the 140 km of river margins already studied, the ten plant communities encountered (both forested and non forested) offered different ecotones which are exploited by specific fish assemblages.

(2) The flooded forest comprised only 40 plant species, *i.e.* less than 10% of the total number of species occurring in this vegetation type in French Guiana. Species of fish associated with flooded forest had a lower biomass than those of the riverine margin of the upland forest.

(3) Conversely, the permanent swamp forest was highly diversified in plant species (close to 25% of the total number of species found in this formation in French Guiana). Species richness of fish in permanent swamp forest was high and comprised the very early developmental stages of several species never found elsewhere.

(4) The riverine margins of the upland moist forest possessed the highest species richness. They were associated with high fish species richness and biomass. The high fish species richness was probably related to the allochthonous food resources that are of major importance to the fish in tropical rivers. Vegetal food sources, such as leaves, fruits, and flowers, were directly exploited by fish as well as insects, especially ants, which were associated with the overhanging vegetation.

Acknowledgements

Thanks to Dr A. Tito de Morais for her help in preparing the manuscript, to C. Charron for the remote sensing treatments, to Dr P. Vauchel for hydrological data, to Dr Lauzanne for field samplings and to Drs Scott Mori and J. J. de Granville for revising the text.

This paper was funded in part by the EDF/Orstom contract n° GP7514.

References

Barmuta L. A. & P. S. Lake, 1982. On the value of the river continuum concept. New Zealand J. mar. freshwat. Res. 16: 227–231.

Bayley, P. B. & H. W. Li, 1992. Riverine fishes. In P. Calow & G. E. Petts (eds), The River Handbook, Vol. 1 – Hydrological and Ecological Principles. Blackwell Scientitic Publications, Oxford: 251–281.

Boujard, T., F. J. Meunier, M. Pascal & J. F. Cosson, 1990a. Les téléostéens d'un haut bassin fluvial guyanais, l'Arataye. 1 – Inventaire des Characoïdes. Cybium 14: 175–182.

Boujard, T., F. J. Meunier, M. Pascal & J. F. Cosson, 1990b. Les téléostéens d'un haut bassin fluvial guyanais, l'Arataye. 2 – Inventaire des 'non-Characoïdes'. Cybium 14: 345–351.

Holland, M. M., 1988. SCOPE/MAB technical consultations on landscape boundaries. Report of a SCOPE/MAB workshop on ecotones. In F. di Castri, A. J. Hansen & M. M. Holland (eds), A New Look at Ecotones: Emerging International Projects on Landscape Boundaries. Biology International, special issue 17. UIBS, Paris: 47–106.

Champeau, A., A. Vaquer & A. Grégoire, 1991. Petit-Saut hydroelectric scheme: vegetal associations of stagnant waters in French Guyana. Hydroécol. Appl. 3: 111–124.

Cremers, G. & M. Hoff, 1990. Inventaire taxonomique des plantes de la Guyane française: I. Les Ptéridophytes. Secrétariat de la Faune et de la Flore, Muséum National d'Histoire Naturelle, Paris, 133 pp.

Cremers, G. & M. Hoff, 1992. Inventaire taxonomique des plantes de la Guyane française: II. Les Orchidaceae. Secrétariat de la Faune et de la Flore, Muséum National d'Histoire Naturelle, Paris, 144 pp.

Cremers, G. & M. Hoff, in press. Inventaire taxonomique des plantes de la Guyane française: III. Les Cyperaceae et les Poaceae. Secrétariat de la Faune et de la Flore, Muséum National d'Histoire Naturelle, Paris, 174 pp.

Ecerex, 1983. Analyse de l'écosystème forestier tropical humide et des modifications apportées par l'homme. CTFT, INRA, MNHN, Orstom, Cayenne, Guyane française, 417 pp.

Granville, J.-J. de, 1979. Végétation. In Atlas des DOM-TOM. IV La Guyane. Orstom, Paris: 12–13.

Granville, J.-J. de, M. Hoff &. Cremers, 1990. La Florule du Haut-Marouni. Enregistrement et exploitation informatique d'une mission botanique en Guyane française: la mission Haut-Marouini 1987. C.R. Soc. Biogéogr. 66: 3–25.

Gregory, S. V., F. J. Swanson, W. A. McKee & K. W. Cummins, 1991. An ecosystem perspective of riparian zones. BioScience 41: 540–551.

Hiez, G. & P. Dubreuil, 1964. Régimes hydrologiques en Guyane française. Orstom, Paris, 119 pp.

Hoff, M., 1991. Etude d'environnement du site du barrage de Petit-Saut et du bassin du Sinnamary. Flore et végétation. Rapport intermédiaire. Convention Orstom/EDF, Orstom, Cayenne, 52 pp.

Hoff, M. & B. Brisse, 1990. Diversité et répartition des formations végétales en Guyane française à partir d'une banque de données sur l'environnement. Atelier sur l'Aménagement et la Conservation de l'Ecosystème forestier tropical humide. Cayenne 12–16 mars 1990: 1–25.

Hoff, M., G. Cremers, C. Feuillet & J. J. de Granville, 1989. La banque de Données 'AUBLET' de l'Herbier du Centre Orstom de Cayenne (CAY). Bull. Jard. bot. nat. Belg. 59: 171–178.

Hoff, M., G. Cremers & J.-J. de Granville, 1990. Carte de la richesse en plantes de la Guyane française à partir d'un observatoire du patrimoine naturel: la Banque de données 'AUBLET' de l'Herbier du Centre Orstom de Cayenne. Nature Guyanaise, Cayenne 4: 12–21.

Humbel, H. X., 1989. Qualité et dynamique des eaux fluviales de Guyane française. Orstom, Cayenne, 35 pp.

Junk, J. J., P. B. Bayley & R. E. Sparks, 1989. The flood pulse concept in river-floodplain systems. In D. P. Dodge (ed.), Proceedings of the International Large River Symposium. Can. Spec. Publ. Fish. aquat. Sci. 106: 110–127.

Le Bail, P. Y., P. Planquette & J. Géry, 1984. Clé de détermination des poissons continentaux et côtiers de Guyane. Rapport INRA CRAAG, Kourou, 63 pp.

Lowe-McConnell, R., 1987. Ecological studies in tropical fish communities. Cambridge Univ. Press, London, 382 pp.

Murphy, M. L., C. P. Hawkins & N. H. Anderson, 1981. Effects of canopy modification and accumulated sediment on stream communities. Trans. am. Fish. Soc. 110: 469–478.

Oldeman, R. A. A., 1972. L'architecture de la végétation ripicole forestière des fleuves et criques de Guyane. Adansonia, ser. 2, 12: 253–265.

Orstom, 1991. Annuaire hydrologique de Guyane. Orstom, Cayenne, Guyane Française, 60 pp.

Planquette, P., J. Géry & P. Y. Le Bail, 1991. Faune characoïde (poissons ostariophysaires) de l'Oyapock, l'Approuague et la rivière de Kaw (Guyane française). Cybium 15 (Suppl.): 1–69.

Planquette, P., R. Rojas-Beltran & P. Y. Le Bail, 1985. Etude d'impact sur le peuplement de Petit-Saut sur le peuplement ichtyologique. Rapport INRA/EDF. Kourou, Guyane française, 76 pp.

Poff, L. N. & J. V. Ward, 1989. Implications of streamflow variability and predictability for lotic community structure: a regional analysis of streamflow patterns. Can. J. Fish. aquat. Sci. 46: 1805–1818.

Puyo, J., 1949. Faune de l'empire français. XII: Poissons de la Guyane française. Orstom, Paris, 280 pp.

Richard, S., 1992. Etat d'avancement des travaux sur la qualité des eaux et des sédiments du bassin versant du fleuve Sinnamary. EDF, Petit-Saut, Guyane Française, 26 pp.

Schiemer, F. & M. Zalewski, 1992. The importance of riparian ecotones for diversity and productivity of riverine fish communities. Neth. J. Zool. 42: 323–335.

Schnell, R., 1965. Aperçu préliminaire sur la phytogéographie de la Guyane. Adansonia 5: 309–355.

Sioli, H., 1984. The Amazon and its main affluents. Hydrography, morphology of the river courses, and river types. In H. Sioli (ed.), The Amazon: Limnology and landscape ecology of a mighty tropical river and its basin. Dr W. Junk Publishers Dordrecht: 127–165.

Tchaplinsky, R. J. & G. F. Hartman, 1983. Winter distribution of juvenile coho salmon (Onchorhynchus kisutch) before and after logging in Carnation Creek, British Columbia, and some implications for overwinter survival. Can. J. Fish. aquat. Sci. 40: 452–461.

Thorpe, J. E., 1991. Review of the introductory session: general considerations. In M. Zalewski, J. E. Thorpe & P. Gaudin (eds), Fish and Land/Inland Water Ecotones. UNESCO MAB. Univ., Lodz, Stirling, Lyon I: 11–15.

Vannote, R. L., G. W. Minshall, K. W. Cummins, J. R. Sedell & C. E. Cushing, 1980. The river continuum concept. Can. J. Fish. aquat. Sci. 37: 130–137.

Wiley, M. J., L. L. Osborne & R. W. Larimore, 1990. Longitudinal structure of an agricultural prairie river system and its relationship to current stream ecosystem theory. Can. J. Fish. aquat. Sci. 47: 373–384.

Welcomme, R. L., 1985. River fisheries. FAO Fish. techn. Pap. 262: 330 pp.

Management aspects of streams and rivers

Hydrobiologia **303**: 183–194, 1995.
F. Schiemer, M. Zalewski & J. E. Thorpe (eds), The Importance of Aquatic-Terrestrial Ecotones for Freshwater Fish.

River restoration concepts – goals and perspectives

S. Muhar, S. Schmutz & M. Jungwirth
Dep. of Hydrobiology, Fishery & Aquaculture, State University of Agriculture Vienna, Austria

Key words: river restoration, revitalization, river management, ecological improvement

Abstract

An analysis of restoration projects carried out both in Austria and abroad shows that most projects are limited to design measures within the river bed with the general intention to increase habitat diversity. The various measures are not based on the type-specific characteristics of the river and thus there is a danger of creating uniform restoration stretches, independent of the specific landscape area, river morphology, discharge regime or site specific biocoenoses. Such proceeding lacks in considering a comprehensive improvement of ecological functioning of the whole river-system. To meet those demands a type-specific guiding view ('Leitbild') has to be developed based on the pristine river conditions.

The comparison of the river pattern including the characteristic features of its unmodified state with the actual situation leads to a comprehensive set of measures. The essential goals of such management-concepts are the integrity of the river habitat, self-regulation and self-regeneration, the preservation of intact resources, as well as the possibility to experience the uniqueness, diversity and beauty of natural river landscapes.

Introduction

River restoration and ecotone research have become inseparable topics since restoration measures were extended beyond immediate low-water channels to include aquatic-terrestrial boundaries. Studies of the land–water ecotone, for example, provide an important basis for river bank design measures. Ultimately, the development of measures such as buffer strips also relies on ecotone research findings.

This paper underscores the call for more comprehensive river restoration projects which encompass the entire river system. During the past few years the term 'ecotone' has also received a wider connotation with regard to space and time (Bretschko, 1995). Both aspects now comprise the interaction between the entire catchment area and the river. Accordingly, the time element has gained importance, as both ecotone research and restoration projects increasingly take the dynamic processes in rivers, the migration of river beds and even wetland dynamics into account.

The challenge for restoration

The ecological situation of numerous rivers – of large streams as well as of rivers or brooks – is characterized by severe human impacts like river engineering, hydro power plants and water pollution.

As an example of the actual state of running waters the situation of the alpine rivers is characterised as follows. A study of Cipra, the International Commission for the Protection of the Alps, shows, that only 10% of the most important rivers of the entire Alpine region can still be call 'pristine' or 'natural like' (Martinet & Dubost, 1992). This degradation especially of the large rivers of the northern Alps is also documented by Müller (1991). On the basis of changes in typical pioneer vegetation of braided rivers, he shows that only minimum areas of intact communities have survived along the River Lech in the Tyrol and the River Isar in Bavaria.

With the heightened awareness of this development over the past few years, river restoration measures in the framework of river engineering have become increasingly important.

The term 'river restoration'

The numerous terms used in the context of river restoration such as 'rehabilitation', 'revitalization', 'renaturation', 'reconversion', 'restructuring', etc., indicate a very wide range of quite different measures. Thus, small-scale measures (e.g. creation of a pool) are subsumed under the same key word as comprehensive restoration projects. Small-scale measures for drain trenches (Newbold et al., 1989) correspond terminologically to large-scale interdisciplinary concepts (Gardiner, 1990).

Although both in practice and in modern literature restoration measures are mainly understood to be design measures, this paper deliberately extends the connotation of the term to describe the most complete 'river-enhancement'.

The following definition of river restoration is proposed:

> 'The totality of measures which change man-induced alterations to rivers (primarily flood control measures, but also diversions, hydro peaking, etc.) in such a manner that the ecological functioning of the new state resembles a more natural river.'

Common river restoration procedures

In recent years the spectrum of different approaches in the field of restoration has widened significantly. Common restoration concepts are based on a wide range of goals and are developed at different levels of planning. In some cases focus has been placed on the importance of river patterns (meandering, braiding, straight river) in developing type-specific concepts.

From an ecological point of view the richness, diversity and age composition of the populations developing after restoration as a result of habitat improvement reveal the degree to which comprehensive concepts were applied. Scientific studies can help to evaluate the ecological success of restoration programmes. Most projects are characterized by an evident lack of such accompanying investigations.

Goals and levels of planning

A common problem is that the set of measures drawn up is based on very vague nature conservation aims or one-sided goals. Most of such restoration programmes are additionally restricted to localized design measures such as structuring the river bed by groynes or small drops, creating pools, widening the profile, creating buffer strips, planting of trees and shrubs etc.

Usually one-sided goals are pursued, such as the enhancement of fish habitat for species of interest to sports fishing. In this context most previous projects can be classified as 'restructuring' projects limited to morphological improvements of the river bed. Even if such goals reflect nature conservation demands, e.g. to restore endangered fish species, the final measures are mostly based on very restricted aims without an overview concept.

In the so-called 'Building Block Model for Stream Restoration', Petersen *et al.* (1992) describe a series of measures (buffer strips, revegetation, horseshoe wetlands, side-slope reduction, meander valley, riffle-pools, riparian wetlands/swamp forest, ponds) designed 'to decrease the transport of nutrients to surface waters' on one hand and 'to restore the animal and plant life along and within the stream channels' on the other. Although these measures improve habitat diversity locally, comprehensive concepts encompassing the entire river system are lacking. The various measures are not based on the type – specific characteristics of the river and thus there is a danger of creating uniform restoration stretches, independent of the specific landscape area, river morphology, discharge regime and site specific biocoenoses.

The River Murr project in Baden-Württemberg, Germany (Landesansalt f. Umweltschutz Baden – Württemberg, 1985) provides another example of a one-sided procedure. Despite an enormous improvement in the habitat conditions, an overall strategy was missing. The poor water quality of River Murr, for example, prevented the appearance of more sensitive species and the development of a balanced fish stock.

Despite emphasizing the importance of changes in the catchment area, other revitalization projects rarely incorporate measures to improve the situation. Changes in discharge and bedload are almost always accepted without being questioned.

Consideration of the river pattern

As mentioned above most previous restoration projects can be classified as 'restructuring' projects. Common measures include the building of groynes, placement of rocks, creation of pools and shelters, etc. which are not specifically adapted to the abiotic and biotic characteristics of the respective river patterns (e.g. White & Brynildson, 1967).

Such a procedure obviously engenders certain deficits in the habitat spectrum for the characteristic range of fish species as well as for other aquatic biocoenoses.

One of the few well-documentated projects is the restoration of the River Melk, a fifth-order potamal river in Austria (Jungwirth *et al.*, 1993). Restructuring measures increased the number of fish species from 10 to 18. Nevertheless, the stagnophile cyprinids, characteristic of the oxbows of the unmodified Melk river system cannot

Fig. 1. Channel widening at the River Isel/Austria (Wiesbauer, 1992).

reestablish themselves. An overall concept including the development of adjacent wetlands is missing; this can be ascribed mainly to the lack of legal regulations for restorations.

The river maintenance project at the River Isel in Austria (Michor, 1990) can be cited as a first step towards a type-specific river restoration. In contrast to other projects the provision of relatively large land areas for the formation of braided channels permits the River Isel to redevelop its formerly characteristic pattern (see Fig. 1).

Desirable river restoration procedure

In more recent projects short-term and use-oriented goals are beginning to be replaced by long-term goals favouring the comprehensive protection of the river ecosystem (Boon *et al.*, 1992). Binder (1992), for example demands the incorporation of dynamic and self-regulating principles of the river ecosystem in the planning stage. Gardiner (1990) underlines river catchment planning as an essential basis of revitalization projects including changes in bedload, discharge, adjacent areas, etc. on a large-scale. At the same time he points out the problems of implementation and overlapping responsibilities of regional and supraregional authorities.

A comprehensive approach to revitalization is increasingly called for (Boon *et al.*, 1992) and public decision-makers in particular are being asked to meet supraregional demands. Mellquist's statement (1992) that 'many administrative bodies must also be forced to act from an overall point of view rather than just tending their own patches', clearly shows the future course.

The prime object of river restoration measures is the reconversion and protection of the river ecosystem with a species composition typical of the site (e.g. Kern *et al.*, 1992). The classical river patterns (straight, braided and meandering rivers) are characterized by distinct differences in slope, flow velocity, substrate composition or the size of the valley occupied by the river. Furthermore, each river pattern is correlated with distinct type-specific biocoenoses. The innumerable river patterns due to the great diversity of landscape types demonstrate that more effort will be required to adapt restoration programmes to specific river patterns (Fig. 2).

Development of a comprehensive restoration concept

The general objective of restoration is the comprehensive improvement of ecological functioning of the river. Primary aims such as:

Fig. 2. River patterns:basis for the 'Leitbild' – concept (Photo: S. Muhar).

* integrity of the river ecosystem,
* self-regulation and regeneration capacity,
* intact resources (water, soil,..), and
* the possibility to experience beauty und diversity of natural river landscapes

must be defined and given precedence over the planning process.

One of the first steps is the determination of the restoration area. Regardless whether the project involves a small stretch or the whole river, the entire system must be taken into account (Fig. 3).

Type-specific 'guiding view'
The next step is the application of general aims to the specific restoration programme. The formulation of site-specific goals requires determining the pristine river conditions. In German these goals are termed 'Ökologisches Leitbild' (ecological guiding view) indicating that all subsequent steps and measures are derived from these principal aims. Kern (1992) and Redl (1990) point out that this 'Leitbild' is based on an ecological view, including cultural and irreversible changes (Fig. 4). In our opinion a separate treatment of the 'Leitbild' on the one hand and the anthropogenic changes on the other has to be considered. We are aware that man is part of the whole river system, but to define the ecological effect of anthropogenic modification in respect to severe impacts as well as restorations we need to have a standardized reference. The determination of irreversible is more sociopolitical than scientific. The 'Leitbild' should be formulated free of such imprecision. The question which changes are to be implemented is automatically answered after the restoration programme has been drawn up. The comparison of the 'Leitbild' and the

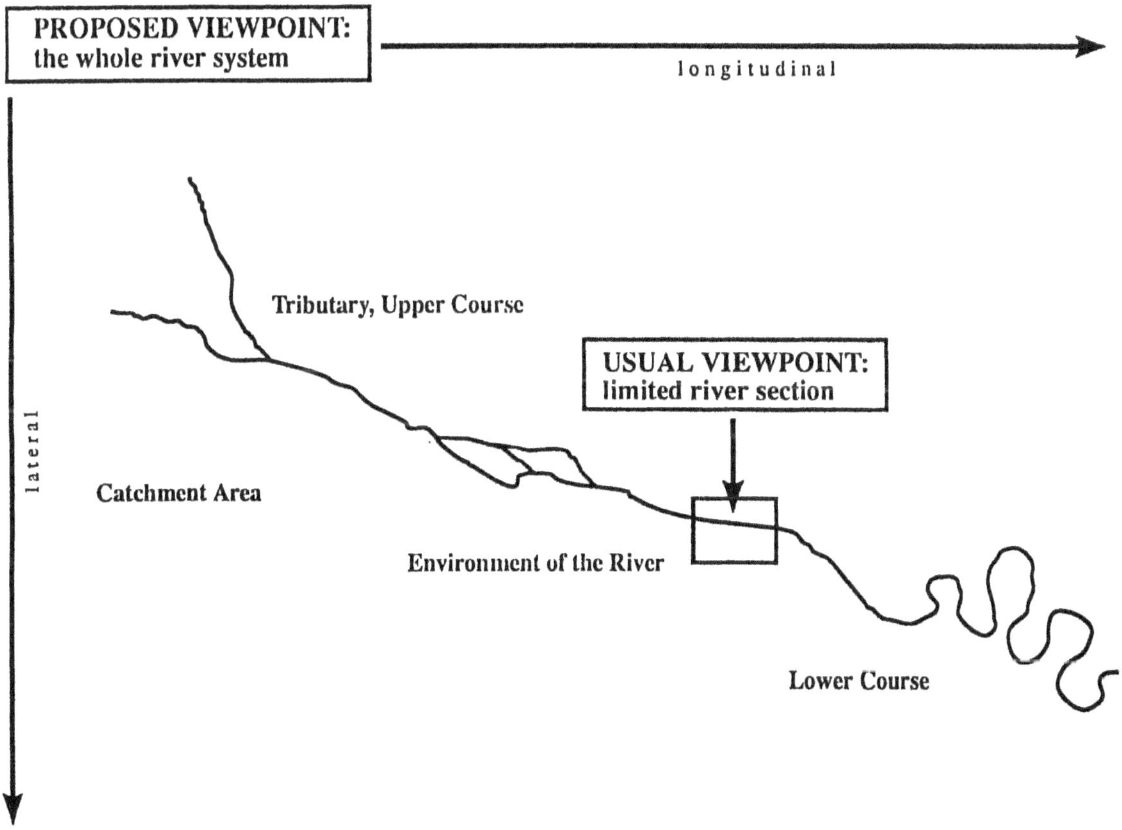

Fig. 3. Different viewpoints of river restoration concepts (A. Muhar, 1991, adapted).

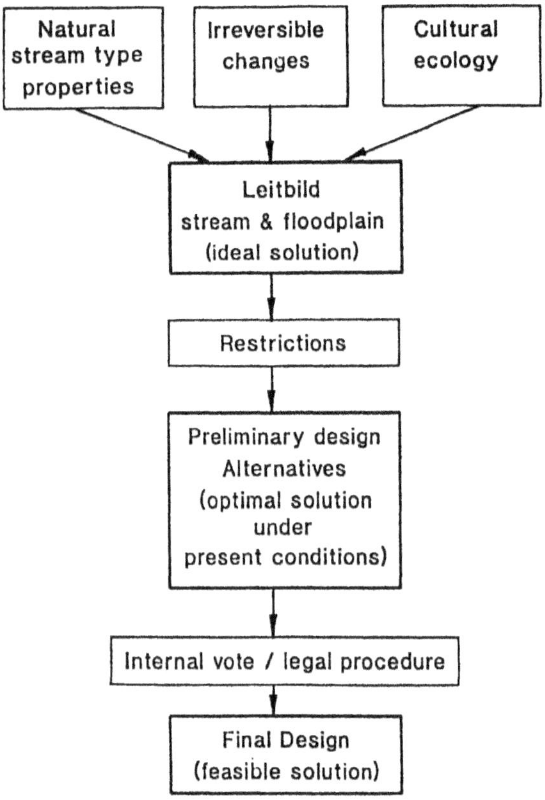

Fig. 4. The 'Leitbild' – concept and steps to the final design (Kern, 1992).

restoration results reveals the degree of success. The function of the 'Leitbild' is not to reestablish an unmodified river system but to show where we are on an ecological scale.

Specification of a type-specific 'guiding view'

The simplest way to obtain a 'guiding view' is to look for reference sites typical for the restoration area. With advanced degradation and correspondingly high demand for restoration, such reference sites may be lacking. This necessitates a more exhaustive examination of the unmodified characteristics of the river.

The problems involved defining the level of modification was pointed out by Moog (1990). 'Can we describe as 'unmodified' the state before of after the first periods of clearing, the rafting of timber and salt, the early industrial exploitation of water power or the first periods of regulation?' The basis for evaluation should not be some far-removed state after the Ice Ages, but rather the time before extensive alterations to the river ecosystem were made – in the form of systematic course straightening, bed stabilization measures or hydro-electric power plants – and before fundamental changes in the essential characteristics of rivers such as channel pattern, flooding dynamics, etc. occurred.

Such a definition certainly comprises landscape modifications caused by agriculture, but only those before the advent of industrial forms of agriculture as pointed out by the 'Rat der Sachverständigen für Umweltfragen' (1985):

> 'Cultivated land, which today seems necessary to protect and preserve, is mainly shaped by man. The plethora of species and communities also relies on a multitude of former agricultural alterations. So the point is not to glorify an original natural landscape, but to preserve and regain the diversity of the cultivated landscape passed down through the centuries. This is the reference system which must form the basis for the maintenance of the landscape, the conservation of the efficiency of natural systems as well as of the protection of species and biotopes'.

Criteria to define the type-specific 'guiding view'

The following list of criteria is a general summary of the most important abiotic and biotic parameters describing the 'Leitbild'; it will have to be modified variously according to the nature of the task. These characteristics enable a rough differentiation of various river patterns and obviously also can be used to describe the actual conditions of the river including man-made alterations.

* nature, geology, valley pattern
* river morphology
  river course
  slope and width
* channel morphology
  profiles of river bed and bank
* hydrology
  regime
  characteristic discharge values

flow characteristics
temperature
* environs of the river
bank and wetland vegetation
river patterns/network of backwaters
* river biocoenoses
fish
benthos
etc.
* faunal communities of river banks and wetlands
amphibians
birds
etc.

The choice of criteria will be determined by their type – specific importance, the availability of data and by financial and time considerations. If the main characteristics of the river are known, an overall picture of the river pattern can be drawn.

Thus records of the river's former course, combined with the zoogeographical classification of the region and basic hydrological and climatic data, provide clues about the fish regions and potential existence of certain fish species. Vice versa, records of the former fish fauna (e.g. records of commercial fisheries) often enable reconstructions of the general river pattern.

For example, historical maps of the entire Austrian territory exist from the 18th century. Historical reports, old surveys, records of old fishing rights etc. help us to reconstruct individual criteria; these, in turn, often permit us to regenerate an overall view of the unmodified situation.

Development of a comprehensive set of measures

Today the range of design-orientated measures includes profile enlargement, shifting of the river course, connection of lateral arms, rebuilding and dismantling of flood dams, vegetation measures, etc. This spectrum must be extended to large-scale measures such as wastewater treatment, reduction of intensive land-use, increase of flooding frequency of the river environs, increase of discharge in stretches affected by diversion, declaration of ecologically valuable or sensitive river stretches as protected areas, etc. In each case, all measures will have to be developed on the basis of the type-specific 'guiding view' in order to guarantee that the river enhancement is being guided toward the unmodified situation.

Restoration of the River Inn with emphasis on river morphology and bedload

The following example described the development of a principal concept to reestablish braided sections in the regulated River Inn.

Present state of the River Inn

The River Inn in the Tyrol, a 6th-order stream, is a 'mountain stream with glacier influences', (Kresser, 1961), and shows the characteristic temperature regime and suspended load. Today the Inn flows in an extremely confined bed due to the intensive exploitation of the valley floor (enormous expansion of urban areas, extension of industrial sites, highways and railway, etc.). Its course was straightened and stabilized in a systematic and uniform way. The construction of run-of-the-river plants, and in particular the continuous regulation of the river, led to a progressive degradation of the river bed. Hydro-peaking in the upper course as well as in the tributaries further severely affect long sections of the river ecosystem.

The Inn and the tributaries show a severe change in bedload, the most important being retentions caused by torrent regulation and power plants, reduction produced by industrial dredging and faster transportation due to river straightening (see Fig. 5). Almost no gravel is transported in many lower courses due to this bedload reduction. The most important consequences for river morphology are river bed degradations, the most severe degradation being caused by the combined effect of bed load reduction and straightening.

From an ecological point of view, the change in bedload has an impact on the habitat of the river

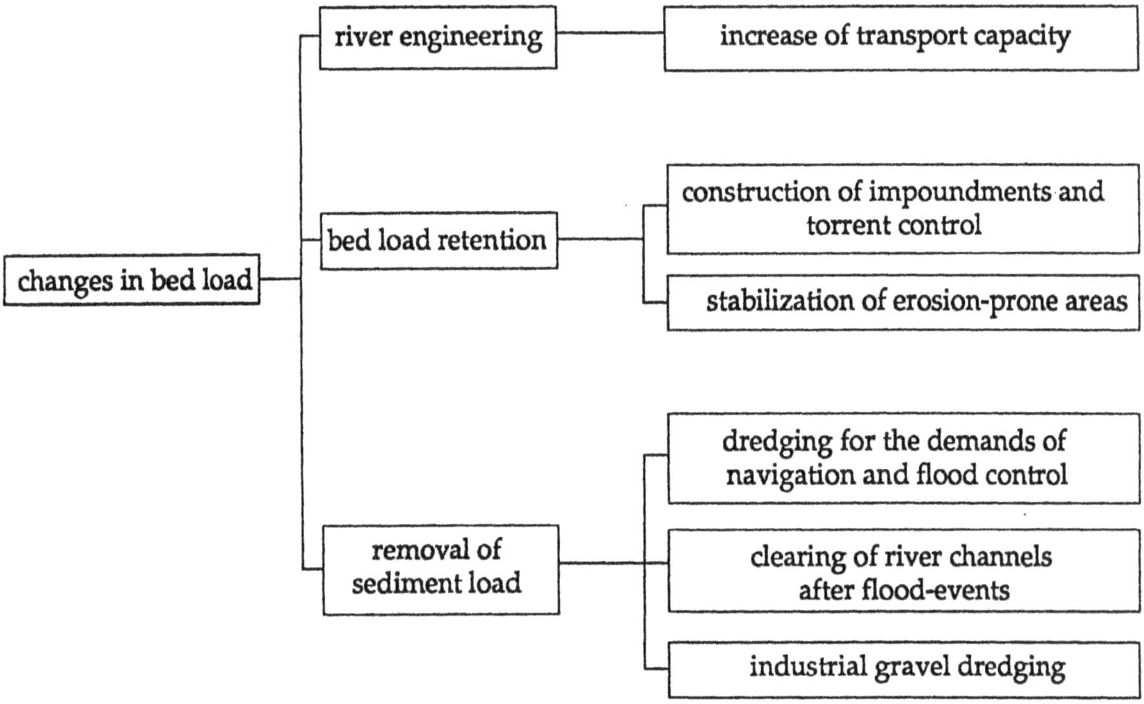

Fig. 5. Changes in bedload (Wiesbauer, 1992, adapted).

bed and on the immediately adjacent areas and on the environs of the river. Degradation of the Inn resulted in loss of braided stretches, lowering of the groundwater level, drying-up of oxbows and reduced or interrupted wetland dynamics caused by lack of floods. This leads to changes in the wetland vegetation and to the loss of the various types of oxbows.

The above alterations are very clearly reflected in fish fauna. Around 1920, before the abovementioned impacts fish stock surveys indicated a range of more than 24 species (Jungwirth et al., 1989). The only indigenous and reproducing species today are the grayling (Thymallus thymallus, L.) and the brown trout (Salmo trutta fario, L.).

The present state of the river, which is only touched on in this paper, shows clearly the prime importance of developing a comprehensive restoration plan. The criteria of section 2 would enable a plan of the original river landscape to be drafted; this would serve both as a guide-line to procedures and as a corrective for misconceived, one-sided restoration projects.

Developing a 'guiding view'

Preliminary evaluations of old fish fauna records, reports of fishing clubs, etc. provide relevant clues for determining the unmodified situation. Nineteenth century literature on this subject indicates that formerly the Inn showed fish species with diverse habitat preferences and needs. The co-existence of Danube salmon (Hucho hucho, L), nase (Chondrostoma nasus, L.) and barbel (Barbus barbus, L.) alongside carp (Cyprinus carpio, L.) tench (Tinca tinca, L.) and bream (Abramis brama, L) demonstrates that the main stream channel was surrounded by a number of different oxbow types. This is further supported by maps dating from the year 1770 (see Fig. 6).

Additional archival material (land-use forms, local history records, etc.) permits further assessment of the former extent of the Inn wetlands and oxbows. Finally, vegetation studies of the last remaining natural like wetlands provide reference data on the communities of the Inn wetland forest.

The above approach enables the gradual estab-

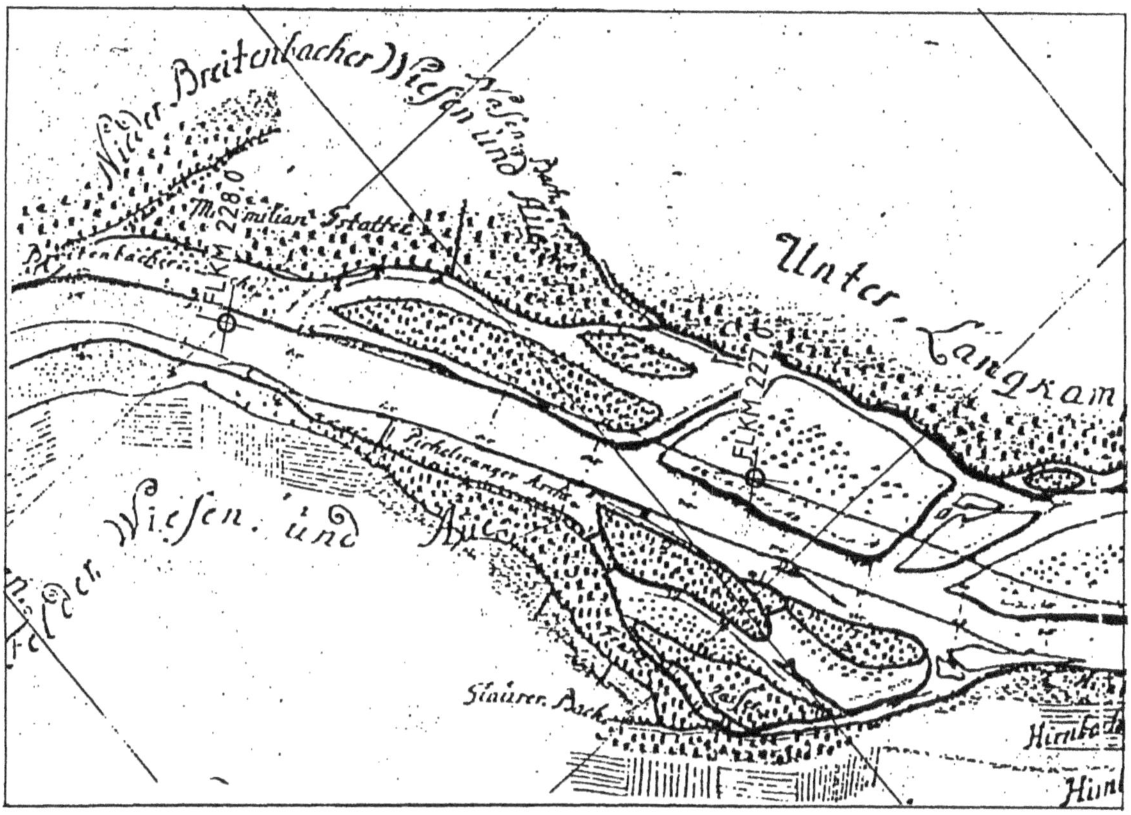

Fig. 6. Historical map of the River Inn (1770).

lishment of original river characteristics; this, in turn, serves as a valuable basis for specifying the 'guiding view'.

Restoration measures

A variety of completed restoration projects shows that design measures specifically aimed at habitat improvements such as profile enlargements, structuring of the bank line, the creation of pools and riffles, flow control by means of transverse structures, etc. have been taken to restore regulated river sections. However, these measures are often effective only on a small scale (see Fig. 7).

According to the criteria of section 2, restoration programmes should not follow a single goal. Sole considerations such as the optimum improvement of the aquatic habitat in order to increase fish stocks, or the enhancement of purely aesthetic aspects of the landscape are to be rejected. The necessity of creating site conditions to establish typical aquatic and terrestrial biocoenoses must be formulated in a general manner and be dealt with in a specific set of measures. In our case study, which deals with an originally braided river, this involves considering not only the immediate river landscape, but the environs in their entirety.

The following restoration measures take the morphological characteristics of the river pattern of the Inn into account and aim at a comprehensive ecological amelioration:

* enlargement of the riverbed;
* integration of adjacent areas into the discharge in order to permit the braiding of the river and the creation of lateral arms, bars, shifting areas and oxbows as sites for a characteristic flora and fauna;

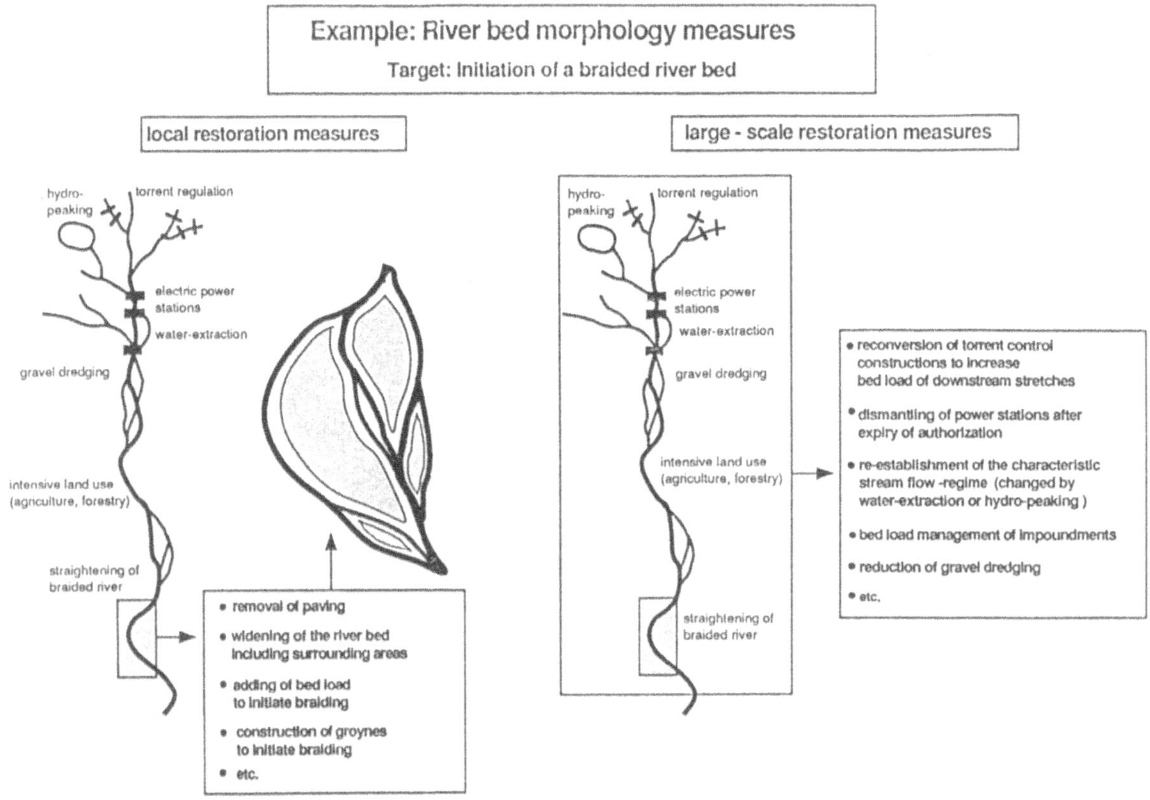

Fig. 7. Local and large-scale restoration measures.

* safeguarding or reconstructing the mouths of tributaries at the same level as the main river (fish migration);
* reconstruction of the originally characteristic oxbows (spawning grounds for fish, development of vegetation characteristic of the site);
* etc.

Basic objective: a comprehensive approach

A comprehensive approach requires the formulation of restoration measures encompassing the entire river system. A number of questions, e.g., will river systems still show their typical ratios of discharge and bedload, are important for establishing detailed morphological and structural measures.

The degradation of the Inn riverbed is characterized by bedload deficit and regulation. In the re-establishment of braided sections, local measures are thus only useful if they guarantee that the river section is resupplied with bedload. This arrests or reverses degradation on the one hand, and secures the dynamic development of shifting areas on the other. An investigation into the causes of changes in bedload therefore serves as a basis for the elaboration of an extensive restoration programme.

Rebuilding or dismantling bedload barriers in the upper course or in tributaries will increase the bedload in the river system again. The bedload accumulated in impoundments should be reintroduced into the natural bedload system. This may involve the programmed rinsing of impoundments or the addition of bedload dredged from impoundments further downstream. After expiry of authorization of power plants also the renewal of the authorization and even the dismantling of power plants must be considered. At the same time, the bedload can also be increased by specifically allowing bank erosion. In these cases industrial dredging should be generally banned or restricted. For reasons of flood control, the clearing of channels after major floods must be reduced to a minimum.

The close interaction between bedload and discharge requires incorporating information on the typical flow regime in the context of restoration measures governing river morphology. An adequate degree of flood dynamics is a prerequisite for initiating braided sections. The shaping power of the river should be retained to guarantee again dynamic processes in braided sections.

The restoration measures presented in this paper are intended to show the scope of a comprehensive restoration plan (see Fig. 7).

Summary

The increasing range of alterations to river ecosystems and their effects on aquatic biocoenoses have increased the importance of river restoration.

Moreover, an analysis of restoration projects carried out both in Austria and abroad shows that most are limited to design measures within the river bed. The main reasons lie in the lack of legal provisions, as is the case in Austria, as well as in one-sided approaches during the planning process. Thus, narrow objectives as well as measures not adequately tailored to the particular type of river tend to produce poorly differentiated restoration projects. The consequences are uniformly revitalized stretches and conversions restricted to structural measures.

Future projects will require a broader understanding of river restoration. The main objective should be a comprehensive ecological improvement. The essential aspects of this objective are the integrity of the river habitat, self-regulation and self-regeneration, the preservation of intact resources, as well as the possibility to experience the uniqueness, diversity and beauty of natural river landscapes.

This approach requires taking the entire catchment area into consideration. In addition to the structural measures carried out on the channel, parameters such as sediment load, discharge, environs and their exploitation, etc. will have to be included in the planning of future restoration projects.

Attaining this goal and drafting specific measures requires the development of a 'guiding view' (German: 'Leitbild') which takes the respective river pattern along with the characteristic features of its unmodified state into consideration.

References

Binder, W., 1992. Renaturierung und Regeneration von Fließgewässern. Naturschutz und Wasser – 21. Deutscher Naturschutztag, Lübeck.

Boon, P. J., P. Calow & G. E. Petts, 1991. Conservation and Management. J. Wiley & Sons, New York, 470 pp.

Bretschko, G., 1995. River/land ecotones: scales and patterns. Hydrobiologia 303 (Dev. Hydrobiol. 105): 83–91.

Gardiner, J. L., 1990. River Projects and Conservation. A Manual for Holistic Appraisal. J. Wiley & Sons, New York, 236 pp.

Jungwirth, M., S. Schmutz & H. Waidbacher, 1989: Fischökologische Fallstudie Inn. Fischerei-Revierausschuß Innsbruck Stadt und Land, Hall, 93 pp.

Jungwirth, M., O. Moog & S. Muhar, 1993. Effects of river bed restructuring on fish and benthos of a fifth order stream, Melk, Austria. Regulated Rivers, Vol. 7, in press.

Kern, K., 1992. Rehabilitation of Streams in South-West Germany. In P. J. Boon, P. Calow & G. E. Petts (eds), River Conservation and Management. J. Wiley & Sons, New York, 321–336.

Kern, K., R. Bostelmann & G. Hinsenkamp, 1992. Naturnahe Umgestaltung von Fließgewässern. Handbuch Wasserbau, Heft 2, 228 pp.

Kresser, W., 1961. Hydrographische Betrachtung der österreichischen Gewässer. Verh. Internat. Verein. Limnol. 14. 417–421.

Landesanstalt f. Umweltschutz Baden – Wüttemberg, 1985. Ökologische Untersuchungen an der ausgebauten unteren Murr. Karlsruhe, 328 pp.

Martinet, F. & M. Dubost, 1992. Die letzten naturnahen Alpenflüsse. CIPRA Kleine Schriften 11/92, Vaduz, in print.

Mellquist, P., 1992. River Management-Objectives and Applications. In P. J. Boon, P. Calow & G. E. Petts, Ed., River Conservation and Management. J. Wiley & Sons, New York, 1–8.

Michor, K., 1990. Revitalisierung verzweigter Gebirgsflüsse – Ausschotterungsbecken an der Isel in Osttirol. In. H. Honsowitz, M. Jungwirth & S. Muhar (eds), Revitalisierung von

Fließgewässern. Wiener Mitteilungen 88, Wien, 331–348.
Moog, O., 1990. Makrobenthologische Aspekte bei der Wiederherstellung naturnaher Flußabschnitte. In H. Honsowitz, M. Jungwirth & S. Muhar (Edr.), Revitalisierung von Fließgewässern. Wiener Mitteilungen 88, Wien, 55–103.
Muhar, A., 1991. Landschaftsplanerische Aspekte bei Fließgewässerrevitalisierungen. In H. Honsowitz, M. Jungwirth & S. Muhar (eds), Revitalisierung von Fließgewässern. Wiener Mitteilungen 88, Wien, 131–148.
Müller, N., 1991. Der Lech. Wandel einer Wildflußlandschaft. Augsburger Ökologische Schriften 2, 174 S. Hrsg. Stadt Augsburg, Referat Umwelt und Kommunales, Amt f. Grünordnung und Naturschutz, Augsburg.
Newbold, C., J. Honnor & K. Buckley, 1989. Nature conservation and the management of drainage channels, Nature Conservancy Council, Peterborough, 108 pp.
Petersen, R. C., L. B.-M. Petersen & J. Lacoursiere, 1992. A Building – Block Model for Stream Restoration. In P. J. Boon, P. Cow & G. E. Petts, Ed., River Conservation and Management. J. Wiley & Sons, New York, 293–310.
Rat der Sachverständigen für Umweltfragen, 1985. Umweltprobleme der Landwirtschaft. Sondergutachten, Verlag Kohlhammer, Stuttgart, 423 pp.
Redl, G., 1990. Moderne wasserwirtschaftliche Planungsansätze. In Landschaftswasserbau 10, Technische Universität Wien, 31–72.
White, R. J. & O. M. Brynildson, 1967. Guide-lines for management of trout stream habitat in Wisconsin. Technical Bulletin Number 39, Dept. Nat. Res. Div. Cons., Madison, 64 pp.
Wiesbauer, K. H., 1992. Störungen des Geschiebehaushaltes – Konsequenzen für die Gewässerumgestaltung und -pflege. Landschaftswasserbau 13, Technische Universität Wien, 389–416.

Hydrobiologia **303**: 195–206, 1995.
F. Schiemer, M. Zalewski & J. E. Thorpe (eds), The Importance of Aquatic-Terrestrial Ecotones for Freshwater Fish.

The effects of recreated instream and ecotone structures on the fish fauna of an epipotamal river

M. Jungwirth, S. Muhar & S. Schmutz
Department of Hydrobiology and Fisheries Management, Universität für Bodenkultur, Vienna, Austria

Key words: river restoration, instream structures, land/water ecotone, fish fauna

Abstract

Investigations of fifteen sections of seven Austrian epipotamal (barbel region) streams between 1981 and 1984 demonstrate the impact of instream river bed structures on fish communities. Reduced spatial heterogeneity due to river straightening resulted in decreasing species number, diversity, stock density and biomass. Reincreased variability of the river bed in the frame of a subsequent restructuring project improved all community-specific values significantly within a 3-year investigation period (1988–1990). Besides the regained habitat variability in form of riffle pool sequences and other instream structures, the newly created riparian zones obviously provided important niches, e.g. as refuge areas during flooding and as nursery grounds for fish fry. The positive effects of the recreated land/water ecotone are discussed with respect to river restoration projects.

Introduction

Structural variety of the habitat is one of the most important conditions for the existence and development of well-balanced aquatic communities (Lelek and Lusk, 1965; Hynes, 1968; Sheldon, 1986; Karr and Schlosser, 1978; Dudley and Karr, 1979; Tarplee *et al.*, 1979; Ward and Stanford, 1979; Schoof, 1980; Meffe and Sheldon, 1988).

After the 2nd World War, Western and Central Europe experienced a boom in the straightening of rivers in many cases with disastrous consequences for the fish fauna. Today more than half of the endemic fish species have either become extinct or are seriously endangered mainly due to straightening and/or damming of rivers. In the 1980s various studies on the ecological impact of reduced river bed structure proved the negative effects of monotonous aquatic environments on fish stocks in Austria (Jungwirth, 1981, 1984). Those studies finally entailed plans for the restructuring of a monotonously straightened section of one of the rivers investigated, the River Melk in Lower Austria. This 5th-order stream with a catchment area of 285 km^2 and a mean annual discharge of 3.5 $m^3 s^{-1}$ showed an absolutely monotonous straightening (Fig. 1), the river bottom being continuously paved. The low flow channel was 10 m wide and about 0.3 to 0.6 m deep along the whole length, its shoreline formed by rip rap.

The most important restructuring measures comprised enlargements of the cross section, structuring of the bottom and shorelines by partial removal of the pavement and rip rap as well as the construction of groynes and bedfalls. Legal restrictions confined the restructuring measures to the low flow channel and its shoreline, but omitted the surrounding landscape (Fig. 2). This

Fig. 1. The straightened Melk River.

limitation explains why the pilot study was designated a 'restructuring' rather than a revitalization project. During the planning stage the correlation between the variance of the maximum depths (as a measure for the structural variety of the river bed, see Fig. 3) and the number of fish species (Jungwirth, 1991) was used to control the project and to forecast the possible development of the fish stock. In addition, a special examination program was elaborated to cover a period of three years after completion of the restructuring measures. It involved investigations of the effects of the various constructions on river morphometry, the fish stock and the benthic community (Jungwirth, 1991). The present paper presents the results of these studies with respect to the instream structures relevant to fish ecology and deals with the newly created land/water ecotone, which, apart from other effects, serves as a nursery niche for fish fry and as a refuge area during floods.

Fig. 2. After restructuring in autumn 1988 the Melk River shows a strongly increased spatial variety of the river bed (compare Fig. 3).

Methods

Fifteen sections of seven Austrian epipotamal streams were investigated with regard to morphometry of the river bed, current and substrate pattern. The widths of at least 25 successive cross sections (distance 5 m) per test site were measured; water depths, flow velocities and substrate types within each cross section were determined at 2 m intervals. Among other parameters, river bed structure was described by the variance of the maximum depths (VMD) and correlated with the number (NFS) and the diversity (according to Brillouin; Pielou, 1975) of fish species (FSD); the latter were calculated from stock estimates by electro-fishing following the De Lury method (Bagenal, 1978). In order to document the development of morphometric and hydraulic conditions as well as of fish community parameters after restructuring, further investigations were conducted using the above-mentioned methods. Finally, the fry and young of the year fishes in shallow littoral zones as well as the adult fishes within the riparian vegetation during floods were sampled by special electric gear (for details see Jungwirth, 1991) following the catch-per-unit-effort method. During the three-year investigation period a still straightened section of the river was sampled as a reference site.

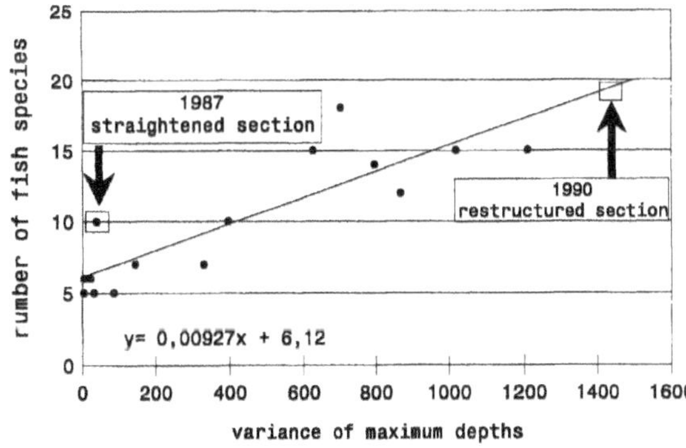

Fig. 3. Variance of maximum depths (VMD) versus number of fish species (NFS) of 15 test sites in 7 Austrian epipotamal streams ($r = 0.861$) investigated between 1981 and 1984. Plotted squares: The formerly straightened and restructured Melk river (compare Fig. 4 and text).

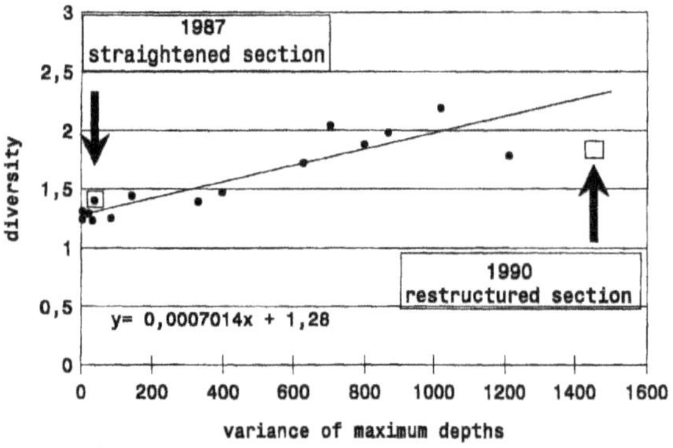

Fig. 4. Variance of maximum depths (VMD) versus fish species diversity (FSD) of 15 test sites in 7 Austrian epipotamal streams ($r = 0.897$) investigated between 1981 and 1984. Plotted squares: The formerly straightened and restructured Melk river (compare Fig. 3 and text).

Results

Instream structures and fish stock

The comparison of 15 straightened and unstraightened epipotamal stream sections of different rivers with great differences in spatial heterogeneity demonstrated clearly the importance of river bed structure for the fish community (Figs 3 and 4). Natural instream structures such as riffles and pools, gravel banks, woody debris etc., were reflected by high VMD values, which in the models corresponded to high numbers and diversities

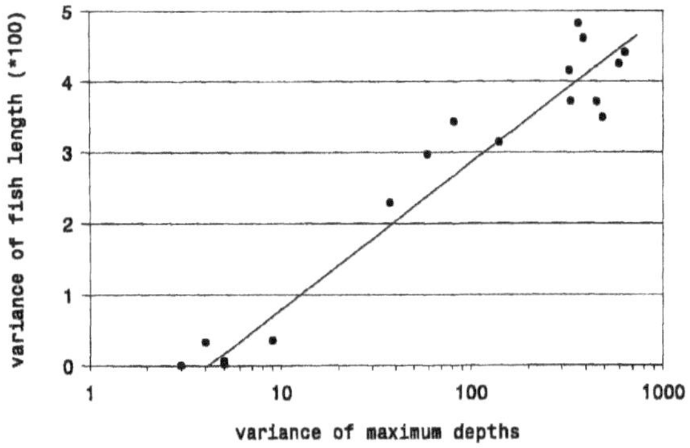

Fig. 5. Variance of maximum depths (VMD) versus variance of fish lengths (brown trout): correlation of 16 headwater test sites of river Ferschnitz, Lower Austria ($p < 0.001$, $r = 0.97$; compare text).

of fish species. In many headwaters of the upper rhithral zone (trout region) the fish community comprised only a few species. Occasionally these waters were inhabited exclusively by brown trout, Salmo trutta. In the latter case the VMD was significantly correlated with the variance of the observed fish lengths, assuming that variable instream structures offer favorable niches for various age classes. Reduced spatial heterogeneity, on the other hand, led to the dominance of only one or few age classes characterized by a low variance of fish lengths (Fig. 5).

The effects of recreated instream structures

Physical effects

Restructuring of a 1.5 km stretch of the Melk River in 1988 resulted in significant changes of the hydrologic and morphometric conditions. The straightened reach only displayed depths between 0.2 and 0.6 m (Fig. 6), whereas depths along the restructured section ranged from <0.1 to 2 m (Fig. 7). The mean current speed of the straightened section showed an 80% dominance of the category 0.4–0.5 m s^{-1}, whereas the restructured section regained a wide range of flow velocities (Figs 6 and 7). The heterogeneous pattern of flow velocities was reflected by more differentiated substrate types as well: both the straightened and the restructured river section were dominated by coarse sediments (Fig. 8), but the latter also exhibited the categories sand and mud.

Fish community

Quantitative sampling of the fish stock following the De Lury method showed that the fish fauna reacted relatively rapidly to the changed instream environment. Two electro-fishings within the straightened section in 1980 and 1987 revealed the same 10 fish species (Table 1). One year after restructuring, NFS already increased to 16. By 1990, 19 species occurred, while NFS in the still straightened reference site remained restricted to 10. Density and biomass of the total fish stock also showed an obvious recovery. Both parameters tripled during the period of investigation

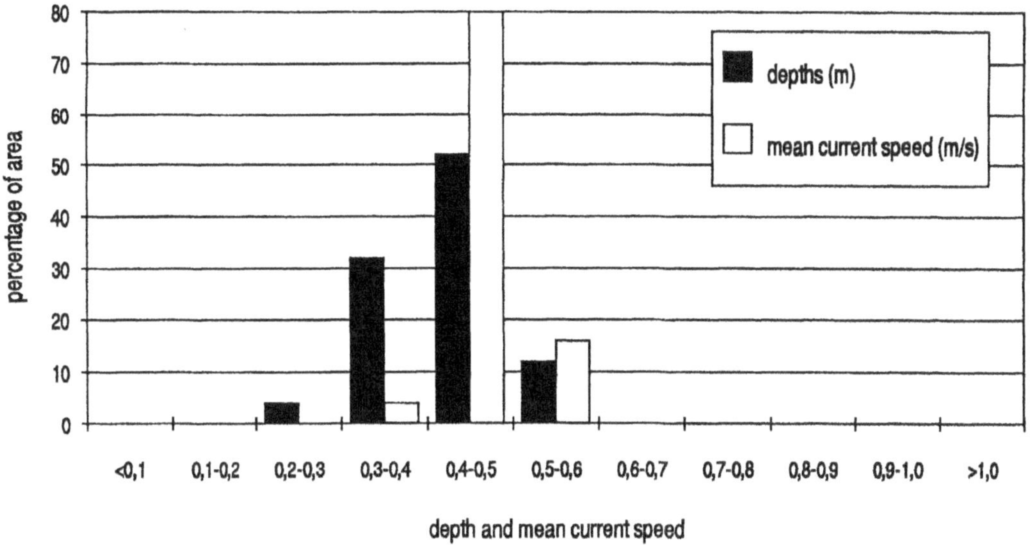

Fig. 6. Depths and mean current speeds in the straightened reference section of Melk River indicate a monotonous environment (compare Fig. 7).

(Fig. 9). At the same time the abundance of the individual species changed considerably. For instance the 66.2% dominance of *Leuciscus cephalus* and the 27.1% dominance of *Gobio gobio* in the straightened section declined by about 50% in the restructured section. This fact, together with the increased NFS demonstrated a more balanced proportion between the individual species; this was also reflected in the diversity index, which increased from an initial 1.40 to 1.87. With re-

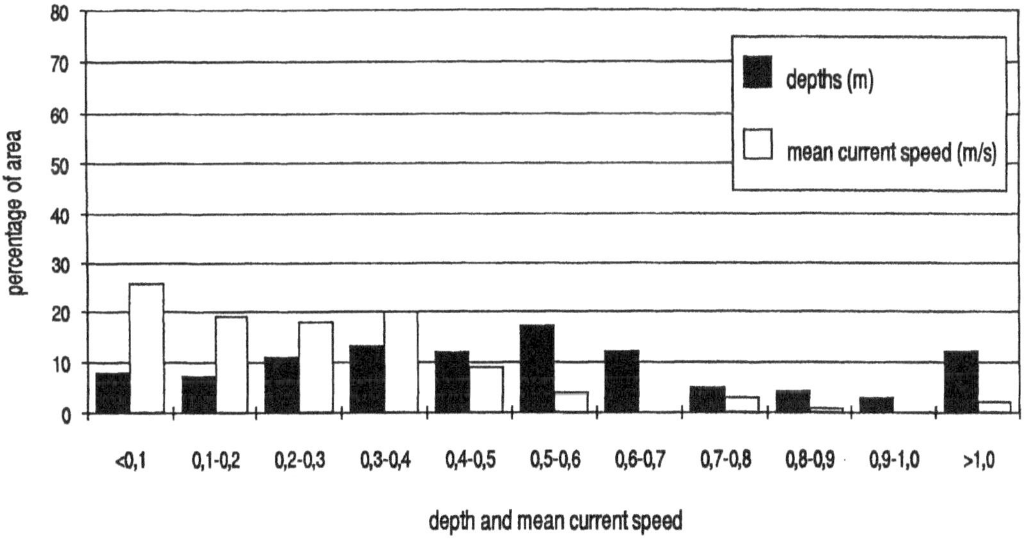

Fig. 7. Depths and mean current speeds in the restructured section of Melk River illustrate a strongly increased range of both parameters (compare Fig. 6).

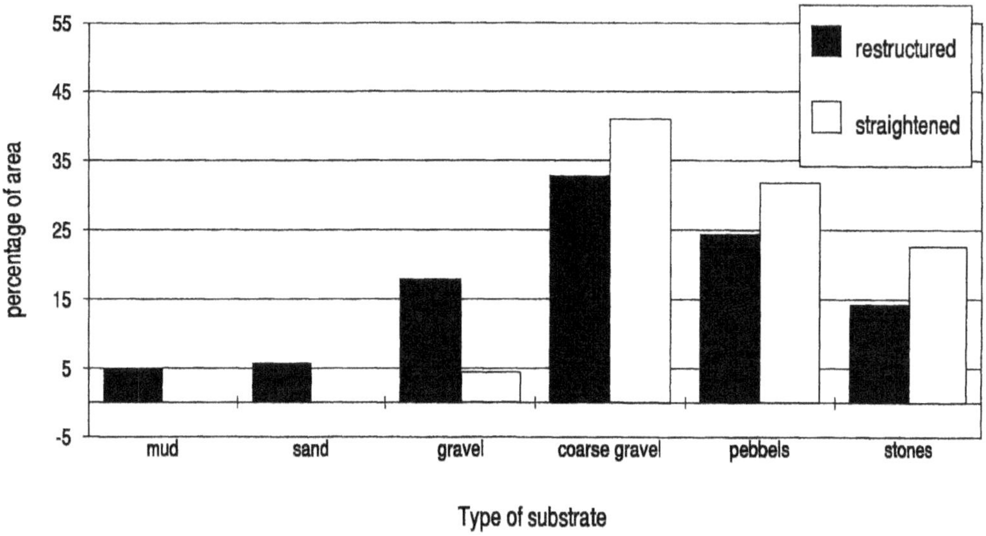

Fig. 8. Coarse substrates dominate in the straightened and in the restructured Melk River, whereas mud and sand are entirely restricted to the restructured sites (for choriotope classification see Moog, 1991).

spect to the VMD/NFS correlation, the restructured river section developed according to the model until 1990 (Fig. 3). The FSD index, however, remained below the value corresponding to the VMD (Fig. 4).

Table 1. Fish species of Melk river sampled in straightened and restructured sections between 1980 and 1990.

Fish species	1980 straightened	1987 straightened	1988 straightened	1988 restructured	1990 straightened	1990 restructured
Salmo trutta forma fario, L.	*	*	*	*	*	*
Salvelinus fontinalis, M.				*		
Hucho hucho, L.			*			
Oncorhynchus mykiss, W.	*	*	*	*	*	*
Thymallus thymallus, L.	*	*	*	*	*	*
Leuciscus cephalus, L.	*	*	*	*	*	*
Barbus barbus, L.	*	*	*	*	*	*
Rhodeus sericeus amarus, B.						*
Phoxinus phoxinus, L.	*	*	*	*	*	*
Gobio gobio, L.	*	*	*	*	*	*
Leuciscus leuciscus, L.	*	*	*	*	*	*
Cyprinus carpio, L.						*
Chondrostoma nasus, L.				*		*
Rutilus rutilus, L.				*		*
Scardinius erythrophtalmus, L.					*	
Alburnoides bipunctatus, B.				*		*
Noemachilus barbatulus, L.	*	*	*	*	*	*
Cobitis taenia, L.				*		*
Cottus gobius, L.	*	*	*	*	*	*
Carassius carassius, L.						*
Number of species	10	10	11	16	10	19

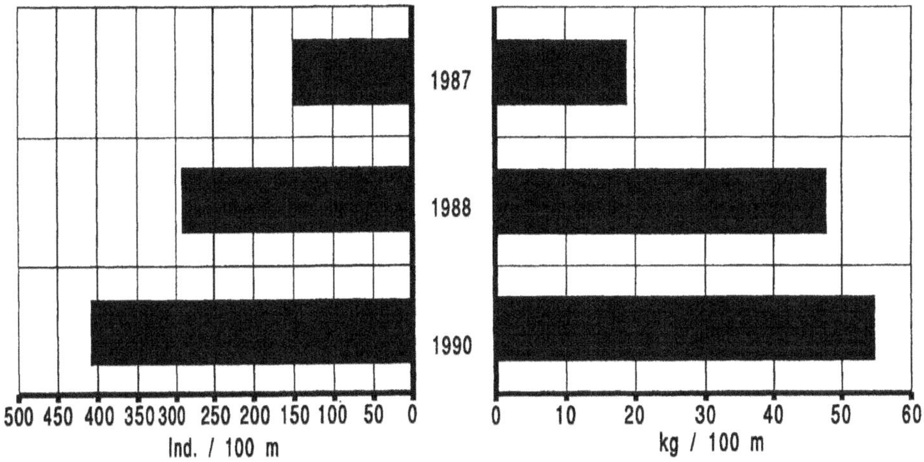

Fig. 9. Density and biomass of the total fish stock (1987: straightened section before restructuring; 1988 and 1990: after restructuring).

Recreated land/water ecotone and fish stock

Physical effects

As mentioned above, legal limitations confined the restructuring measures of the Melk River to the low flow channel and its adjacent riparian zones. Ecotone-relevant structures therefore only comprised the shoreline and their adjoining vegetation but not the surrounding landscape.

The formation and dynamic development of natural littoral structures characterizing a complex land/water interface presuppose variable width conditions of the river bed, which can be described by the variance of the widths (VW) of successive cross sections. The straightened Melk river section showed a VW of 5310 at a mean width of 867 cm, while a still unmodified reference site had a more than five-fold greater VW at a mean width of 1146 cm (Table 2). After restructuring, the mean width increased to 1226 cm and the VW even exceeded that of the nearly pristine reference site by about 50%. The same held true 2 years later, indicating that the restructuring measures maintained the high variability of the environment.

Table 2. Mean widths and variances of widths of different test sites of the Melk river (distance of cross sections 5 m)

	Mean width (cm)	Variance of widths
Almost unmodified section 1981	1,146	29,700
Straightened section 1981	867	5,310
Restructured section 1988	1,226	47,600
Restructured section 1990	1,220	44,906

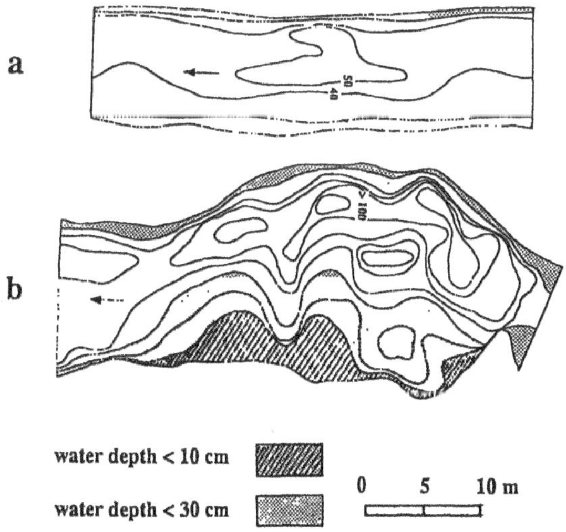

Fig. 10. Distribution of depths (a: still straightened and uniform reference site; b: restructured subsection.Compare Tab. 3).

Table 3. Percentages of dominating substrate types and area of shallow water (<30 cm depth) in % of total surface area of a still straightened versus a restructured Melk River section (compare Fig. 10)

	Shallow water area in % of tot. surface	% mud	% sand material	% coarse
Straight. sect.	14.5	0	36	64
Restruct. sect.	37.4	50	18	32

Due to channelization the straightened reference site lacked variable shallow areas, depths <30 cm only representing about 15% of the total surface area (Fig. 10). On the other hand, a characteristic subsection of the restructured Melk River showed a comparatively large shallow water area of about 37% (<30 cm depth). Within this zone, areas shallower than 10 cm, with very low currents and/or backwaters constituted about 30%, their bottom being mainly formed by sandy or muddy material (Table 3).

The smooth slope of the slip-off banks and their littoral zones enabled the development of a complex land/water boundary. Typical pioneer vegetation was found mainly at the transition from the low flow to the mean flow channel, where inundation and/or erosion or sedimentation occurred several times a year. On the other hand, higher bushes and trees of the adjacent vegetation belt, such as willow, *Salix sp.* and alder, *Alnus sp.*, were restricted to the upper parts of the cross section (above the mean flow level) inundated only during floods (Fig. 11).

Fish community

The fish community of the Melk River showed an overlap of rhithral and epipotamal elements (e.g.

Fig. 11. Restructured Melk River: A gravel bank develops within a widened cross section downstream of a groyne and offers shallow water areas with smooth and inverse currents. The adjacent pioneer vegetation is restricted to the upper parts of the low flow channel; higher bushes and trees grow above the mean flow level.

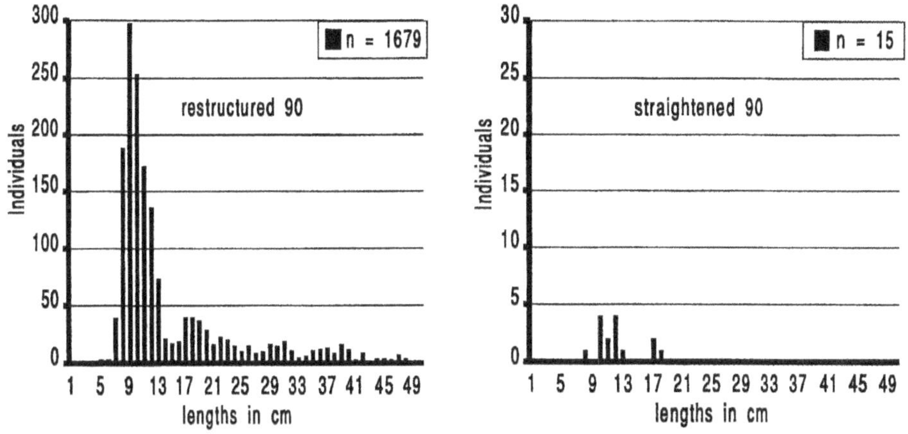

Fig. 12. Length/frequency diagram of barbel within the still straightened and restructured Melk River sections in 1990.

trout; bullhead, *Cottus gobio*; grayling, *Thymallus thymallus;* barbel *Barbus barbus;* nose, *Chondrostoma nasus;* chub, *Leuciscus cephalus*; compare Table 1). Successful reproduction and development presuppose a wide range of species-specific spawning places and nursery grounds for fry and juveniles. With regard to the spawning requirements of both rhithral and epipotamal species, the newly created riffle pool sequences, gravel banks, bays and backwaters of the restructured Melk River obviously offered adequate niches (comp. Figs 2 and 11). Since successful reproduction of both ecological groups generally led to an increased number of fish species (Table 1), most of the typical riverine species (e.g. barbel), now show a much more balanced population structure (Fig. 12). The survey of juveniles and small fish species (e.g. minnow, *Phoxinus phoxinus* and loach, *Noemacheilus barbatulus*) within the shallow water areas provided further evidence for the enhanced supply of adequate niches for fry and young of the year fishes. Comparing the mean catches within the littoral zones of 3 straightened and 9 restructured river sections during different seasons showed clearly the generally higher stocks along restructured shorelines (Fig. 13). Densities

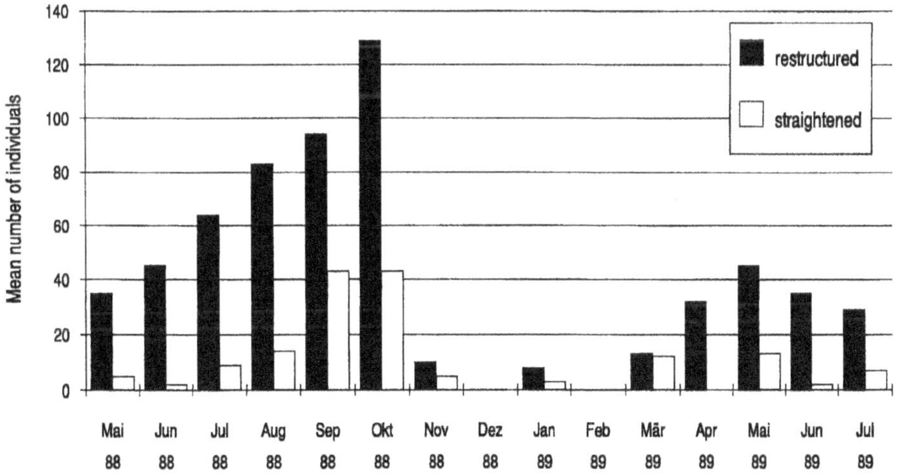

Fig. 13. Mean number of juveniles and small fish species sampled within the shallow water zones of 9 restructured and 3 still straightened Melk River sections during different seasons.

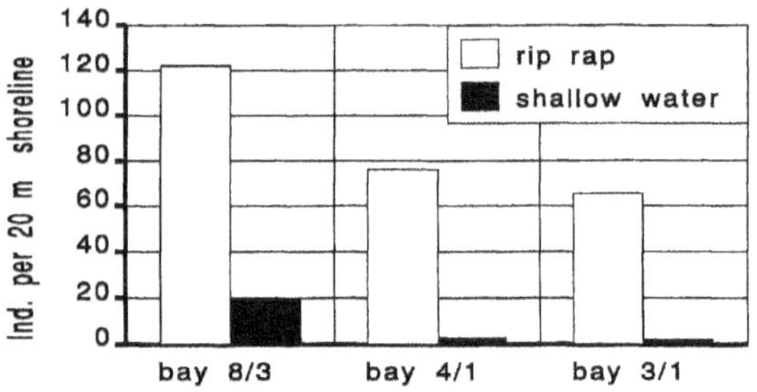

Fig. 14. Juveniles and small fish species sampled within rip rap and shallow water of 3 bays in winter.

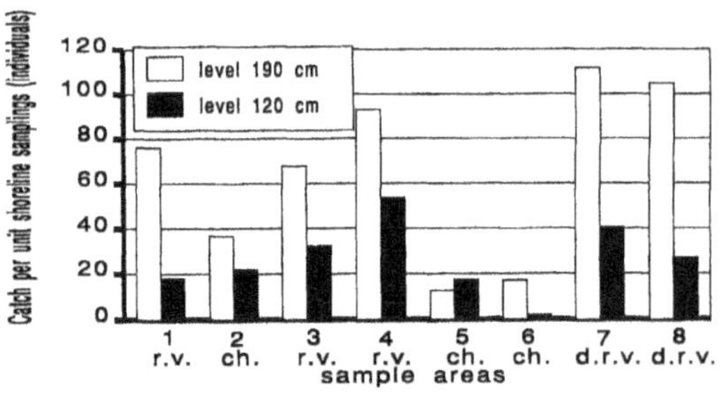

Fig. 16. Fish stock at 8 differently structured littoral sites at mean (level = 120 cm) and high (level = 190 cm) water discharge (ch = channelized, r.v. = riparian vegetation, d.r.v. = dense riparian vegetation).

of juveniles and small fish species increased in these flat water areas from spring to late summer, but the fish disappeared abruptly in late autumn. The results of samplings along the rip rap of 3 bays versus those from 3 shallow water areas in January (water temperature <1 °C) showed that the young of the year and/or small fish species preferred stony or woody structures in winter (Fig. 14). Current stress was obviously a main reason for this demand for shelter during the cold season. Comparison of the fish samplings of all 12 sections in October 1988 according to 3 different classes of current speeds, showed that more than 80% of the fishes concentrated in areas with flow velocities below 15 cm s^{-1} (Fig. 15).

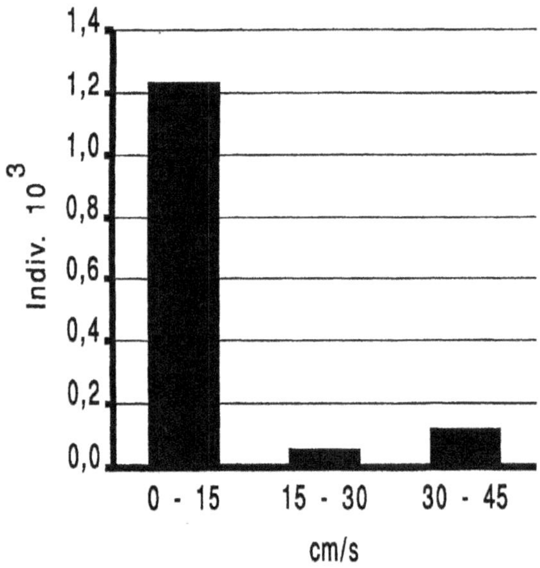

Fig. 15. Juveniles and small fish species sampled within the shallow water zones of 9 restructured and 3 still straightened Melk River sections in October 1988 according to flow velocities.

Another important function of the land/water interface is the supply of adequate shelter during floods. Such shelter consists largely of woody debris and/or higher vegetation within the upper parts of the cross sections. The comparative catch-per-unit-shoreline samplings in eight differently structured littoral test sites at mean and high water discharge (Fig. 16) showed that fish stocks generally increased in littoral areas during flood periods ($p<0.05$; student's *t*-test). This applied especially to littoral test sites 7 and 8 with dense riparian vegetation. On the other hand, less structured test sites typical for the channelized Melk River (sections 2, 5 and 6), showed low densities (independent of water level) or only a small increase at high water discharge.

Discussion

Numerous authors have reported that high natural variety of the river bed is one of the most important conditions for the development of well-balanced aquatic communities (Lelek & Lusk, 1965; Hynes, 1968; Sheldon, 1986; Karr and Schlosser, 1978; Dudley and Karr, 1979; Tarplee *et al.*, 1979; Ward & Stanford, 1979; Schoof, 1980; Jungwirth & Winkler, 1983; Zerz & Moog, 1991). Unmodified rivers with type-specific riffle pool sequences, gravel banks, woody debris and dynamic development of the littoral zones contain a broad spectrum of instream habitats as well as a highly complex land/water interface. Channelization results in monotonous aquatic habitats and a quantitative and qualitative decrease of the fish stock. The present results demonstrate that the variability of instream structures can be described fairly well by the variance of maximum depths (VMD) of successive cross sections. This measure showed a highly significant correlation ($p<0.001$) with both the number of fish species (NFS) and fish species diversity (FSD); it is therefore a practical instrument for the evaluation of water management projects. The VMD/NFS model can also be used to forecast the effects of river restoration plans. The Melk River project showed that regained spatial heterogeneity of the river bed led rapidly (1987–1990) to a significant recovery of the fish stock. Due to the improvement of the aquatic environment with respect to variable depths, flow velocities and substrate types, the number of fish species developed exactly in correspondence to the VMD/NFS model and both the density and biomass of the total fish stock tripled. Species diversity increased as well, but did not reach the value predicted by the VMD/FSD model; this was an indication that fish communities need more time to regain balanced proportions.

In addition to instream structures complex land/water interfaces are key features for the existence and development of natural fish communities. Essential functions of these ecotone zones include nursery grounds for fry and/or young of the year fishes and shelter during floods. Since the location of the land/water boundary changes strongly with the magnitude of discharge, the seasonal availability of relevant ecotone structures differs in quantity and quality. In the restructured Melk River, the variance of the maximum widths described fairly well the increased variability of the width conditions. High values corresponded to a high supply of shallow water areas with low flow velocities and fine substrates.

The importance of the recreated shallow water zones as nursery grounds for young stages and as a habitat for small fish species (e.g. minnow, loach) was obvious from spring to late summer. During the warm season, fishes occurred here in high densities. In autumn, however, these fishes (stages) disappeared abruptly from shallow littoral zones and migrated to other habitats. During winter the main concentrations were found in stony or woody shelters with current speeds $<15\ \mathrm{cm\ s^{-1}}$ and low hydraulic stress. Other critical ecotone-specific structures were bushes and trees or woody debris above the mean water level of the restructured Melk River. In contrast to straightened river sections, such living or dead woody materials prevented fry and adults from being washed off during floods. The combined results presented here prove that heterogeneous instream structures are particularly important for fish communities dominated by rheophilic species. This appears to be valid for headwaters exclusively inhabited by brown trout and/or bullhead, where VMD showed a significant correlation with population-(age-) structure parameters (see Fig. 5), as well as for streams corresponding to the barbel region, where VMD is strongly correlated with the community parameters NFS and FSD (see Figs 3 and 4). In small headwaters, however, the instream structures overlap strongly with the land/water interface. A differentiation in instream and ecotone structures therefore only makes sense in larger running waters of increased stream order.

Most of the rheophilic fish species accomplish their whole life cycle within the river itself. Instream structures oftering a broad spectrum of adequate species- and stage-specific niches (e.g. for spawning, embryonic development, etc.) are therefore of utmost importance. On the other hand the instream structures alone cannot provide adequate niches for all species and/or stages and their habitat requirements. This appears to be particularly true with respect to hydraulic stress (Bretschko, 1995), which increases with decreasing stream order. Fry and young of the year of rheophilic species as well as small and/or stagnophilic species therefore require additionally the typical elements of the land/water interface (shallow water zones, backwaters, seasonally inundated littoral vegetation, woody debris etc.) as nursery grounds and for shelter. The adjacent vegetation above mean flow level for all species and stages serves as a vital refuge area during floods.

In potamal streams the mean size of instream areas with uniform hydraulic stress increases parallel with increasing stream order (Bretschko, 1995). These areas alone represent a comparatively monotonous environment for fish communities.

On the other hand, regular inundation of the surrounding landscape becomes a stream-specific phenomenon. Therefore the large and complex ecotone zones of potamal regions represent the classical environment for high diversity fish communities.

Based on this knowledge, planning of river restorations should take into account that every stretch of running water possesses its unique features. Consequently, restoration measures should orient themselves on the original situation of the unmodified river and not be confined to restructuring. From an ecological point of view the ideal strategy is a set of measures which allows the river to regain its characteristic instream and ecotone structures by dynamic processes corresponding to the type-specific conditions.

Acknowledgements

We thank P. Parasiewitcz for the graphics and Dr M. Stachowitsch for comments and improving the language. Financial support was provided by the Austrian Fonds zur Förderung der wissenschaftlichen Forschung, the Ministry of Agriculture and the Government of Lower Austria. This paper is dedicated to Prof. Dr F. Berger, Biologische Station Lunz, on the occasion of his 90th birthday.

References

Bagenal, T., 1978. Methods for Assessment of Fish Production in Fresh Waters, 3rd edn. Blackwell Scientific Publications Ltd.

Bretschko, G., 1995. River/land ecotones: scales and patterns. Hydrobiologia 303 (Dev. Hydrobiol. 105): 83–91.

Dudley, D. R. & J. R. Karr, 1979. Reconsiling streambank erosion control with waterquality goals. Env. Impact of Land Use on Water Quality: Final Rep. on the Black Creek Project, Vol. 4.

Hynes, H. B. N., 1968. Further studies on the invertebrate fauna of a Welsh mountainstream. Arch. Hydrobiol. 65: 360–379.

Karr, J. R. & I. J. Schlosser, 1978. Water Resources and the Land-Water Interface. Science 201: 229–234.

Jungwirth, M., 1981. Auswirkungen von Fließgewässerregulierungen auf Fischbestände am Beispiel zweier Voralpenflüsse und eines Gebirgsbaches, Teil I, Wasserwirtschaft Wasservorsorge. BMLF, Wien, 104 pp.

Jungwirth, M. & H. Winkler, 1983. Die Bedeutung der Flußbettstruktur für Fischgemeinschaften, Österr. Wasserwirtschaft 35: 229–234.

Jungwirth, M., 1984. Auswirkungen von Fließgewässerregulierungen auf Fischbestände, Teil II. Wasserwirtschaft und Wasservorsorge. BMLF, Wien, 188 pp.

Jungwirth, M., 1991. Auswirkungen von Fließgewässerregulierungen auf Fischbestände, Teil III: Das Restrukturierungsprojekt Melk. Wasserwirtschaft und Wasservorsorge, BMLF, Wien, 388 pp.

Lelek, A. & S. Lusk, 1965. Occurrence of Fishes in Relation to Formation of the Bed. Zool. Listy 14/3, 255 pp.

Meffe, G. K., A. L. Sheldon, 1988. The Influence of Habitat Structure on Fish Assemblage Composition in Southeastern Blackwater Streams. Am. Midl. Nat. 120: 225–240.

Moog, O., 1991: Makrobenthologische Aspekte bei der Wiederherstellung naturnaher Flußabschnitte. Wr. Mitt. 88: 56–103.

Pielou, E. C., 1975. Ecological Diversity. J. Wiley, New York, 165 pp.

Schoof, R., 1980. Environmental Impact of Channel Modification. Wat. Res. Bul. 16: 697–702.

Sheldon, A. L., 1986. Species diversity and Longitudinal Succession in Stream Fishes. Ecology 49: 193–198.

Tarplee, W. H., D. E. Louder & A. J. Weber, 1979. Evaluation of the effects of channelisation on fish populations in North Carolinas coastal plain streams. N. Carolina Wildlife Res. Comm., Raleigh.

Ward, J. V. & J. A. Stanford, 1979. The Ecology of Regulated Streams. Plenum Press, New York & London, 398 pp.

Zerz, H. J.& O. Moog, 1991. Auswirkungen von Fließgewässerregulierungen auf Fischbestände/Makrozoobenthos. In: Jungwirth, M. (ed.), Auswirkungen von Fließgewässerregulierungen auf Fischbestände, Teil III. Wasserwirtschaft und Wasservorsorge, BMLF, Wien, 388 pp.

Hydrobiologia **303**: 207–210, 1995.
F. Schiemer, M. Zalewski & J. E. Thorpe (eds), The Importance of Aquatic-Terrestrial Ecotones for Freshwater Fish.

Effects of removal and regeneration of bankside vegetation on fish population dynamics in the Warta River, Poland

Tadeusz Penczak
Department of Ecology and Vertebrate Zoology, University of Łódź, 90-237 Łódź, 12/16 Banacha Str., Poland

Key words: lowland river, ecotones, fish density, standing crop

Abstract

In March 1989 both banks of a 450 m section of the Warta River, 2 km above a reservoir, were cleared of their riparian vegetation (osiers, willows and alders). In October 1988 17 fish species had been recorded there, with a standing crop (SC) of 31.9 kg ha^{-1}. In April 1989, the number of species decreased to 11, and their SC to 7.9 kg ha^{-1}.

Over the next three years new alder trees appeared and osiers regrew to 3.5 m. The SC in October 1990, October 1991 and May 1992 was recorded at 36.5, 66.2 and 40.9 kg ha^{-1}, respectively. The primary effect was on phytophils, whose biomass was directly proportional to the height of the trees and osiers ($r = 0.87$, $p = 0.05$). The litho-phytophils appeared to occupy the habitat vacated by the phytophils soon after bank clearance ($r = -0.87$, $p = 0.05$), but then declined as the osiers grew.

Introduction

Stream animals benefit from 'the shade provided by riparian ecotones' (Petersen *et al.*, 1987). Schiemer & Zalewski (1992) considered that the 'benefit' was limited in certain situations and that optimal fish diversity and production may be expected at an intermediate level of ecotone complexity down to the fifth stream order of a river.

By contrast, in the third to fifth-order sections of the alluvial, shallow Pilica River the most abundant and diverse populations of fish occur close to banks that are compactly overgrown by shrubs and trees (Penczak, 1989; Penczak & Mann, 1990), while fish are absent from the exposed mid-current sections of this river (Mann & Penczak, 1984). These different opinions may result from the inadequacy of classifying streams by order (Matthews, 1986).

The investigated section of the Warta River was a fifth order stream. This alluvial river was twice as deep as the Pilica, although of similar width and velocity. It had a densely vegetated riparian zone, which was cleared, and the effect of this on the fish is the subject of the present monitoring investigations.

Study area, material and methods

The Warta, 808 km long, is the largest tributary of the Odra (Oder) River. The study site was 315 km downstream from the source, and 2 km upstream of the Jeziorsko Reservoir (Penczak, 1992a). The 450 m long site (width: 40–55 m, mean depth in the current zone: 1.4 m, area: 20080 m^2) has a predominantly sandy bottom (75%), with additional components of mud, stones and gravel (Penczak, 1992a). Detailed morphological and water quality data are given by Galicka *et al.* (1991), Penczak *et al.* (1991) and Penczak (1992a). Between May 1985 and October 1988 (two years after impoundment) the left bank had a compact, continuous ecotone boundary with osiers, willows and alders, while 50% of its right bank was not forested. At the end of winter 1989 both banks were cleared for many kilometres. By May 1992, young alders had grown 2 m high, and osiers to 3.5 m.

The site was sampled 12 times, once or twice a year, between July 1985 and May 1992, by successive electrofishing from a boat, (Penczak & Romero, 1990). Twenty two fish species were recorded, phytophils and litho-phytophils being dominants on every occasion. Data on their density and biomass to October 1989 are published elsewhere (Penczak *et al.*, 1991).

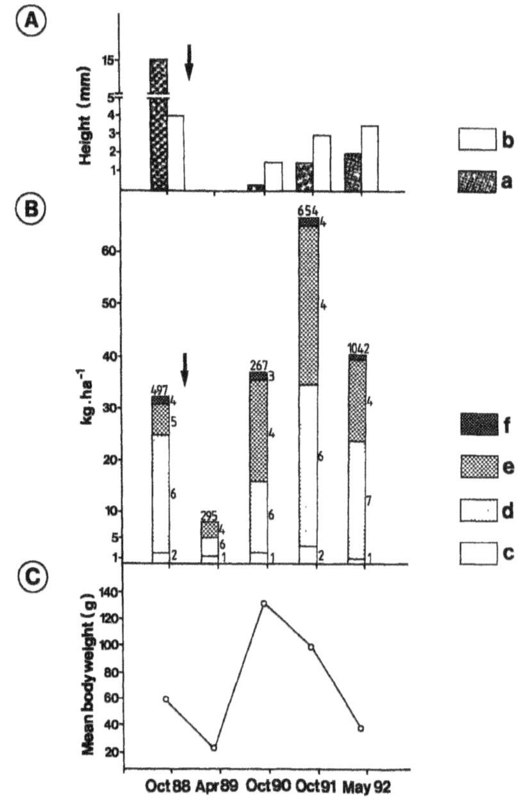

Fig. 1. A: Height of riparian vegetation, a = alders, b = osiers. The black arrow indicates the period of the clearing. B: Standing crop (kg ha^{-1}) of c = lithophils, d = phytophils, e = litho-phytophils, f = others (psammophils, partially pelagophils and eel). Density (indiv. ha^{-1}) is marked at the top of a bar. The number of species in a given ecological group is indicated on the right side of bars. C: Mean weight of fish against dates of sampling.

Results

In October 1988, 17 species were recorded, with a standing crop (SC) of 31.9 kg ha^{-1}. The mean SC from 7 samplings for the period July 1985–October 1988 was 22.5 kg ha^{-1}, and on average 15 species were identified.

In April 1989, one month after the clearing, 11 species were recorded, but their SC had declined to 7.9 kg ha^{-1}, but their mean individual weight to 26.8 g (Fig. 1). This indicates that large specimens are more seriously affected by the removal of the riparian zone, despite submerged roots being only partially destroyed during the levelling and strengthening of the banks. Also, psammophilous and pelagophilous species (Fig. 1) were absent.

In October 1990, the osiers had regrown to 1–1.5 m, other shrubs and trees remained almost absent, but the fish fauna increased to 14 species (Fig. 1). SC and mean body weight had increased, but density had not. In October 1991 the number of species increased to 16, the density by 2.5 times and the SC doubled, while the mean individual weight decreased.

Over the study period the water quality improved, being only slightly alkaline (pH 7–8), while the levels of suspended solids and phosphates decreased considerably from their 1988–89 levels, which may have benefitted recruitment of young-of-the-year.

In May 1992 the highest density of fish since 1985 was recorded, yet SC declined to post-clearance 1989 values and mean individual weight was also low. This might have been related to a serious decrease in water quality in that year, and to emigration of young fish from the polluted reservoir.

During riparian regrowth the percentage of litho-phytophils in the fish fauna increased at the expense of phytophils (Fig. 2). An inverse statistically significant correlation between the proportion of phytophils and litho-phytophils was found both for the percentages of number and biomass. In the last year of investigations, when osiers had regrown almost to their preclearance height, litho-phytophil biomass declined but density still increased a little bit (Fig. 3). This may testify to age composition having changed in favour of young individuals.

Discussion

Petersen *et al.* (1987) noted 'the shade which cools the river' among numerous benefits supplied by riparian vegetation. This factor, always of impor-

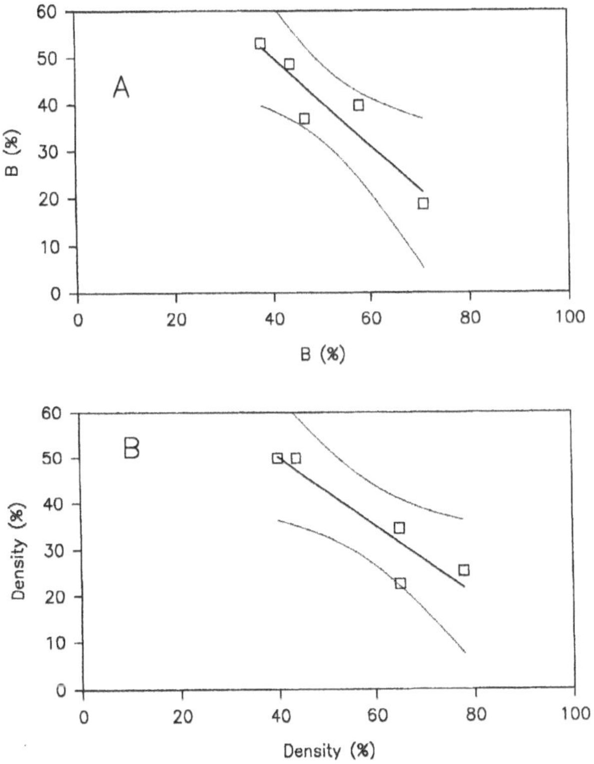

Fig. 2. A: Correlation between the percentage biomass of lithophytophils (ordinate) and of phytophils (abscissa) ($r = -0.93$, $p = 0.02$). B: Correlation between the percentage density of lithophytophil (ordinate) and of phytophils (abscissa) ($r = -0.91$, $p = 0.03$).

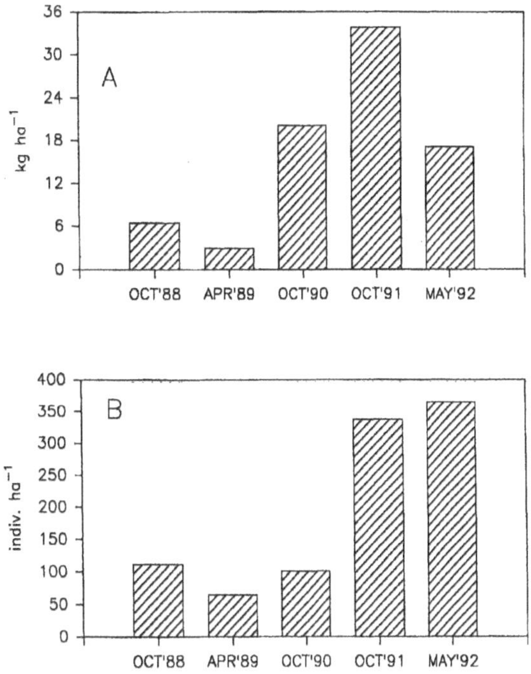

Fig. 3. Biomass (A) and density (B) of litho-phytophils against dates of sampling.

tance in tropical and subtropical zones, may become so intermittently in temperate zones, during heat waves such as occurred in Central Europe in summer 1992. That year, during tagging fish in the Pilica River, 95% of fish were aggregated, due to low water levels, into pools under cover of osiers and other trees. Without the riparian zone, lithophils and part of litho-phytophils would not have survived and the regeneration of this species-rich fish fauna would have taken several years.

The complete lack of large specimens in the first sampling after riparian clearance (Fig. 1) is attributed to shortage of hiding places (Swales, 1982), whose lack in large alluvial rivers is particularly harmful for fish (Penczak, 1989). Transparent, shallow water and large areas of sandy bottom (abovc 75%) arc also poor in food (Penczak & Grzybkowska, 1991; Grzybkowska *et al.*, 1990). However, the reduction of hiding places to roots only is probably a more important factor (Swales, 1982), emphasizing the importance of overhanging bank vegetation. The numerically large sample in October 1991 may have been increased by upstream migration from the reservoir (Fig. 1).

Petersen *et al.* (1987) recorded that West Africa village fishermen 'understand the link between riparian protection and continued fish production'. In contrast, Europeans do not and by complete removal of riparian vegetation, small and large rivers are changed into unstable and environmentally poor habitats (Naiman *et al.*, 1988).

Acknowledgements

Thanks are extended to an anonymous referee for extensive critical comments, including improvements to the English, F. Schiemer for modifications of figures, Ł. Glowacki for help in preparing the English version of the draft manuscript, and M. Ułańska and K. Galicka for making the figures. The study was financed from Grant No. 2661/6/91 awarded by KBN.

References

Balon, E. K., 1975. Reproductive guilds of fishes: a proposal and definition. J. Fish. Res. Bd Can. 32: 821–864.

Galicka, W., Ł. Głowacki & T. Penczak, 1991. Water pollution in the upper Warta River in 1985–1989. Acta Univ. Lodz., Folia limnol. 5: 5–20.

Grzybkowska, M., J. Hejduk & P. Zieliński, 1990. Seasonal dynamics and production of Chironomidae in a large lowland river upstream and downstream from a new reservoir in Central Poland. Arch. Hydrobiol. 119: 439–455.

Kryzhanovsky, S. G., 1949. Ecomorphological principles of development among carps, loaches, and catfishes. II. Ecological groups of fishes and patterns of their distribution. Tr. Inst. Morfol. Zhivotn. Severtsova, L.: 237–331. [Translated from Russian in Fish. Res. Bd. Can., Transl. Ser. No. 2945 in 1974].

Mann, R. H. K. & T. Penczak, 1984. The efficiency of a new electrofishing technique in determining fish numbers in a large river in Central Poland. J. Fish Biol. 24: 173–185.

Matthews, W. J., 1986. Fish faunal 'breaks' and stream order in the eastern and central United States. Envir. Biol. Fishes. 17: 81–92.

Naiman, R. J., H. Décamps, J. Pastor & C. A. Johnston, 1988. The potential importance of boundaries to fluvial ecosystems. J. N. Am. Benthol. Soc. 7: 289–306.

Penczak, T., 1989. Ichtiofauna dorzecza Pilicy. Część II. Po utworzeniu zbiornika [The ichthyofauna of the Pilica River drainage basin. Part II. Post impoundment study]. Sci. Annual Pol. Anglers Assoc. 2: 116–186. [In Polish with English summ.]

Penczak, T., 1992a. Fish production in the Warta River, Poland: a preimpoundment study. Hydrobiologia 237: 117–129.

Penczak, T., 1992b. Fish production in the Warta River: postimpoundment study. Hydrobiologia 242: 87–93.

Penczak, T., W. Galicka & H. Koszaliński, 1991. Fish populations density and biomass in the Warta River: pre- and post-impoundment study. Acta Univ. Łódź., Folia limnol. 5: 59–77.

Penczak, T. & M. Grzybkowska, 1991. Fish and invertebrates *vs* riparian ecotones in a large river (the Warta River, Central Poland). In M. Zalewski, J. E. Thorpe & P. Gaudin (eds), Proceedings of Symposium 'Role of Land/Inland Water Ecotones in Landscape Management and Restoration', UNESCO MAB, University of Łódź, Łódź: 41–44.

Penczak, T., R. H. K. Mann, 1990. The impact of stream order on fish populations in the Pilica drainage basin, Poland. Pol. Arch. Hydrobiol. 37: 243–261.

Penczak, T. & T. Romero, 1990. Accuracy of a modified catch-effort method for estimating fish density in a large river (Warta River, Poland). In I. G. Cowx (ed.), Developments in Electric Fishing, Fishing News Books, Blackwell Scientific Publications, Oxford: 191–196.

Petersen, R. C. Jr., B. L. Madsen, M. A. Wilzbach, Ch. H-. D. Magadza, A. Paarlberg, A. Kullberg & K. W. Cummins, 1987. Stream Management: Emerging Global Similarities. Ambio 16: 166–179.

Schiemer, F. & M. Zalewski, 1992. The importance of riparian ecotones for diversity and productivity of riverine fish communities. Neth. J. Zool., 42: 323–335.

Swales, S., 1982. Impact of weed-cutting on fisheries: an experimental study in a small lowland river. Fish. Mgmt 13: 125–137.

Hydrobiologia **303**: 211–219, 1995.
F. Schiemer, M. Zalewski & J. E. Thorpe (eds), The Importance of Aquatic-Terrestrial Ecotones for Freshwater Fish.

Effects of siltation on stream fishes and the potential mitigating role of the buffering riparian zone

Charles F. Rabeni & Martin A. Smale
Missouri Cooperative Fish and Wildlife Research Unit, The School of Natural Resources, University of Missouri, 112 Stephens Hall, Columbia, Missouri 65211, USA

Key words: sediment, guilds, fish community metrics, riparian zones, ecological health

Abstract

Riparian buffer zones serve several important roles in linking a stream to its watershed. A main function is controlling the dynamics of sedimentation. This paper documents how siltation impacts fish communities and how proper riparian management can mitigate the negative effects of sedimentation. Two studies examined the relation between stream siltation and fish community characteristics. Community responses to siltation were poorly described by common structural indices. Community level responses to varying siltation were most consistently described by changes in functional characteristics of the resident fish species using a guild-based analysis. Herbivores, benthic insectivores and simple lithophilous spawners were most sensitive to siltation while other guilds were not. These results were repeatable in both intraregional comparisons among sites of similar size and character, and in interregional comparisons of streams which varied in characteristics besides siltation. This suggests the index may be useful in separating the effects of siltation from other environmental variables. A discussion of bufferstrip characteristics important in mitigating against, or preventing, excess siltation is presented.

Introduction

Riparian ecotones connect a stream with its watershed and control within-stream biological communities by influencing solar radiation and water temperature, regulating nutrient budgets, stabilizing stream banks and providing a source of organic energy (Schiemer & Zalewski, 1992). When buffer strips are degraded or removed, a common result is a change in sediment loads within the stream channel. Streamside vegetation plays an important role in the regulation of sediment dynamics within the channel by a variety of mechanisms. The most commonly assumed function is to minimize overland transport (Williams & Nicks, 1988), but the ability of vegetation to stabilize stream channels (Beschta & Platts, 1986) and to act as sediment traps during overbank flow (Wilkin & Hebel, 1982) are also important. Sedimentation has been linked, usually indirectly, with impacts on fish communities. There is a need to more quantitatively evaluate the effects of siltation on fishes and to evaluate mitigative strategies.

Settlement of the central United States has resulted in extensive agriculturalization of the original prairie and mixed forest-prairie ecoregion over the past 150 years. Agricultural development coincided with considerable changes in the original fish fauna, which some researchers attributed to increased siltation from soil erosion (Larimore & Smith, 1963; Menzel *et al.*, 1984; Karr *et al.*, 1985). However, most conclusions implicating siltation in the decline of fish populations are primarily subjective and are based largely on the observations of changes in the fish fauna as the scope of agricultural activity increased (Smith, 1971; Menzel *et al.*., 1984). Determining the extent to which siltation has been responsible for altering midwestern fish communities is an important issue for resource managers concerned with anticipating and planning for changes in stream biota which might result from improved farming practices and reduced soil erosion. But this determination has been inhibited both by the incompleteness of historical records of presettlement fish communities and by the difficulty of isolating siltation effects from those due to numerous other changes in stream characteristics which usually accompany agriculturalization.

Various mechanisms by which siltation affects fishes have been described, including physiological stress from clogged gills and other sensitive surfaces, smothering of eggs and larvae, and disruption of normal feeding and other activities dependent on vision. Secondary effects due to changes in the trophic character of a stream and to altered habitat conditions brought about by channel siltation are also considered likely effects of siltation (Cordone & Kelley, 1961; Muncy *et al.*., 1979). Although silt-fish interactions have often been examined at the species level, our focus in this paper is on effects at the community level. Relations between siltation and community level metrics such as richness, diversity, stability, biotic integrity and composition should generalize to more situations than would research into effects on individual species. We also expected that community level analyses would enable predictions about the sensitivity of individual species based on common features of their life history strategies.

Community metrics such as diversity, richness and Index of Biotic Integrity (IBI) scores (Karr *et al.*., 1986) are fairly generic indices sensitive to most types of stream degradation rather than specific to siltation effects. Therefore our research projects focused on developing a biotic index for fish community data which was specific to and diagnostic of siltation effects independent of other factors. This siltation index was based on functional classifications of species into both trophic guilds and reproductive guilds (Balon, 1975; Muncy *et al.*., 1979). Since 1982, we have undertaken two research projects with the common objective of determining how stream siltation altered fish communities in the central United States. In both studies, guild compositions of communities from sites with varying levels of siltation were contrasted, and we expected that guilds which were most likely to be sensitive to siltation effects would diminish as levels of silt increased.

The objective of this paper is to delineate how the effects of siltation on fish communities can be distinguished from other related effects of buffer strip degradation, and how buffer strip restoration might most efficiently be carried out to ameliorate siltation.

Approach

Species were classified into one of four feeding guilds and one of four reproductive guilds, based on life history information available from various sources (Starrett, 1950; Cross, 1967; Pflieger, 1975; Muncy *et al.*., 1979). Trophic guilds included herbivores, omnivores, general insectivores and benthic insectivores (piscivorous fish were absent or scarce in these samples) and assignments were based on modes of feeding and types of food consumed. Reproductive guilds were based on the degree of nest preparation and spawning site defense, the degree of parental care given the eggs and young and the type of substrate typically used. Guilds included simple lithophilous spawners which require a coarse substrate, simple miscellaneous spawners which utilize other substrates such as vegetation or sand, complex spawners without parental care which may build and occupy nests that are abandoned following spawning and complex spawners with parental care which continue to guard sites after spawning occurs.

Intensive study

The first study, which we will term 'intensive', was conducted in a localized geographic area of northeast Missouri (Fig. 1) on similar sized streams (Berkman & Rabeni, 1987). Fish communities were sampled at 12 sites on three second or third order streams of approximately 10000 ha in drainage area. Streams and sites were chosen to minimize physiographical, hydrological and morphological differences between them while still varying distinctly in the degree of siltation from site to site.

Fish sampling was stratified by three different habitat types: pools, riffles and the intermediate run habitat. The first step in the analysis consisted of ordinating communities from each site and habitat using Detrended Correspondence Analysis (DCA) to assess the degree of dissimilarity among the communities sampled. Both pool and run communities were clustered tightly together in this ordination, indicating that commu-

Fig. 1. The location of study streams and their drainage networks in Missouri.

nity compositions from these two habitat types did not respond to differences in characteristics, including siltation, among the different sites. Riffle communities on the other hand, were widely dispersed across the ordination plot, indicating a much greater sensitivity of riffle communities to physical differences among the sites. For fish communities from riffles, associations between both the relative abundance and absolute abundance of each guild and the percent silt by weight in each of the 10 riffles were analyzed using Spearman's rank correlation coefficients (r_s) (Table 1). Absolute abundance estimates of two trophic guilds, herbivores and benthic insectivores, were significantly ($P \leq 0.05$) and negatively correlated with increasing riffle siltation. Absolute abundances of general insectivores and of omnivores were not correlated with siltation, but as siltation increased the frequency of these two guilds in the communities tended to increase because of the diminished numbers of herbivores and benthic insectivores present. Of the four reproductive guilds, absolute abundance estimates for only simple lithophilous spawners were negatively correlated with siltation ($r_s = -0.80$, $P \leq 0.01$). The abundance estimates of the three remaining guilds were uncorrelated with siltation, but the frequencies of simple miscellaneous spawners and, to a lesser extent, of complex-no parental care spawners tended to increase with increasing siltation. These results indicated that the numbers of herbivores, benthic insectivores and simple lithophilous spawners diminished with increased siltation levels in riffles while other guilds remained relatively constant in number, becoming more frequent in the community only because of the diminished abundance of sensitive guilds.

Table 1. Spearman's rank correlation coefficients for the relationship between the percent silt and abundance estimates for each of four trophic and reproductive guilds of fishes in riffle communities at 10 sites on 3 Missouri streams in 1983 (values from Berkman & Rabeni, 1987).

	Absolute abundance	Relative abundance
Trophic guilds:		
Benthic insectivores	−0.70	−0.82
Herbivores	−1.00	−0.89
General insectivores	0.00	0.70
Omnivores	0.26	0.54
Reproductive guilds:		
Simple lithophilous	−0.80	−0.84
Simple miscellaneous	0.14	0.58
Complex with parental care	0.18	0.05
Complex without parental care	0.15	0.35

This intensive study also examined relations between siltation levels and community metrics commonly used to assess general stream biotic health. Neither species richness nor species diversity, as measured by the Shannon-Wiener diversity index, were associated to any degree with the percent silt. This absence of correlation was seen in analyses of both riffle communities alone and for data combining all habitats at each site. Siltation varied from 0.0 to 7.5 percent by weight over the study sites, which was apparently not sufficient to result in consistent differences in these two structural metrics. IBI scores were, however, correlated with siltation levels, particularly in riffle habitats alone ($r = -0.44$, $0.2 < P < 0.4$). Though not significant, this correlation between siltation and IBI scores was strong enough to suggest that this index, which is a composite of several structural and compositional metrics, can be sensitive to siltation when other differences between sites were minimized. But on the whole, this study indicated that siltation affected fish community composition but not overall diversity or richness.

Extensive study

A second, more extensive, study examining siltation's effects on fishes was carried out from 1986 to 1990 (Smale & Rabeni, 1991). Starting in 1986, we sampled fish communities and benthic sediment composition at a total of 21 sites on 13 streams distributed throughout Missouri. Core samples of benthic sediments taken over the 3–5 year duration of the study showed no evidence of temporal change in sediment composition in most of the study streams. Fish communities in many of the streams changed considerably during the study. Variation in overall abundance of fishes exceeded 2 orders of magnitude at some sites, and extensive changes in species composition occurred at most. These changes in fish communities at individual sites were attributable to factors such as severe dehydration, oxygen depletion and high temperatures due to drought conditions during 3 of the 5 years of this study (Smale & Rabeni, 1991). Siltation was the only influential physical stream characteristic which remained stable at each individual site throughout the study.

This study provided an opportunity for additional comparisons and contrasts of communities in streams of similar size along a gradient of siltation to further test our initial hypotheses about compositional changes in fish communities in response to siltation. Study streams were located in three regions of Missouri: the prairie region, consisting of level terrain, deep, impermeable and fine-grained soils and intense agriculture; the Ozark region, consisting of steeper terrain, coarser and shallower soils and a mixture of pasture and woodland usage; and the Ozark border region, which was intermediate in character. These physiographic differences among regions resulted in streams which varied considerably in character, particularly in the composition of their benthic sediments. Prairie region streams were characterized by primarily fine substrates, with clay, silt and fine sand fractions comprising from 18 to 75 percent of the streambed. Similarly, coarse sediments were found in both Ozark and Ozark border study sites, where clay, silt and fine sand comprised from 0.5 percent to, at most, 8.1 percent of the streambed. This enabled a comparison of communities from streams with a much steeper siltation gradient than in the intensive study, and also enabled us to test whether compositional responses to siltation were robust to the numerous other differences in stream characteristics among the three regions sampled.

Fish communities were not stratified by habitat type during the extensive study. Reaches of 200–250 m in length were subdivided into segments, sampled by repeated hauls with minnow seines and considered as whole sites in all analyses. Fish were sampled two or three times each year, and species compositions averaged over the last 4 years of the study were used for most of the analyses following.

Ordination of the study communities indicated that levels of siltation were not a principal factor influencing species composition (Fig. 2). There was a general gradient of community types from prairie communities at the extreme left to Ozark communities at the right, with Ozark border communities in the middle. This X-axis separation does correspond somewhat with the gradient of decreasing siltation from prairie through Ozark streams. But, although Ozark border sites were virtually identical to Ozark sites in their low levels of siltation, Ozark border communities tended to cluster together with communities from the highly silted prairie sites. Secondly, although the four Ozark sites sampled were similar in sediment composition, they separated widely along the ordination Y-axis corresponding to longitudinal position in their respective drainages. On the whole, the degree of similarity between communities in overall species composition, as tested by both ordination and percent similarity calculations, did not strongly correspond with the degree of

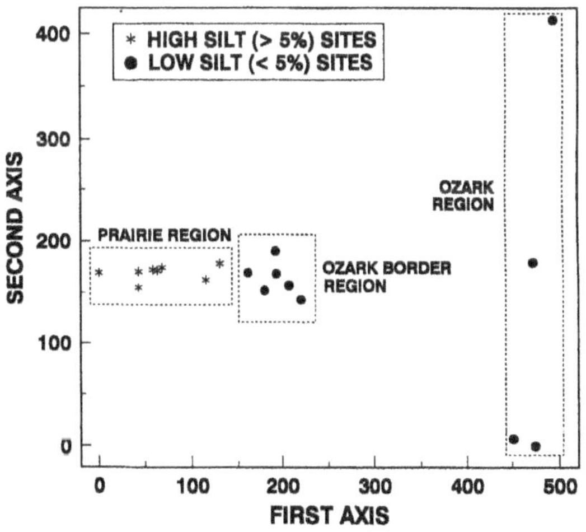

Fig. 2. Detrended correspondence analysis ordination of mean community species composition data from 1987–90 samples at 18 study sites throughout Missouri. Communities are designated by region (prairie, Ozark border or Ozark) and by the degree of siltation at their sites.

similarity in sediment composition. Low silt communities varied considerably in composition, and several low silt communities were more similar to high silt communities than they were to other low silt communities.

Relationships between siltation levels and community guild composition were stronger than those with simple species compositions, as our intensive study would predict. Frequencies for each of the three guilds (herbivore, benthic insectivore and simple lithophilous spawners) which were sensitive to siltation in the intensive study were also negatively correlated with increased siltation in the extensive study. Correlation coefficients for the relationships between percent silt and the percent of each guild in the community ranged from $r = 0.41$ for simple lithophilous spawners to $r = 0.61$ and 0.62 for herbivores and benthic insectivores, respectively. The most sensitive and temporally stable index of siltation effects was obtained by summing the frequencies of these three guilds together. As shown in Fig. 3, averages for the summed frequencies of silt sensitive guilds were negatively correlated to the percent siltation ($r = 0.77$, $P < 0.01$). This index was

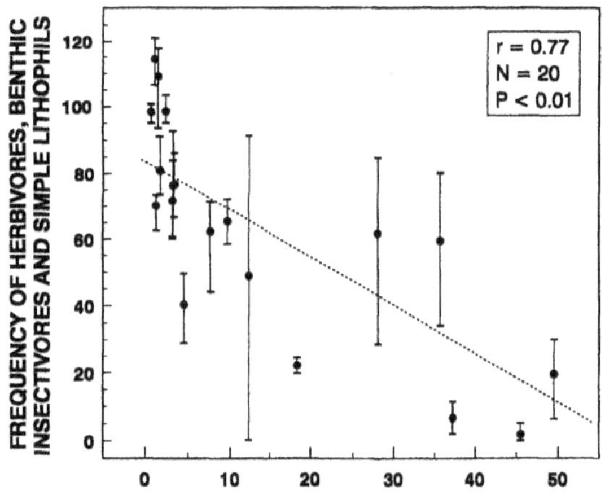

Fig. 3. The relationship between the combined frequencies of three silt-sensitive guilds (herbivores, benthic insectivores and simple lithophilous spawners) and the mean percent streambed silt at each of 20 sites. Points represent the mean frequencies, and error bars represent the minimum and maximum frequencies obtained during sampling 1987–90.

also relatively stable over the duration of sampling at most sites. Error bars in Fig. 3 represent the range of frequencies for the three guilds over several samples, and in most cases these ranges were typically smaller than differences among sites. Most temporal variation in guild frequencies was due to surges of recruits of particular species, an effect which could be minimized by more careful selection of sampling periods.

As with the intensive study, community structural metrics in the extensive study were also poorly related to siltation levels. Mean community Shannon-Wiener species diversity was uncorrelated with siltation ($r = 0.01$) and mean species richness was weakly correlated ($r = 0.29$). Diversity and richness metrics were more related to other characteristics of the study sites, particularly the size of the drainage areas, the degree of dehydration occurring during extended dry periods and dissolved oxygen and temperature regimes. IBI scores were, however, significantly correlated ($r = 0.58$, $P < 0.05$) with silt percentages. Since several IBI metrics reflect community features which overlap features of community guild composition, sensitivity of IBI scores to siltation should be expected. But compared to the guild composition index, temporal variation in IBI scores was relatively high at most sites, suggesting that IBI scores responded most to temporally varying factors other than siltation.

Discussion

The two studies which contrasted community characteristics across gradients of varying siltation produced several complementary results. First, we found that community responses to siltation were at best poorly described by common structural indices such as diversity and species richness. Although species composition of communities was influenced by siltation in both studies, it was also apparent that variation in stream conditions due to longitudinal and geophysical factors independent of siltation could result in very different species compositions in streams with essentially identical sediment characteristics.

Community level responses to varying siltation were most consistently described by changes in functional characteristics of the communities. In both studies, herbivores, benthic insectivores and simple lithophilous spawners were most sensitive to siltation while other guilds were not. Species from each of these three guilds possess life history traits which make them especially sensitive to the composition of benthic sediments for various reasons, and the association between their frequency in communities and siltation gradients was present in both studies. These frequencies were reliable indicators of siltation effects in both intraregional comparisons among sites of similar size and character, and in interregional comparisons of streams which varied in characteristics besides siltation, suggesting that this index may be robust to variation from other conditions.

The studies reported here show the effects of siltation on fish communities to be significant and predictable and to have a functional basis. Once effects are measurable, mitigation strategies can be evaluated. It is important that any bufferstrip restoration be efficiently carried out to achieve the goals of a reduction of siltation input, movement, and deposition within the channel.

A simple, but not trivial, question in any bufferstrip program is 'where is the silt coming from?' It is clear from our studies that under particular conditions, the channel itself can be a major source of sediment. Measurements from our extensive study on prairie headwater streams indicated that more than 2000 kg of sediment per year is eroded from each meter in length of the streambed and associated banks (Smale & Rabeni, 1991). Additionally, half the study streams drained watersheds which received intensive management to reduce soil erosion from primarily non-riparian upland areas during the 3–5 year study period. The absence of any detectable change in streambed sediment composition in both treated and untreated streams along with these extremely high measurements of bank and bed erosion indicated that riparian management can influence the degree to which streams behave as sources or sinks of fine sediment. In some, perhaps many, cases, riparian management may have a more substantial effect on sediment budgets and their ecological consequences than would management of more diffuse upland sources. Using a bufferstrip to stabilize channels (Erman *et al.*, 1977) and to collect extra sediment during high flows (Wilkin & Hebel, 1982) should be primary objectives.

Considerable experimentation has been carried out to delineate important variables of the source area affecting transport of eroded soil to a stream. While distance traveled, slope, soil type, geology and macro- and microrelief are all important, they are variables the stream manager has little control over. Characteristics of the bufferstrip are of more practical importance. Most emphasis has been given to calculating an appropriate bufferstrip width (see Osborne & Kovacic, 1993, for a review), but research at the University of Missouri has shown width to be of relatively minor importance (Gough, 1988). Qualitative characteristics are more important, and the factor of most concern should be microrelief. The bufferstrip must prevent formation of rills, gullies or any other pathways that offer easy access of silt to the stream. Continuity is the key: even the finest bucket is useless if a hole is poked through its bottom. Once continuity is achieved, concern can be turned to the vegetative covering. Vegetation should contribute organic matter to the forest floor, increase hydraulic roughness, bind both the litter layer and mineral layer and be able to respond quickly to changes on the surface caused by deposited sediment (Gough, 1988). Herbaceous plants meet these criteria more than woody plants – if siltation is the main focus. Transport of silt through a bufferstrip is also dependent on ground slope and infiltration capacity of the soil, two characteristics that are not amenable to management.

Besides the features of the source area and the bufferstrip, characteristics of the stream channel are also important because siltation affects fish mainly when it comes out of suspension and is deposited on the stream bottom. Particle size distribution is the most important within-channel feature. A stream in a physiographic region where silt-sized particles dominate will possess a fauna typical of high siltation rates regardless of whether excess siltation is occurring. Bufferstrips preventing siltation will only be useful where the mean particle size of the channel is sufficiently large to be impacted (*i.e.*, interstices filled). A final factor relating to channel characteristics is stream power, especially the aspects of slope and depth. Sufficient power moves the majority of suspended sediments downstream and can result in a 'clean substrate' even with a moderately high sediment input. Of course, the sediment will ultimately be deposited, although it may be many kilometers downstream from the source.

Conclusions

Our focus on siltation represented only one of numerous possible relations between land use practices and the biological health of streams in which proper riparian management plays an important role. Our research illustrates some of the issues which must be addressed regardless of whether the concern is siltation, nutrient loading, shading and thermal dynamics, or channel morphology and instream habitat characteristics.

First, it must be established that riparian degradation is affecting the ecological health of a stream. Our research shows that fish community guild analysis can be sensitive to siltation when other biological metrics were insensitive or inconsistent; a result which we suspect could be developed for application to other relationships. Secondly, riparian management is essential regardless of other activities in a watershed. In headwater reaches of streams draining a region of highly erodible soils, we found that even substantial efforts to reduce erosion in the watershed were ineffectual in improving instream conditions without good riparian conditions.

Acknowledgements

This is a contribution from the Missouri Cooperative Fish and Wildlife Research Unit (U.S. Fish and Wildlife Service; Missouri Department of Conservation; The School of Natural Resources, University of Missouri; and Wildlife Management Institute cooperating). Eric Nelson assisted in field collections. Partial funding was from The Missouri Department of Natural Resourses Nonpoint Source Program and the U.S. Environmental Protection Agency.

References

Balon, E. K., 1975. Reproductive guilds of fishes: a proposal and definition. J. Fish. Res. Bd Can. 32: 821–864.

Berkman, H. E. & C. F. Rabeni, 1987. Effect of siltation on stream fish communities. Envir. Biol. Fishes 18: 285–294.

Beschta, R. L. & W. H. Platts, 1986. Morphological features of small streams: significance and function. Wat. Res. Bull. 22: 370–379.

Cordone, A. J. & D. W. Kelley, 1961. The influences of inorganic sediment on the aquatic life of streams. Calif. Fish Game 47: 189–228.

Cross F. B., 1967. Handbook of fishes of Kansas. Museum of Natural History, University of Kansas Misc. Publ. 45, 357 pp.

Erman, D. C., J. D. Newbold & K. B. Roby, 1977. Evaluation of streamside bufferstrips for protecting aquatic organisms. Calif. Wat. Resour. Center, University of Calif., Davis, USA: 1–50.

Gough, S. C., 1988. Stream water quality protection using vegetated filter strips: structure and function related to sediment control. MS Thesis, University of Missouri, Columbia, USA, 92 pp.

Karr, J. R., L. A. Toth & D. R. Dudley, 1985. Fish communities of midwestern rivers: a history of degradation. BioScience 35: 90–95.

Karr, J. R., K. D. Fausch, P. L. Angermeier, P. R. Yount & I. J. Schlosser, 1986. Assessing biological integrity in running waters: a method and its rationale. Illinois Nat. Hist. Surv. Spec. Publ. 5.

Larimore, R. W. & P. W. Smith, 1963. The fishes of Champaign County, Illinois as affected by 60 years of stream changes. Illinois Nat. Hist. Surv. Bull. 28.

Menzel, B. W., J. B. Barnum & L. M. Antosch, 1984. Ecological alterations of Iowa prairie-agricultural streams. Iowa State J. Res. 59: 5–30.

Muncy, R. J., G. J. Atchinson, R. V. Bulkley, B. W. Menzel, L. G. Perry & R. W. Summerfelt, 1979. Effects of suspended solids and sediment on reproduction and early life of warmwater fishes: a review. EPA-600/3-79-0. Corvallis Environmental Research Laboratory, Corvallis, OR.

Osborne, L. L. & D. A. Kovacic, 1993. Riparian vegetated buffer strips in water quality restoration and stream management. Freshwat. Biol. 29: 243–258.

Pflieger, W. L., 1975. The fishes of Missouri. Missouri Dep. Conserv., Jefferson City, MO.

Schiemer, F. & M. Zalewski, 1992. The importance of riparian ecotones for diversity and productivity of riverine fish communities. Neth. J. Zool. 42: 323–335.

Smale, M. A. & C. F. Rabeni, 1991. The effects of special area land treatment (SALT) agricultural management practices on the biological health of headwater streams. Completion Rep. to Missouri Dep. Nat. Resour., Jefferson City, MO, 283 pp.

Smith, P. W., 1971. Illinois streams: a classification based on their fishes and an analysis of factors responsible for disappearance of native species. Illinois Nat. Hist. Surv., Biol. Notes 76.

Starrett, W. C., 1950. Food relationships of the minnows of the Des Moines River, Iowa. Ecology 31: 216–233.

Wilkin, D. C. & S. J. Hebel, 1982. Erosion, redeposition and delivery of sediment to midwestern streams. Wat. Resour. Res. 18: 1278–1282.

Williams, R. D. & A. D. Nicks, 1988. Using CREAMS to simulate filter strip effectiveness in erosion control. J. Soil & Wat. Conserv. Jan–Feb.: 108–112.

Evolutionary aspects

Hydrobiologia **303**: 223–228, 1995.
F. Schiemer, M. Zalewski & J. E. Thorpe (eds), The Importance of Aquatic-Terrestrial Ecotones for Freshwater Fish.

Evolutionary adaptations by fish to ecotonal complexity in spatially variable landscapes – a perspective

M. M. Coelho [1] & M. Zalewski [2]
[1] *Departamento de Zoologia & Antropologia, Faculdade de Ciências, Bloco C-2, 3° Piso 1700 Lisboa, Portugal;* [2] *Department of Applied Ecology, University of Łódź, Banacha 12/16, PL-90-237, Łódź, Poland*

Key words: ecotones, environmental heterogeneity, genetic diversity

Abstract

In most types of freshwater ecosystems fish diversity depends greatly on land/inland water ecotones. So, to maintain biodiversity of fish communities in inland waters, management and restoration of aquatic terrestrial ecotones will be an important tool. However, to provide a scientific background for such conservation activities, it will be desirable to test the importance of different types of ecotones in structuring and maintaining the genetic diversity of fish populations. The relevance of population genetics data to ecotone studies can only be understood in an ecological context as evolution is a function of environment. We suggest that as ecotone complexity increases opportunities for survival of individuals, improving trophic conditions and spatial habitat heterogeneity, so the population size and variation increase with increased genetic diversity and vulnerability to environment changes decreases.

Introduction

A central problem in population genetics is the maintenance of genetic variation.

The interpretation of spatial and temporal patterns of genetic variation requires an integration of information from several different areas, including population ecology and genetics, and biogeography. Comparative studies of spatial genetic patterns and their relations with abiotic and biotic features of the environment are needed to understand and predict the importance of environmental heterogeneity in structuring and maintaining genetic diversity in natural populations.

In most types of freshwater ecosystems fish diversity depends greatly on land/inland water ecotones (Zalewski *et al.*, 1991). These boundary environments, as usually the most diversified element of aquatic ecosystems, could be of great importance in maintaining population genetic diversity, even in those populations subject to anthropogenic pressures. Also these interface zones are important for maintaining landscape biodiversity because they harbour a large number of species and diverse genomes (Salo *et al.*, 1986; Petts, 1990; Risser, 1990) and are sites for active microevolution (Naiman *et al.*, 1988).

A good starting point for understanding local genetic diversity is analysis of between-habitat differences in fish populations. The interactions between population density and the scale of environmental heterogeneity determines the exposure of alternative evolutionary strategies to selection (Brown & Pavlovic, 1992).

Relationships between ecological and genetic profiles

The amount of genetic variation in a population is a fundamental parameter in evolutionary studies, because genetic variation determines the evolutionary potential of populations. Natural selection, as well as random drift, can only take place in populations that possess genetic variability. Loss of genetic variation may occur through small population size, population bottlenecks, genetic drift, inbreeding, selection and introgression (Meffe, 1986).

In the last decade the classical selectionist/neutralist controversy, in interpretation of genetic variation, has been under debate in spite of the enormous accumulation of data on the molecular basis of genetic change. In general, selectionists claim that observed variation is to a great extent a consequence of selection and neutralists claim that the majority of amino-acid substitutions are neutral. The treatment of the scientific status of the competing theories for the maintenance of genetic variation has been discussed recently by Gillespie (1991).

Although the importance of parameters that are supposed to explain heterozygosity (mutation and migration rates, genetic drift, dynamics of effective population sizes and selection) is emphasized differently by different authors, remains the idea that demography and genetics are closely intertwined, and it may be unwise to think of them separately. The maintenance of genetic variance requires that the effective population size be an order of magnitude greater that would be required to avoid an effect of inbreeding and retard the effects of drift. Otherwise, selection has a predictable influence on genetic variation and operates in concert with mutation and genetic drift. Selection can lead to the retention of heritable variation in natural populations; these include antagonistic pleiotropic effects of genes, high fitness for heterozygotes, and fluctuations in the environment that prevent equilibrium being maintained (Cockburn, 1991).

More than six hundred references were summarized and reanalyzed by Nevo *et al.* (1984) demonstrating that environmental heterogeneity is a major factor in maintaining and structuring genetic diversity in natural populations. In fact, fluctuations in relative fitness among genotypes in heterogeneous environments has been suggested as a mechanism for the maintenance of enzyme polymorphism in natural populations (Gillespie & Langley, 1974; Gillespie, 1991). Mitton & Lewis (1989) emphasised the role of environmental variation in the control of genetic variation in bony fishes although Waples (1991a) considered that those data were heavily weighted by freshwater species (Catastomidae, Centrarchidae and Cyprinidae).

Life history features have been used as indicators of similarity in the operation of certain kinds of selection, as referred to by Mitton & Lewis (1989) who demonstrated the significant association of genetic variation and four life-history variables: fecundity, body size, rate of maturation and egg size. Waples (1991a) criticised the inferring of causal relationships between environmental variation and genetic variation from a correlation between genetic variation and a third class of variables (life-history traits). However, Mitton & Lewis (1992) considered that the relationships between heterozygosity and controlling variables were obviously subtle, but could be studied most effectively across the broadest possible range of independent variables and with large sample sizes.

Many authors accept that the genetic-environmental correlations demonstrate inferentially the adaptive nature of biochemical polymorphism in fish mainly if it occurs at many loci, in many populations and unrelated species in a repetitive and consistent pattern (Nevo, 1988). A series of examples referred to as microadaptations, are presented in Gillespie (1991), where one or a few amino-acid changes in a protein are implicated in an adaptation.

A review of the literature point to the conclusion that fitness in natural populations is a positive function of heterozygosity and that any decrease in genetic variation will be paralleled by a diminution of fitness (viability, vigour, fecundity, fertility, etc.) (Frankel & Soulé, 1981). The opinion that more heterozygous individuals display higher fitness with respect to characters as body length, growth rate, survival, fecundity and developmental stability was expressed by several authors, specially in Cyprinidae, Poecilidae and Salmonidae (e.g. Mitton, 1978, Feder *et al.*, 1984, Leary *et al.*, 1984, 1985; Dazmann *et al.*, 1986, Leary *et al.*, 1987; Quattro & Vrijenhoek, 1989; Kirpichnikov, 1992). However, we know little about actual mechanisms of heterozygote superiority and we do not know whether all species or populations are affected similarly by heterozygosity (Meffe, 1987; Meffe & Vrijenhoek, 1988). Vrijenhoek *et al.* (1992) suggest that some form of overdominance is necessary to maintain polymorphism in small, dynamically fluctuating populations that inhabit temporally and spatially varying environments.

Genetic patterns in spatially variable landscapes

Genetic differentiation is the result of a balance between homogenizing effects of gene flow and divergent effects such as drift, selection (e.g. Slatkin, 1987; Meffe, 1990) and DNA turnover mechanisms (Hoelzel & Dover, 1991). Gene flow may either constrain evolution by preventing adaptations to local conditions or promote evolution by spreading new genes and combinations of genes throughout a species range. Mutation, genetic drift due to finite population size, and natural selection favouring adaptations to local environmental conditions will all lead to the genetic differentiation of local populations (Slatkin, 1987).

About the effects of migration on population diversity, little mixing is necessary to overcome the effects of genetic drift and to maintain genetic homogeneity between populations. In practical terms if one or more reproductively active individuals migrate between subpopulations per generation, then there will be little local differentiation (Cockburn, 1991; Hoelzel & Dover, 1991). The effect will be less pronounced in species that tend to disperse over a short range, because neighbouring subpopulations will tend to be genetically similar. However, strong selection can produce adaptations to local conditions in the presence of substantial gene flow (Slatkin, 1987).

One of the most distinctive characteristics of riverine fish populations is their spatial seasonal dynamics, which to a great extent is reflected in their evolution (Thorpe, 1987) and genetic changes (Waples, 1991b).

Distance and geographic barriers reduce the movement of breeding individuals and therefore the transfer of genes (Hoelzel & Dover, 1991). About the first aspect McClenaghan *et al.* (1985) considered that the spatial genetic subdivision of mosquitofish populations appeared to be largely independent of their proximity in space. Also Brown & Chapman (1991) observed in *Fundulus heteroclitus* that populations separated by distances greater than 2 km accumulated significant genetic differences. However, Sheldon (1988) considered that speciation rates may be greater in larger drainages if isolation by distance and intervening high order streams reduce gene flow among populations.

Some barriers are consequences of environmental or anthropogenic effects in aquatic ecosystems allowing the fragmentation of natural populations, which may alter their genetic structure and evolutionary potential. Size-selective fishing gear, habitat elimination, pollution stress, introduction of exotic species and other such activities can impose new selection pressures or alter the existing ones (Thorpe *et al.*, 1981). Alterations in migration rates, occurrence of bottlenecks and the acceleration of genetic drift can be consequences of loss or disruption of connectedness by habitat alteration, and can greatly alter population structure (Meffe, 1990). However, the responses of a species to stress from human activities will depend to a large degree on its genetic and phenotypic variability (Thorpe *et al.*, 1981).

Sometimes, temporal variation can confound our ability to predict microgeographic spatial patterns (Smith *et al.*, 1983). Fluctuations in stream flow sometimes results in drought in some sectors, and can influence the survival of fishes and the genetic composition of local populations by the occurrence of temporary bottlenecks. The consequences of these possible genetic changes depend largely on the time and period at which they occur. Also, differential dispersal potentially may alter the genetic and demographic constituency of resident groups. Differential dispersal by fish from different age and sex classes among subpopulations was observed in *Gambusia affinis* (Brown, 1985). Differential dispersion, breeding behaviour, age-specific changes and recruitment of offspring that differ in genotypic composition can lead to temporal genetic variation within populations (Scribner *et al.*, 1991).

So, the study of adaptive evolution must include consideration of fitness differences among individuals and differences among populations in space and time (Barker & Thomas, 1987). Taylor (1991) showed that adaptation to local environments is a major factor that maintains and promotes genetic variation among salmonid populations. Also, we found electrophoretically detectable differences in the population structure of the Iberian cyprinid *Chondrostoma lusitanicum*, presumed related to different local abiotic and biotic factors (Alves & Coelho, *in press*). Some subpopulations lived in habitats which differred in fish assemblages (some local predominance of exotic species), riparian vegetation, and levels of oxygen content related to lower depth and lower current velocity. These local differences in genetic and ecological characteristics should be tested further to assess the potential ecological genetic consequences of the different habitats.

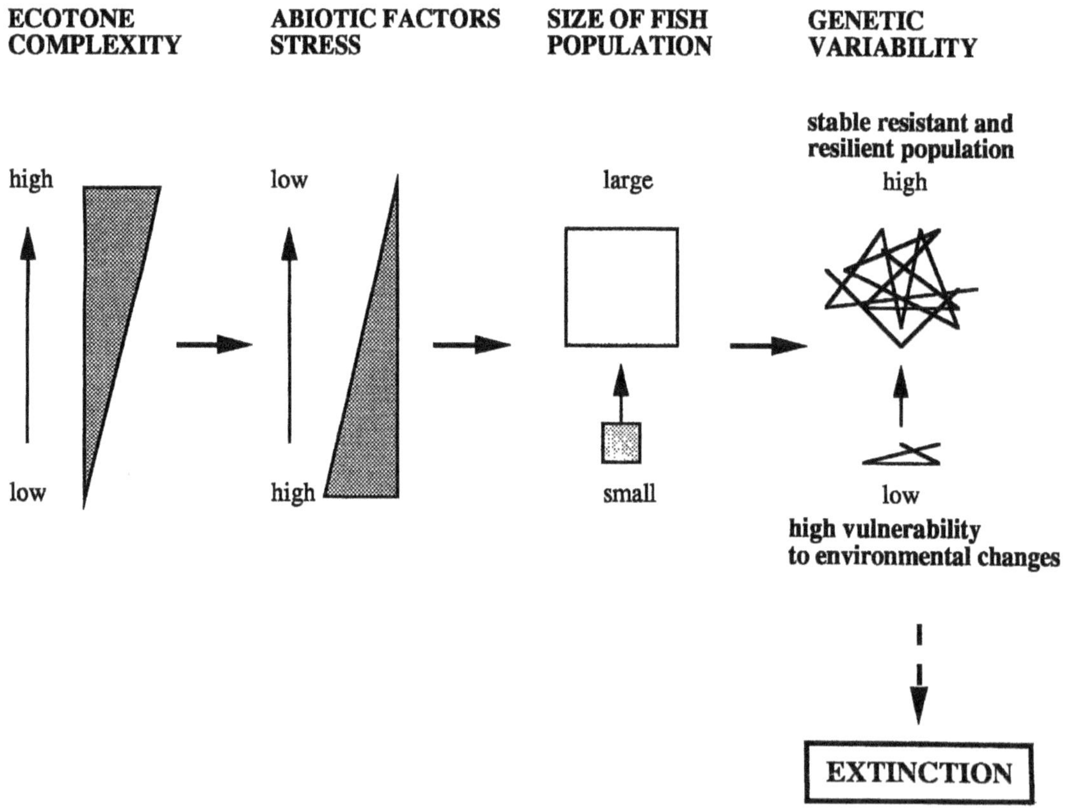

Fig. 1. The hypothetical effect of restoration in land/water ecotone complexity on population size, genetic variability and increase vulnerability to environmental changes.

Perspectives of the research on ecotones and genetic patterns

The increasing intensity of land use is reflected in the degradation of land/water ecotones, which in turn affects aquatic system dynamics. According to Zalewski (1990) and Schiemer and Zalewski (1995) fish biodiversity in most freshwater ecosystems is directly or indirectly dependent on land/inland water ecotone complexity. Consequently, to maintain biodiversity of fish communities in inland waters, management and restoration of aquatic terrestrial ecotones will be an important tool.

However, to provide a scientific background for such conservation activities, it will be desirable to test the importance of different types of ecotones in structuring and maintaining the genetic diversity of fish populations. The relevance of population genetics data to ecotone studies can only be understood in an ecological context as evolution is a function of environment. The questions to answer are: how and why will the maintained or restored high complexity of ecotonal habitats enchance the genetic diversity of a fish population, and thus the biodiversity of its community? We suggest that as ecotone complexity increases opportunities for survival of individuals, with improved trophic conditions and spatial habitat heterogeneity (Benke *et al.*, 1984; Zalewski *et al.*, 1990), so the population size and variation increase with increased genetic diversity and vulnerability to environment changes decreases. (Fig. 1).

Acknowledgements

We thank John Thorpe and Gunther Gollman for their valuable comments on the early draft of the manuscript. We are also indebeted to John Thorpe for stimulating conversations and helpful suggestions. This paper was partially supported by fnict (Portugal).

References

Alves, M. J. & M. M. Coelho, *in press*. Genetic variation and population subdivision of the endangered Iberian cyprinid *Chondrostoma lusitanicum*. J. Fish Biol.

Barker, J. S. F. & R. H. Thomas, 1987. A quantitative genetic perspective on adaptive evolution. In Genetic Constraints on Adaptive Evolution, Springer-Verlag, Berlin: 3–23.

Benke, A. C., T. C. Van Arsdall Jr., D. M. Gillespie & F. K. Parrish, 1984. Invertebrate productivity in a subtropical blackwater river. Ekol. Monogr. 54: 25–63.

Brown, B. L. & R. W. Chapman, 1991. Gene flow and mitochondrial DNA variation in the killifish, *Fundulus heteroclitus*. Evolution 45: 1147–1161.

Brown, J. S. & N. B. Pavlovic, 1992. Evolution in heterogeneous environments: effects of migration on habitat specialization. Evol. Ecol. 6: 360–382.

Brown, K. L., 1985. Demographic and genetic characteristics of dispersal in the mosquitofish, *Gambusia affinis* (Pisces: Poeciliidae). Copeia: 597–612.

Cockburn, A., 1991. An Introduction to Evolutionary Ecology. Blackwell Scientific Publications, Oxford, 370 pp.

Danzmann, R. G., M. M. Ferguson, F. W. Allendorf & K. L. Knudsen, 1986. Heterozygosity and developmental rate in a strain of rainbow trout (*Salmo gairdneri*). Evolution 40: 86–93.

Feder, J. L., M. H. Smith, R. K. Chesser, M. W. Godt & K. Asbury, 1984. Biochemical genetics of mosquitofish. II. Demographic differentiation of populations in a thermally altered reservoir. Copeia: 108–119.

Frankel, O. H. &. E. Soulé, 1981. Conservation and Evolution. Cambridge University Press, Cambridge, 327 pp.

Gillespie, J. H., 1991. The Causes of Molecular Evolution. Oxford Series in Ecology and Evolution. Oxford University Press, 336 pp.

Gillespie, J. H. & C. H. Langley, 1974. A general model to account for enzyme variation in natural populations. Genetics 76: 837–884.

Hoelzel, A. R. & A. Dover, 1991. Molecular Genetic ecology. Oxford University Press, 75 pp.

Kirpichnikov, V. S., 1992. Adaptive nature of intrapopulational biochemical polymorphism in fish. J. Fish Biol. 40: 1–16.

Leary, R. F., F. W. Allendorf & L. Knudsen, 1984. Superior developmental stability of heterozygotes at enzyme loci in salmonid fishes. Am. Nat. 124: 540–551.

Leary, R. F., F. W. Allendorf & L. Knudsen, 1985. Developmental instability as an indicator of reduced genetic variation in hatchery trout. Trans. Am. Fish. Soc. 114: 230–235.

Leary, R. F., F. W. Allendorf & L. Knudsen, 1987. Differences in inbreeding coefficients do not explain the association between heterozygosity at allozyme loci and developmental stability in rainbow trout. Evolution 41: 1413–1415.

McClenaghan Jr., L. R., M. H. Smith & M. W. Smith, 1985. Biochemical genetics of mosquitofish. IV. Changes of allele

frequencies through time and space. Evolution 39: 451–460.
Meffe, G. K., 1986. Conservation genetics and the management of endangered fishes. Fisheries 11: 14–23.
Meffe, G. K., 1987. Conserving fish genomes: philosophies and practices. Envir. Biol. Fishes 18: 3–9.
Meffe, G. K., 1990. Genetic approaches to conservation of rare fishes: examples from North American desert species. J. Fish Biol. 37A: 105–112.
Meffe, G. K. & R. C. Vrijenhoek, 1988. Conservation genetics in the management of desert fishes. Conservation Biology 2: 157–169.
Mitton, J. B., 1978. Relationship between heterozygosity for enzyme loci and variation of morphological characters in natural populations. Nature 273: 661–662.
Mitton, J. B. & W. M. Lewis Jr, 1989. Relationships between genetic variability and lifehistory features of bony fishes. Evolution 43: 1712–1723.
Mitton, J. B. & W. M. Lewis Jr, 1992. Response to Waples's comment on heterozygosity and life–history variation in bony fishes. Evolution 46: 576–577.
Naiman, R. J., H. Décamps, J. Pastor & C. A. Johnston, 1988. The potential importance of boundaries to fluvial ecosystems. J. N. Am. Benthol. Soc. 7: 289–306.
Nevo, E., 1988. Genetic diversity in nature. Patterns and theory. In Max K. Hecht & B. Wallace (eds), Evol. Biol. 23. Plenum Press, New York: 217–246.
Nevo, E., A. Beiles & R. Ben-Shlomo, 1984. The evolutionary significance of genetic diversity: ecological, demographic and life history correlates. In G. S. Mani (ed.), Evolutionary Dynamics of Genetic Diversity. Liss, New York: 69–92.
Petts, G. E., 1990. The role of ecotones in aquatic landscape management. In R. J. Naiman & H. Décamps (eds), The Ecology and Management of Aquatic-Terrestrial Ecotones. Man and the Biosphere Series, 4. The Parthenon Publishing Group, UNESCO: 227–261.
Quattro, J. M. & R. C. Vrijenhoek, 1989. Fitness differences among remnant populations of the endangered sonoran topminnow. Science 245: 976–978.
Risser, P. G., 1990. The ecological importance of land-water ecotones. In R. J. Naiman & H. Décamps (eds), The Ecology and Management of Aquatic-Terrestrial Ecotones. Man and the Biosphere Series, 4. The Parthenon Publishing Group, UNESCO: 7–21.
Salo, J., R. Kalliola, I. Hakkinen, Y. Makinen, P. Niemela, M. Puhakka & P. D. Coley, 1986. River dynamics and the diversity of Amazon lowland forest. Nature 322: 254–258.
Schiemer, F. & M. Zalewski, 1992. The importance of riparian ecotones for density and productivity of riverine fish communities. Neth. J. Zool. 42: 323–335.
Scribner, K. T., M. H. Smith, R. A. Garrott & L. H. Carpenter, 1991. Temporal, spatial, and age-specific changes in genotypic compositions of mule deer. J. Mamm. 72: 126–137.
Sheldon, A. L., 1988. Conservation of stream fishes: patterns of diversity, rarity, and risk. Conservation Biology 2: 149–156.
Slatkin, M., 1987. Gene flow and the geographic structure of natural populations. Science 236: 787–792.
Smith, M. W., M. H. Smith & R. K. Chesser, 1983. Biochemical genetics of mosquitofish. I. Environmental correlates, and temporal and spatial heterogeneity of allele frequencies within a river drainage. Copeia (1): 182–193.
Thorpe, J. E., 1987. Smolting versus residency: developmental conflict in Salmonids. Am. Fish. Soc. Symp. 1: 244–252.
Thorpe, J. E., J. F. Koonce, D. Borgeson, B. Henderson, A. Lamsa, P. S. Maitland, M. A. Ross, R. C. Simon & C. Walters, 1981. Assessing and managing man's impact on fish genetic resources. Can. J. Fish. aquat Sci. 38: 1899–1907.
Taylor, E. B., 1991. A review of local adaptation in Salmonidae, with particular reference to Pacific and Atlantic salmon. Aquaculture 98: 185–207.
Vrijenhoek, R. C., E. Pfeiler & J. D. Wetherington, 1992. Balancing selection in a desert stream-dwelling fish, *Poeciliopsis monacha*. Evolution 46: 1642–1657.
Waples, R. S., 1991a. Heterozygosity and life-history variation in bony fishes: an alternative view. Evolution 45: 1275–1280.
Waples, R. S., 1991b. Genetic interactions between hatchery and Salmonids: Lessons from the Pacific Northwest. Can. J. Fish. aquat Sci. 48: 124–133.
Zalewski, M., 1990. The ecotone concept in conservation and fisheries management of riverine ecosystems. Proc. UNESCO/MAB Int. Seminar: 80–94.
Zalewski, M., P. Frankiewicz, M. Przybylski, J. Banbura & M. Nowak, 1990. Structure and dynamics of fish communities in temperate rivers in relation to the abiotic-biotic regulatory continuum concept. Pol. Arch. Hydrobiol. 37: 151–176.
Zalewski, M., J. Thorpe & P. Gaudin, 1991. Fish and Land/Inland Water Ecotones. UNESCO, MAB. Universities of Łódź, Stirling and Claude Bernard Lyon I, 102 pp.

Hydrobiologia **303**: 229–234, 1995.
F. Schiemer, M. Zalewski & J. E. Thorpe (eds). The Importance of Aquatic-Terrestrial Ecotones for Freshwater Fish

The genetic structure of roach populations in two contrasted large rivers

Y. Bouvet, M. Bobin, J. L. Maslin & E. Pattee
Laboratoire de Biologie Animale et Ecologie, Université Claude Bernard Lyon-I, 43 bd du 11 Novembre, F-69622 Villeurbanne Cedex, France (U.R.A. 1451 Ecologie des Eaux Douces et des Grands Fleuves)

Key words: genetic diversity, enzyme electrophoresis, French Upper Rhone, Saone

Abstract

Overall heterozygosity was high in all roach populations, and so was gene flow between them. However, genetic diversity was greater in the fast-flowing and more diverse French Upper Rhone than in the uniformly lentic River Saone. Rhone backwaters may be considered as ecotones and reservoirs of genetic diversity, but such is not the case in the Saone owing to the uniformity of the environment.

Introduction

Roach, *Rutilus rutilus* L. 1758, is one of the most common cyprinid fish species in Europe. It inhabits mostly lentic and often warm waters such as those of the River Saone, or lakes and reservoirs. The Saone belongs to the bream zone. However, roach is present along the banks and in reservoirs of the fast-flowing River Rhone itself – the French Upper Rhone between the cities of Geneva and Lyon belongs to the barbel zone – and some of its backwaters and main tributaries.

Genetic diversity of the populations of the French Upper Rhone has been investigated for the last six years as recorded by Soewardi (1988), Bouvet *et al.* (1991) and Bouvet *et al.* (1992).

Our present aim was to include the Saone population in our investigations, to compare genetic diversity in the two fluvial systems and to stress possible ecotone effects.

The two contrasted large rivers (Fig. 1)

The southern outskirts of Lyon, in southeastern France, are the meeting point of two large rivers. Both flow through the city, the Saone (mean annual discharge 430 $m^3 s^{-1}$) from the north and the Upper Rhone (mean annual discharge 600 $m^3 s^{-1}$) from the east.

Over 200 km, the Saone flows slowly in the plains with a slope of 0.04‰, across what was a cenozoic lake owing to the threshold where it meets the hills that it cuts through in entering Lyon. Floating and emerged macrophytes occupy its banks. The district investigated was the confluence of the Saone and one of its small eastern tributaries, the Seille, that meet 5 kilometres to the south of the city of Tournus. A weir with a lock for letting boats through closes the tributary, the mean discharge of which is 7 $m^3 s^{-1}$. Roach thrives in every part of this system. Other species are, by order of decreasing abundance after roach, *Abramis brama*, *Blicca bjoerkna*, *Alburnus alburnus*, *Leuciscus cephalus*, *Scardinius erythrophthalmus*, and many other limnophilic species. During high water that occurs at least once a year in winter, the whole plain becomes a shallow lake several kilometres wide, through which limnophilic fish move about freely.

The French Upper Rhone springs from Lake Geneva and cuts its way to Lyon through the Jura mountains. The district between the Alps and Lyon was covered by the Quaternary Würm glacier, and hence the present course of the Upper Rhone is much more recent than that of the Saone and the Lower Rhone, but indiscriminate fish stocking has been carried out for years in all these rivers.

Owing to the slope of 0.4‰, trout (*Salmo trutta fario*), grayling (*Thymallus thymallus*) and rheophilic

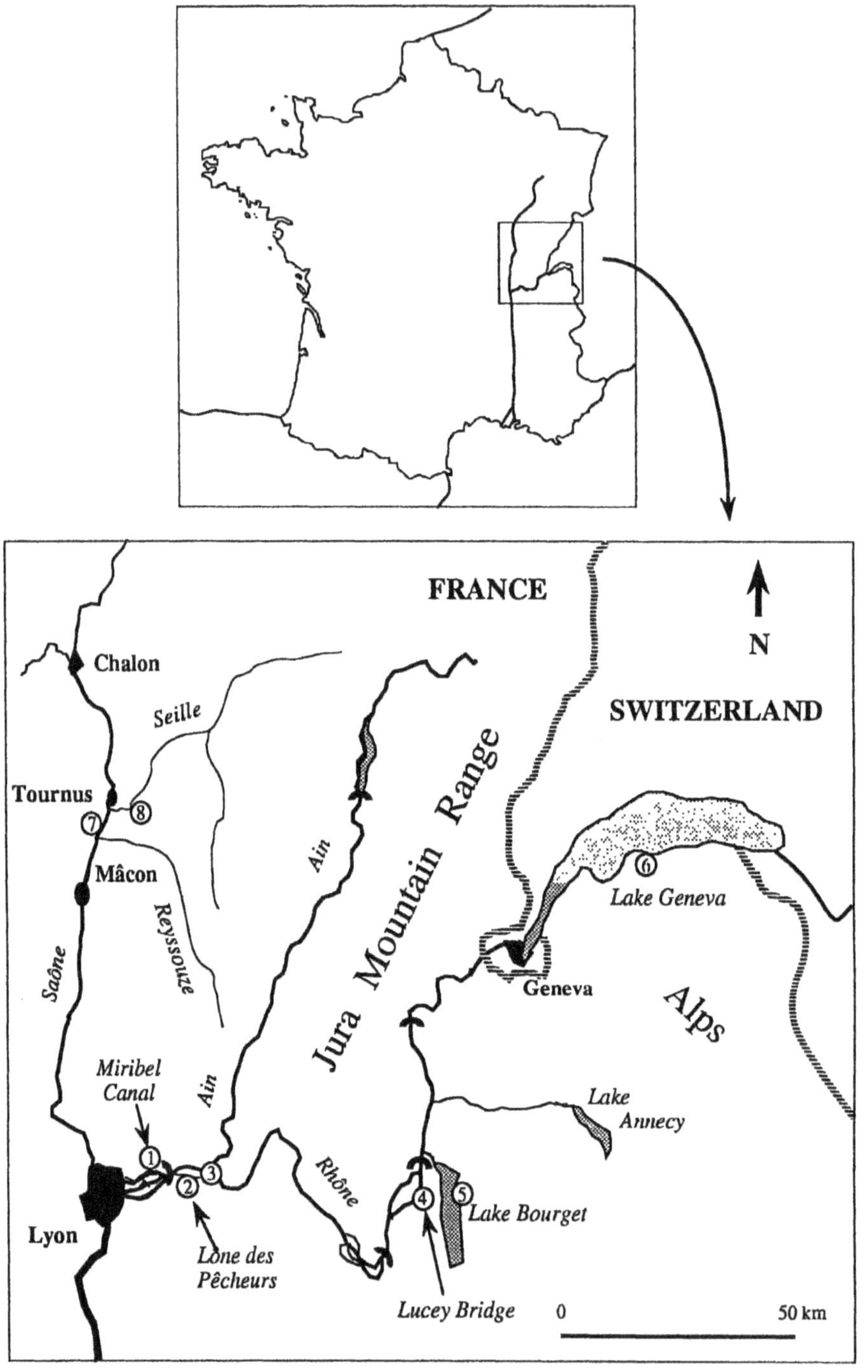

Fig. 1. Location (n ° 1 to n ° 8) of the roach populations investigated in the Rhone and the Saone hydrosystems.

cyprinids (*Leuciscus leuciscus*, *Leuciscus cephalus*, *Chondrostoma nasus*) occur as well as roach. Dams have been erected, the first in 1948 and most others since 1981 (list in Pattee, 1988). However the main channel has preserved three kinds of lateral appendices: fast-flowing mountain tributaries, large lakes such as Lake Bourget, and more or less connected former channels of the meander or braided types. Though roach occurs in most parts of this system and indeed along the banks of the main channel itself, it is more abundant in the reservoirs, lakes and backwaters. Bouvet *et al.* (1982) listed the following fish species in these backwaters in 1976–1980, by order of decreasing abundance: *R. rutilus, Ictalurus melas, Lepomis gibbosus, Tinca tinca, Perca fluviatilis, Esox lucius, Leuciscus cephalus, Abramis brama*, followed by 15 other species. The spawning grounds are invaded in spring by successive shoals of roach that migrate up the dead arms, leave fry behind them and migrate down again. Where they go after spawning is still unclear: during an 8-months study, 25 marked specimens were recaptured in the same arm, at a maximum distance 900 m from the marking spot (Bouvet *et al.*, 1982, Bouvet *et al.*, 1991).

Material and methods

Roach were caught by electrofishing or by gillnets (30 specimens per station) in 1993 in the tributary 2 km above its confluence, and in three stations of the Saone extending 4 km downstream from the confluence. Only live specimens were brought back to the laboratory or frozen on the spot. Then they were preserved at −80 °C. Tissue extracts were prepared and zymograms obtained by migration on starch gels according to Pasteur *et al.* (1987). The results were compared with those of Bouvet *et al.* (1992) on the Rhone. Thus 16 enzymes and 27 loci were considered in the populations of both rivers (Table 1). The dates of collection are also shown in this table. Genetic diversity was evaluated by the degrees of heterozygosity and polymorphism as defined by Pasteur (1985), Nei's (1972) genetic distances on an UGPMA dendrogram (Sneath & Sokal, 1973) and multiple correspondence factorial analyses of the alleles. Gene flow between the Saone and the Seille was calculated according to Slatkin (1985).

Results

9 enzymes proved to be polymorphic, each with 2 to 4 alleles (Table 1). Gene flow, i.e. the number of individuals in each generation that was estimated to have migrated within the 6-km long district in the Saone was Nm = 13.9.

Though no valid statistic test could be applied, heterozygotic within-population diversities (Fig. 2) seemed lower in the Saone than in the Rhone, in which 89% of the heterozygotic variability was within the populations and only 11% between the populations.

Multivariate factorial analysis also stresses a greater diversity between populations in the Rhone than in the Saone (Fig. 3). Allele diversity within the

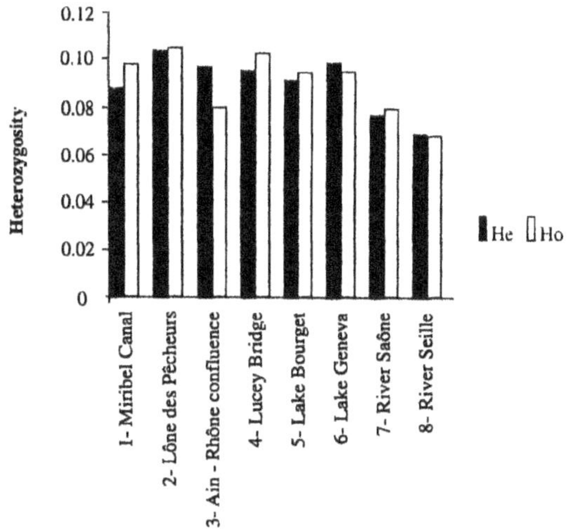

Fig. 2. Roach heterozygosity in the different stations. He = estimated heterozygosity. Ho = observed heterozygosity.

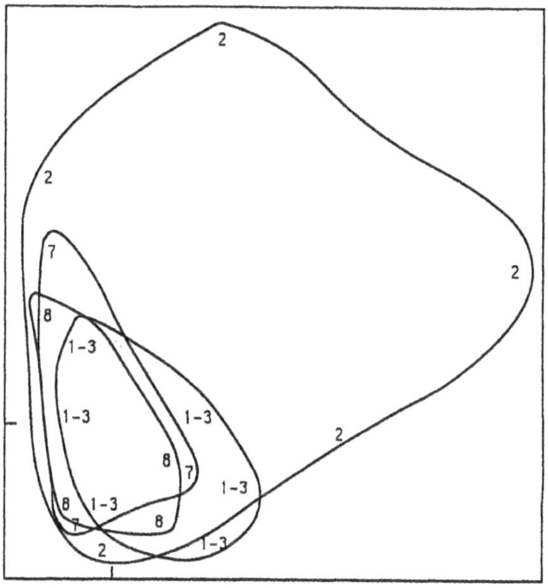

Fig. 3. A comparison of genetic diversity in the Rhone and in the Saone: projection on plane F1–F2 of the dots obtained by multivariate factorial analysis on roach individual genotypes. 1–3 (shaded area) = Rhone channel downstream (Miribel Canal + Ain-Rhone confluence pooled) 2 = Lône des Pêcheurs (sidearm of the Rhone) 7 = Saone channel (3 stations pooled) 8 = River Seille (Saone tributary).

Table 1. List of loci and alleles investigated. Stations and dates of collection: Mir. Can. = Miribel Canal. - A. Confl. = Confluence between the Ain and the Rhone. Lucey = Lucey Bridge on the Rhone. - L. Bo. = Lake Bourget. - L. Ge. = Lake Geneva

Enzymes	1984 IUBNC n°	Number of loci	Loci	Alleles Observed	Mir. Can. 07/85-10/86	Side Arm 11/84-04/85	A. Confl. 10/85-04/86	Lucey 10/88	L. Bo. 04/88	L. Ge. 03/86	Saone 09/91	Seille 06/92
AAT	2.6.1.1	2	Aat-1	100	+	+	+	+	+	+	+	+
			Aat-2	100	+	+	+	+	+	+	+	+
				110	0	+	0	0	0	0	0	0
				90	+	+	+	+	+	+	+	+
ADH	1.1.1.1	1	Adh1	100	+	+	+	+	+	+	+	+
AK	2.7.4.3	2	Ak-1	100	+	+	+	+	+	+	+	+
			Ak-2	100	+	+	+	+	+	+	+	+
CK	2.7.3.2	2	Ck-1	100	+	+	+	+	+	+	+	+
			Ck-2	100	+	+	+	+	+	+	+	+
EST	3.1.1.1	3	Est-1	100	+	+	+	+	+	+	+	+
				95	+	+	+	+	+	+	+	+
			Est-2	100	+	+	+	+	+	+	+	+
				98	+	+	+	+	+	+	+	+
				90	+	+	+	+	+	+	+	+
			Est-3	100	+	+	+	+	+	+	+	+
FH	4.2.1.2	1	Fh-1	100	+	+	+	+	+	+	+	+
G6PDH	1.1.1.49	1	G6pdh 1	100	+	+	+	+	+	+	+	+
				110	0	+	0	0	0	0	0	0
				90	+	+	+	+	+	+	+	+
				70	+	+	+	+	+	+	+	+
GDP	1.1.1.8	1	Gdp-1	100	+	+	+	+	+	+	+	+
GPI	5.3.1.9	1	Gpi-1	100	+	+	+	+	+	+	+	+
			Gpi-2	100	+	+	+	+	+	+	+	+
				115	0	+	0	+	+	+	+	+
IDHP	1.1.1.42	2	Idhp-1	100	+	+	+	+	+	+	+	+
			Idhp-2	100	+	+	+	+	+	+	+	+
LDH	1.1.1.27	2	Ldh-1	100	+	+	+	+	+	+	+	+
				10	0	0	0	+	0	0	0	0
			Ldh-2	100	+	+	+	+	+	+	+	+
MDH	1.1.1.37	2	Mdh-1	100	+	+	+	+	+	+	+	+
				60	0	+	0	0	0	+	0	0
			Mdh-2	100	+	+	+	+	+	+	+	+
				120	0	0	0	0	0	0	+	0
MEP	1.1.1.40	2	Mep-1	100	+	+	+	+	+	+	+	+
			Mep-2	100	+	+	+	+	+	+	+	+
MPI	5.3.1.8	2	Mpi-1	100	+	+	+	+	+	+	+	+
			Mpi-2	100	+	+	+	+	+	+	+	+
				88	+	+	+	+	+	+	+	+
PGM	2.7.5.1	1	Pgm-1	100	+	+	+	+	+	+	+	+
				110	0	+	0	+	+	+	0	0
SOD	1.15.1.1	1	Sod-1	100	+	+	+	+	+	+	+	+

sidearm of the Rhone was much greater than that in the other stations and included it. The dots of individual specimens could not be shown in the figure, but 25 out of 138 individual dots (18%) from the sidearm (area 2) were located outside the other areas (areas 1, 3, and 4). Allele diversity within the Saone tributary (area 4) was very similar to that of the nearby channel (area 3).

Between-population genetic distances are shown in Fig. 4. The dendrogram identifies several groups: one includes the populations of Lake Geneva and the channel of the Rhone 50 km downstream ($D = 0.0087$), another includes those that were found in the side arm and the neighbouring parts of the main river within a distance of 5 km ($D = 0.0072$ and 0.0121). The Saone channel and its tributary, the River Seille ($D = 0.0097$), appear as a very distinct group, that is somewhat closer to the population of Lake Bourget than to those of the Rhone.

Discussion

Backwaters and small tributaries may be considered as ecotones between the main river and its floodplain. The high genetic diversity (heterozygosity) discovered in roach from the Rhone is confirmed in the Saone, and thus appears as a general characteristic of this species in Southeastern France.

However, within-population heterozygosity was somewhat lower in the Saone than in the Rhone. Multivariate analysis also pointed out slightly greater genetic diversity in the channel and much greater genetic diversity in the backwaters of the Rhone.

In both rivers, gene flow was high (14.6 in the Rhone according to Soewardi (1988) and 13.9 in the Saone), indicating the exchange of individuals between stations.

The Saone population lives in lentic water with occasional beds of macrophytes, the most uniformly favourable environment for roach. Gene flow is high and genetic diversity is lower. Here we have one broadly-distributed population, occupying as well the main river as its local tributary. The same favourable conditions occur in the shallow reed-beds of Lake Bourget, in which the roach were captured, and this may explain the slightly lower genetic distance between these two populations in the dendrogram despite the absence of any geographic connection between them. Lake Bourget is only connected with the Rhone by the narrow 3 km-long Savières Canal that now ends in a sluice-gate.

Living in the Rhone implies for roach the exploitation of marginal habitats (ecotones), with migrations within and perhaps between these habitats. Overall genetic diversity is greater in such a harsher and more diverse environment. Greater genetic diversity certainly means better possibilities of adaptation and survival.

The Rhone flows through Lake Geneva, on the rocky shores of which the roach were sampled. According to genetic distances, the lake and the river downstream had closely-related populations, and so had the sidearm called Lône des Pêcheurs and the neighbouring parts of the river. This stresses both longitudinal and transversal connections.

High gene flow along the river in a lentic species despite the current and the dams might seem paradoxical in comparison with the more distinct rheophilic grayling populations of the Rhone, that are firmly attached to their local biotopes (Bouvet *et al.*, 1992). One explanation may be the great number of roach offspring produced, which invade the whole system and drift down the Rhone from Lake Geneva, thus passing through the rapids and the dams.

The sidearm of the Rhone (ecotone) offers habitat conditions far closer to the ecological optimum of the species than the fast-flowing channel. Bouvet *et al.* (1992) considered it as a reservoir of genetic diversity for the main river itself. The outflow of excess larvae and juveniles, combined with possible adult spawning migrations, may explain the existence of common roach genotypes in the sidearm and main river. The greater diversity in the arm may be due to behavioural (only some genotypes migrate out of the arm), selective (only some genotypes can survive along the banks of the fast-flowing channel) and/or anthropic (possible unnecessary coarse fish stocking in the backwaters) factors.

In contrast, the Seille and the Saone have similar and stable conditions and one cannot be considered a reservoir for the other.

The topographical distance between the Rhone and Saone populations is difficult to evaluate. The actual distance is about 110 km, including 90 km along the lentic Saone and 20 km along the lotic Upper Rhone. The course of the rivers is channelized inside the city of Lyon, so that current is faster and shelter is scarce. It seems hardly possible for roach to swim up the rivers through the city, and only downward drift seems possible. The degree of biological connection between both rivers and the Lower Rhone is being investigated presently.

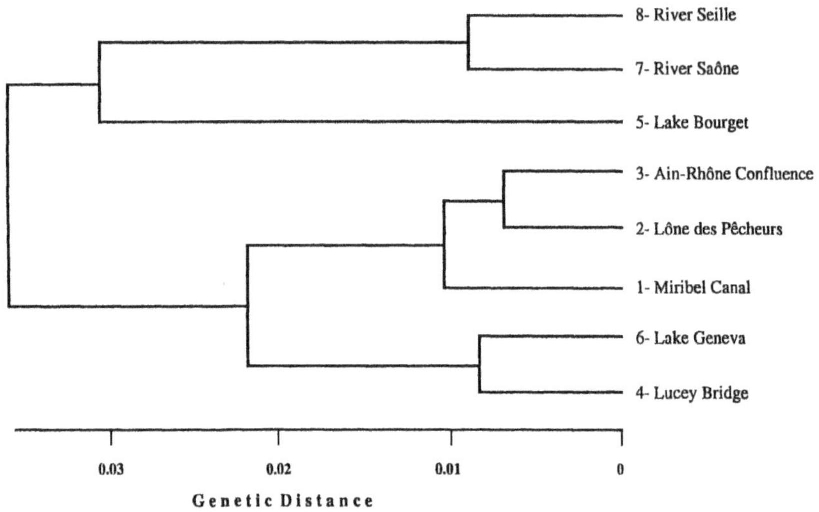

Fig. 4. Genetic distances between roach populations, according to Nei (1973).

Conclusions

1. Several genetic indices are available for describing diversity. Some describe one aspect of this diversity and others another aspect. It is still difficult to associate each of these genetic indices with precise ecological adaptations.
2. However genetic data can be used as ecological markers inasmuch as they indicate flow or separation between populations.
3. The general law of biological diversity coinciding with environmental diversity applied in the present comparison between the Rhone and the Saone.
4. In the harsher environment, the ecotones appeared as reservoirs of genetic diversity. In the more favourable environment, this was not the case.
5. Roach is opportunistic: it seems to adapt to various types of environment and this may be a cause of its growing success and ubiquitous presence in our inland waters.

References

Bouvet, Y., I. Cristau-Quost & N. Thiero Yatabary, 1982. Structure et fonctionnement des écosystèmes du Haut-Rhône français. XII: Contribution à l'étude des poissons d'un bras mort, la Lône des Pêcheurs. Bull. mens. Soc. linn. Lyon 51: 134–147.

Bouvet, Y., E. Pattee & J. L. Maslin, 1992. Comparaison de la variabilité génétique de deux espèces de poissons, l'ombre commun et le gardon, dans un fleuve aménagé. Bull. fr. Pêche Piscic. 324: 26–35.

Bouvet, Y., K. Soewardi & E. Pattee, 1991. The discrimination of roach, *Rutilus rutilus* (Linnaeus, 1758) populations in different parts of a river system. An investigation using biochemical markers. Hydrobiologia 209: 161–167.

Nei, M., 1972. Genetic distance between populations. Am. Nat. 106: 283–292.

Nei, M., 1973. Analysis of gene diversity in subdivided populations. Proc. natn. Acad. Sci. USA 70: 3321–3323.

Pasteur, G., 1985. Les paramètres statistiques communément utilisés dans l'exploitation des résultats de l'électrophorèse des protéines et leur avenir en systématique. In Goyffon, M. & J. L. d'Hondt, (eds), Electrophorèse et Taxonomie. Soc. zool. Fr., Mémoire 42, Paris, 141–180.

Pasteur, N., G. Pasteur, F. Bonhomme, J. Catalan & J. Britton-Davidian, 1987. Manuel technique de génétique par électrophorèse des protéines. Tec & Doc, Lavoisier, Paris, 217 pp.

Pattee, E., 1988. Fish and their environment in large European river ecosystems. The Rhone. Sci. Eau 7: 35–74.

Slatkin, M., 1985. Rare alleles as indicators of gene flow. Evolution 39: 323–335.

Sneath, P. H. & R. R. Sokal, 1973. Numerical taxonomy. Freeman, San Francisco CA, 573 pp.

Soewardi, K., 1988. Caractérisation des populations de *Thymallus thymallus* L. (Poissons, Salmonidés) et de *Rutilus rutilus* L. (Poissons Cyprinidés) dans le réseau fluvial du Rhône. Approche par l'étude du polymorphisme enzymatique. Thèse Univ. Lyon-I, 125 pp.

Résumé

La valeur élevée des flux géniques et de l'hétérozygotie du gardon est confirmée. Cependant, cette diversité génétique est plus grande dans le réseau du Haut-Rhône Français au courant rapide et aux habitats contrastés que dans celui de la Saône, milieu lentique et uniforme. Les annexes de ces fleuves (en l'occurrence bras mort et cours inférieur d'un affluent), se présentent comme des écotones entre le chenal et sa plaine alluviale. Ces écotones fonctionnent comme réservoirs de diversité génétique pour les populations du Rhône, mais non pas pour celles de la Saône du fait de l'uniformité des conditions de milieu.

Hydrobiologia **303**: 235–245, 1995.
F. Schiemer, M. Zalewski & J. E. Thorpe (eds), The Importance of Aquatic-Terrestrial Ecotones for Freshwater Fish.

The land/inland-water ecotones and fish population of Lake Valley (West Mongolia)

Yurij Yu Dgebuadze
Russian Academy of Sciences, A. N. Severtzov Institute of Evolutionary Animal Morphology & Ecology, Leninsky, 33, Moscow V-71, 117071, Russia

Key words: closed basin, desert waters, Mongolia, fish, life history strategies, *Oreoleuciscus humilis*

Abstract

New data on fish populations of a closed desert watershed of Mongolia were obtained in 1990 and 1991. For this region periodic droughts, with the accompanying disappearance of lakes and some parts of rivers, are typical. Two forms of a Cyprinid species *Oreoleuciscus humilis* (dwarf Altai osman) occur in this region during wet periods which usually last for 10–30 years. The dwarf form, is characterized by a maximum SL of 200 mm and early maturation (SL = 70 mm, four years of age). It inhabits small desert rivers in dry periods which last for 3–5 years and both rivers and the riparian zone of lakes during wet periods. The larger lake form occurs only in lakes during the wet periods. It can attain a maximum size of 450 mm and matures in six years, SL = 200 mm. These two forms of *O. humilis* differ in feeding habits, rates of growth, and morphology. The dwarf form feeds mainly on insect larvae and on plants. The lake form consumes the same food items until it reaches 180 mm SL and then becomes piscivorous. Populations of *O. humilis* in lakes are restored after a dry period, originating anew from river populations of the dwarf form.

Currently there is a transition from a dry period to a wet one. Orog-Nur (one of the lakes of Lake Valley) has been filling with water since 1990. In July 1991 the depth of this lake reached 0.5–1.0 m and fish were found in the lake. The large individuals of dwarf form which came to the lake from the Tuyn-Gol River became cannibals, and their growth rate increased rapidly. The homogeneous environment and low food supply in the restored lakes are suggested to be the main causes of these phenomena.

Introduction

Ecological studies often have been very fruitful when it was possible to observe the response of populations and of individuals to natural environmental fluctuations. Lake Valley in Mongolia presents a good example of such fluctuations. The environments of Mongolia are rather variable and sometimes unique. The landscapes of the Central Asian closed watershed are especially peculiar. This region is separated from others by the Hangai mountains in the North and by the chain of the Mongolian Altai in the South. In the Western part of this territory is the so-called Lake Valley (45–47°N, 98–105°E).

The ichthylogical group of the Joint Soviet-Mongolian biological expedition has investigated the Lake Valley waters since 1975. Beginning in 1978, the gradual drying up of the waters has been observed. Lakes Orog-Nur and Taytzyn-Tzagan-Nur have practically disappeared since the middle of eighties. As a result of periodical observations

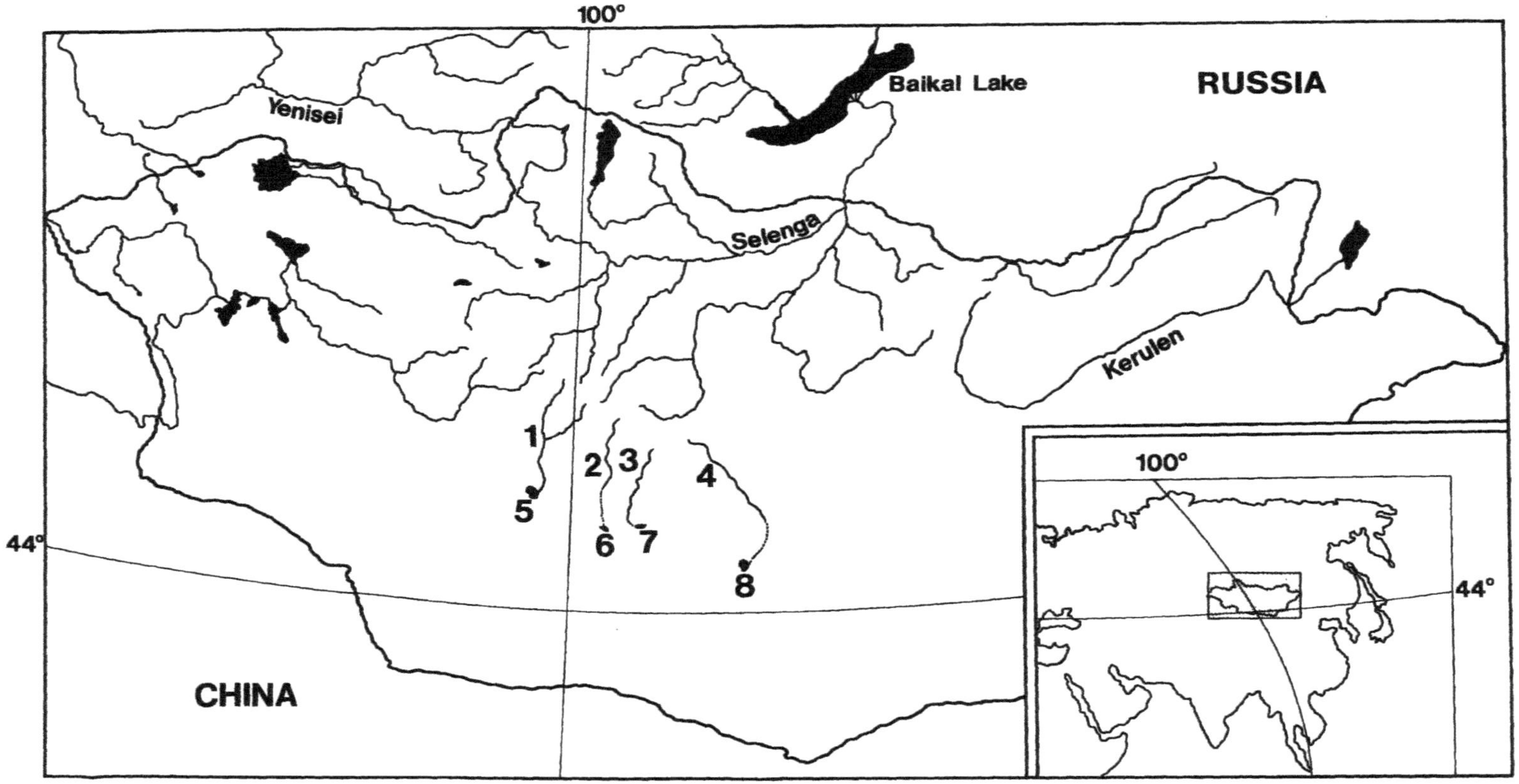

Fig. 1. Map of Mongolia. 1 = Baydrag-Gol River, 2 = Tuyn-Gol River, 3 = Taytzyn-Gol River, 4 = Ungijn-Gol River, 5 = Lake Bon-Tzagan-Nur, 6 = Lake Orog-Nur, 7 = Lake Taytzyn-Tzagan-Nur, 8 = Lake Ulan-Nur.

during the last 16 years, space photo analysis, using literature data and information received from local people, we have come to the conclusion that Lake Valley is characterized by long-term climatic fluctuations. These fluctuations are expressed in alternation of dry and wet periods and they influence first of all the water supply of four main lakes of the valley. Thus the drought observed in Lake Valley now is a phase of a long-term natural climatic fluctuation.

The aim of the present paper is to describe the influence of the land/inland-water ecotones on fish populations during the long-term environmental alterations in Lake Valley

Description of the region studied

Lake Valley in Central Asia is a part of a closed watershed without any connection with other river basins or with the ocean. There are four rivers in this valley: the Ungijn-Gol, the Taytzyn-Gol, the Tuyn-Gol, and the Baydrag-Gol, which flow down from the Hangai mountains into Lakes Ulan-Nur, Taytzyn-Tzagan-Nur, Orog-Nur and Bon-Tzagan-Nur (Fig. 1).

The climate of Lake Valley is very extreme. In summer temperature reaches +35 °C, but in winter it falls to −50 °C. This region is situated in the desert steppe zone with a deficit in water. Precipitation usually does not exceed 200 mm per year (Vostokova, 1985). The density of human population is very low, with no more than 2–3 people per km^2, employed in cattle ranching. In recent times river water has been used for irrigation.

Bon-Tzagan-Nur, the largest lake, exceeds 250 km^2, with average and maximum depths of 9 m and 16 m respectively. The mineral content of the lake fluctuated from 3.0 to 5.7 g l^{-1}. The main tributary of Bon-Tzagan-Nur is the Baydrag-Gol. The length of the river is about 300 km, its width (in the wet period) 20–40 m, depth 0.3–1.5 m, and current speed 1.0–1.5 m s^{-1} (Kuznetzov, 1959). Bon-Tzagan-Nur is the only lake in the valley which has never dried up, although the area of the lake decreases substantially during the dry periods. For example in 1986 and 1988 the shoreline in the northern part of the lake had shifted by 5–10 m in comparison with its position in 1983.

Lake Orog-Nur is smaller than Bon-Tzagan-Nur, with an area of 140 km^2 during the wet period, average depth of 3 m, and maximum depth of 5 m. The main tributary of the lake is the Tuyn-Gol River. The length of the river is 243 km, its

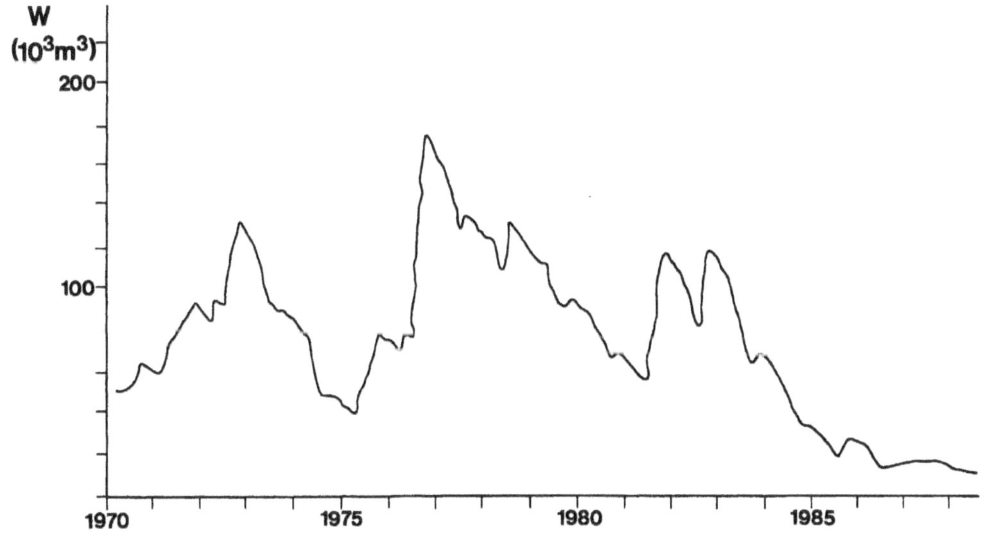

Fig. 2. Water volume (million m^3) fluctuations in Lake Orog-Nur during the last 20 years (data of LENGIDROVODKHOZ).

width in the wet period 10–20 m, and depth 0.5–1.5 m (Kuznetzov, 1959). Orog-Nur periodically dries up partially and may disappear completely (Potanin, 1893; Kozlov, 1905; Murzaev, 1952). Hearsay evidence from local people and analysis of space photo showed that periodical drying up and restoration of Orog-Nur has taken place during the last 40–50 years. In July 1986 the shoreline of the lake had shifted approximately 300 m, the depth of the lake was only 30–40 cm in its western part and 50–60 cm in the eastern part. There were no fish in the lake. Lake Orog-Nur completely dried up in 1988–1989. Only a small part of the lake was filled by water in the beginning of the summer during the flood. The retrospectively assessed fluctuation of water volume is presented in Fig. 2.

The area of lake Taytzyn-Tzagan-Nur reaches only 10 km during the wet period; the average depth of the lake is 1 m and its maximum depth 3.5 m. The mineral content of water varied in 1979, 1980 and 1982 between 3 and 4 g l^{-1}. The Taytzyn-Gol River flows into Taytzyn-Tzagan-Nur. The length of the river is 197 km, width 2–15 m, depth 0.3–1.2 m, and its water velocity 0.5–1.2 m s^{-1} (Kuznetzov, 1959). As Orog-Nur, the Taytzyn-Tzagan-Nur completely dried up during the dry period. In 1982 the area of this lake was reduced by 30% since 1980, and the Taytzyn-Tzagan-Nur River reached by the lake only in spring. Taytzyn-Tzagan-Nur completely dried up in 1986 and has not restored to date.

The area and depth of Ulan-Nur, the eastern most lake in Lake Valley has fluctuated widely

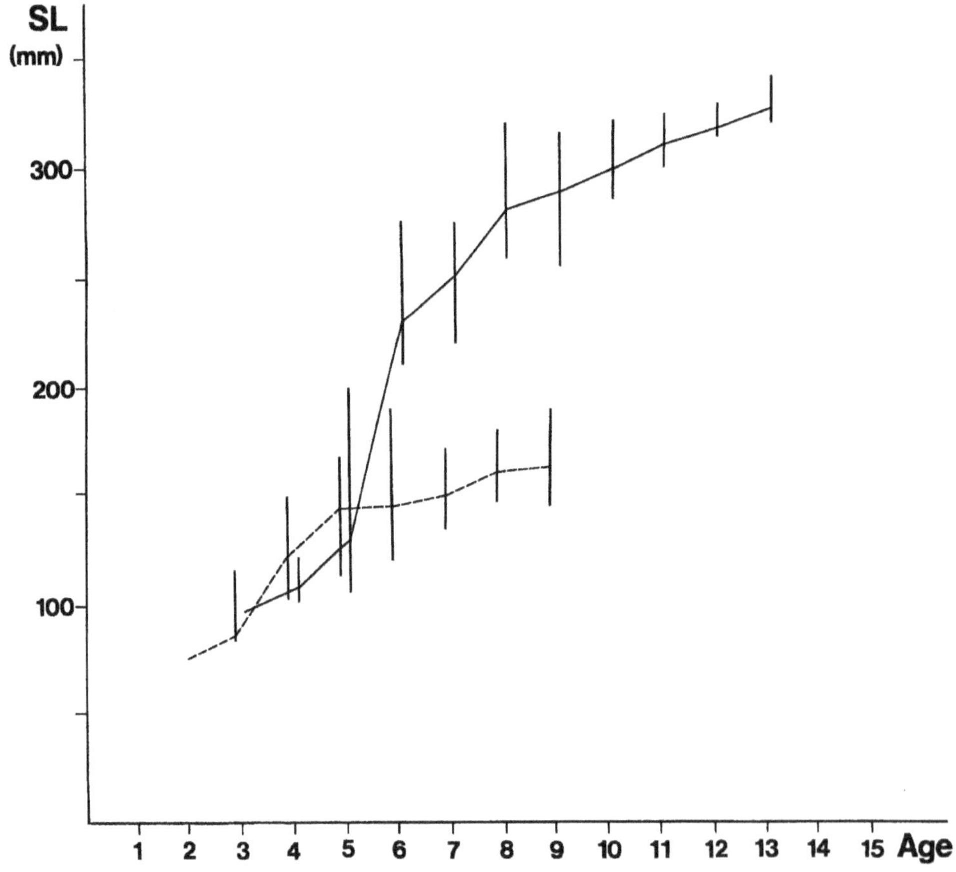

Fig. 3. Growth curves of *Oreoleuciscus humilis* in Bon-Tzagan-Nur Lake (means and limits of variations). Data of 1983. A broken line – the dwarf form; a solid line – the lake form.

due to the water regime of different years. The shores of the lake are very swampy. The tributary of Ulan-Nur is the Ungijn-Gol River, 435 km long, 15–30 m wide, 0.4–1.7 m deep, with a water velocity of 1.2–1.8 m s^{-1} (Kuznetzov, 1959). Ulan-Nur periodically dries up completely. For example, Tzerensodnom (1971) noted the drying up of Ulan-Nur in 1952–1953.

Materials and methods

The investigations of fish populations of Lake Valley were carried out during wet and dry periods. Observations and sampling were made in summer time of the wet period in Lake Bon-Tzagan-Nur (1975, 1982 and 1983), Lake Orog-Nur (1975), Lake Taytzyn-Tzagan-Nur (1979, 1980 and 1982), the Baydrag-Gol River, the Tuyn-Gol River and the Ungjin-Gol River (1975, 1979, 1980, 1981 and 1982); and in summer time of the dry period in Lake Bon-Tzagan-Nur (1986, 1988), Lake Orog-Nur (1986), Lake Taytzyn-Tzagan-Nur (1986, 1988), the Baydrag-Gol River (1988) and the Taytzyn-Gol River (1988). In addition, observations were carried out in Orog-Nur and the Tuyn-Gol and Taytzyn-Gol during the summer of 1990 and 1991, which was probably the transitional period between the dry and the wet ones.

Fish were collected from gill nets (10–55 mm mesh size) in the lakes. Rivers were sampled using frame nets, hook and lines, and electrofishing. The last method was used to study distribution of fish in the rivers. Before electrofishing sampling, barrier nets (3 mm mesh size) were put at the upstream and downstream boundaries of the site. A site usually was 10–25 m and correspond to the physical habitat of dwarf Altai osman, including pool and raceway. For each site 3 or 4 successive electrofishing runs were made. The fish of age 0 + were underestimated because the spanning period of *O. humilis* lasts from June to August. All fish were counted, measured (SL) and weighed. Opercular bones and dorsal fin rays were taken for age determination. Growth rate was estimated for fish from Bon-Tzagan-Nur (about 300 specimens) and Orog-Nur (about 200 specimens sampled in 1975 and 41 specimens collected in 1991). The gut contents of more than 500 fish from the Taytzyn-Gol River (1979, 1988, 1990), 50 individuals from the Tuyn-Gol River (1975), about 300 specimens from:the Baydrag-Gol River (1980, 1981, 1988) and 67 from Lake Orog-Nur (1991) were analysed qualitatively. For analysis of growth of fish from Orog-Nur (sampled in 1975) measurements of the opercular bones were used. We could not use observed or back-calculated data of fish lengths because we have not got young individuals of the dwarf form or of the lake form.

Results

The fish fauna of Lake Valley is extremely poor. Only two species of fish: stone loach (*Nemacheilus barbatulus toni* (Dybowski)) and dwarf Altai osman (*Oreoleuciscus humilis* Warpachowsky) live in this area. The former prefers upstream parts of rivers, while the latter is more abundant in the middle and downstream parts.

There are different views concerning subdivisions of the genus *Oreoleuciscus*. The problem is caused by great morphological variability of Altai

Table 1. Species of the genus *Oreoleuciscus* inhabiting waters of Lake Valley according to different authors.

Species	Author
(1) *Oreoleuciscus potanini*	
O. altus	Nichols, 1930
(2) *O. pewzowi*	Dashdorzh *et al.*, 1969
O. humilis	Vasiljeva, 1985
(3) *O. potanini*	
O. pewzowi	Dulma, 1974
O. humilis	
(4) *O. potanini*	Svetovidova, 1965
(5) *O. potanini*	Basanzhav *et al.*, 1983
A. lake form	
B. darf form	
(6) *O. humilis*	Bogutzkaya, 1988
(7) *O. humilis*	Our present opinion
A. lake form	
B. darf form	

osmans (Svetovidova, 1965; Basanzshav *et al.*, 1983; Borisovetz *et al.*, 1985). Six different points of view concerning species composition of Altai osmans inhabiting waters of Lake Valley exist (Table 1). Our long-term observations (more then 10 years) revealed substantial changes in the pattern of morphological diversity of Altai osmans in Lake Valley. To describe this diversity we proposed to use a neutral term 'form' (according to Mayr *et al.*, 1953), assuming that morphological and ecological differentiation proceeds within a single species. Now we believe that the dwarf form and the lake form of *Oreoleuciscus humilis* inhabit the waters of Mongolian Lake Valley.

The dwarf form is characterized by a relatively small maximum size (SL about 200 mm) and by early maturation (SL = 70 mm, at four years of age). The fecundity of the dwarf osman varies between 2400 and 12200 eggs. It inhabits small desert rivers in drought periods and both rivers and the riparian zone of lakes (ecotones) during wet periods. The diet of the dwarf form consists of invertebrates (mainly insect larvae) and plants. The dwarf form in rivers prefers sections with slow water: raceways and pools.

In the middle and low parts of the rivers the dwarf form of *O. humilis* reaches very high density and biomass. In some downstream pools of the Taytzyn-Gol River the biomass of Altai osman exceeds 600 kg ha^{-1}. Electrofishing data for 1988 showed that the total average biomass for this site was 26.8 kg ha^{-1} and the density was 3114 ind ha^{-1}. These estimates are rather close to Naiman's data for American desert rivers (1976) and those of Penczak & Molinski (1984) for a North African small river.

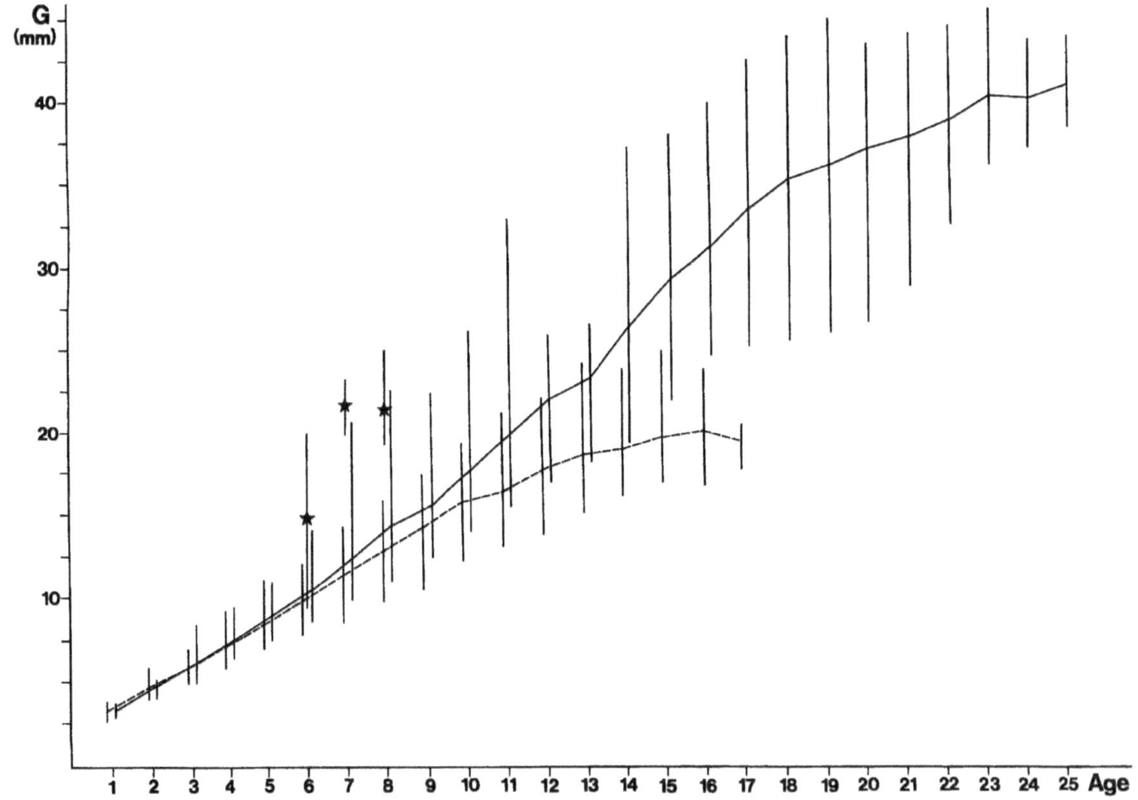

Fig. 4. Growth curves of opercular bones (G = length of opercular bone in mm) of *Oreoleuciscus humilis* in Orog-Nur Lake (means and limits of variations). The curves are plotted using measurements from a sample taken in June 1975 (wet period). A broken line – the dwarf form; a solid – the lake form; ★ – data of July 1991 (beginning of wet period).

The individuals of the lake form reach SL 450 mm and during the early years of life use the same foods as the dwarf form. After their SL exceeds 180 mm the individuals of the lake form become piscivorous. The fecundity of the lake form varies from 33720 to 105300 (Basanzsah *et al.*, 1985). There are some morphological differences between the dwarf and the lake forms, particularly in the external morphology of the head: relative size of eyes, length of mandible, length of operculum and interorbital distance (Borisovetz *et al.*, 1985a, 1985b).

The differences between the dwarf and the lake forms of *Oreoleuciscus humilis* in life history strategy are shown for the Bon-Tzagan-Nur population (Fig. 3).

According to our observations, during the wet period (in 1975, 1979, 1980, 1981 and 1982) both the dwarf and the lake forms inhabit all the four main lakes of Lake Valley. In the dry period (our observations in 1986 and 1988) fish populations of Lakes Orog-Nur and Taytzyn-Tzagan-Nur were extinguished.

Now we can observe a transition from the dry period to the wet one. Orog-Nur has been filling with water since 1990. In July 1991 this lake was 0.5–1.0 m deep and the Tuyn-Gol River reached the lake. The dwarf form of *O. humilis* came to the lake from the river.

The analysis of growth of fish colonizing the lake gave us unexpected results. The larger individuals of the dwarf form, which came to the lake from the Tuyn-Gol River, became cannibals after about 1 year in the lake, and their growth rate increased drastically (Fig. 4). Also, part of the dwarf population after maturation sharply changed its life history, the guts of most individuals > 186 mm SL contained small fish.

Discussion

The situation in Mongolian Lake Valley helps us to understand mechanisms of intraspecific life history differentiation. During the dry period both the dwarf and the lake forms of *O. humilis* disappear from the lakes. Only the river populations of the dwarf form remain, which use rivers as a refuge. The lake populations of *O. humilis* are restored after the dry period, originating anew from

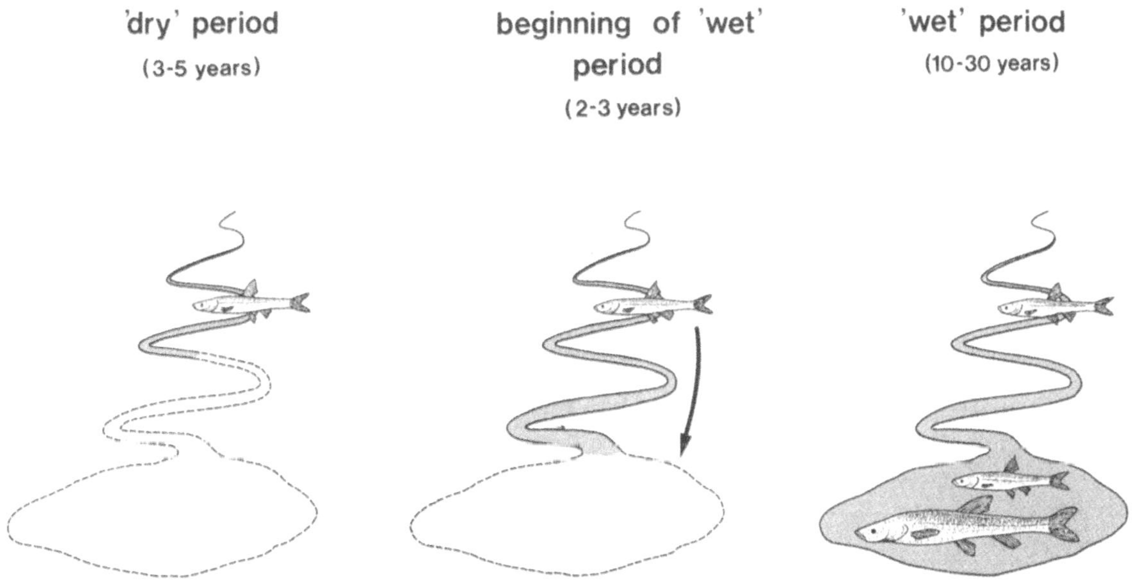

Fig. 5. Scheme illustrating changes in fish population structure in Lake Valley waters in connection with climatic fluctuations (On the basis of data from Orog-Nur and Taytzyn-Tzagan-Nur Lakes.) The way of fish migration is indicated by a pointer.

the river populations (Fig. 5). Initially we assumed that differentiation of the forms takes place during the first summer of life of fish originated from the eggs deposited by the dwarf river fish in the lake. However the data obtained during the transition period from the dry to the wet period in July of 1991 showed that differentiation of adult individuals was feasible. This is consistent with Skreslet's (1973) hypothesis about Arctic charr (*Salvelinus alpinus* (L.)) of the Nordlaguna on Jan Mayen Island. Skreslet explained the appearance of large charrs by pronounced increase of their growth rate after maturation, due to cannibalism.

According to our data *Oreoleuciscus humilis* in the lakes become piscivorous only after reaching SL = 180 mm. In the rivers, perhaps due to bioenergetic reasons, only a small amount of individuals exceed SL = 160 mm. Still the main factor of differentiation is the ecotonal structure of the waters. The peculiarity of the Mongolian Lake Valley rivers is the existence of well delimited habitats: raceways, pools and riffles. Such environmental heterogeneity favours segregation of fish in different habitats. The observations in downstream parts of the Taytzyn-Gol River confirm this assumption. As illustrated by Fig. 6 the bigger individuals of Altai osman preferred pools while the smaller ones occupied mainly shallow raceways. The average standard length of Altai osman in pools was 103.2 mm whereas in raceways their average standard length was 65.1 mm. Such a distribution probably allows the small fish to avoid the biggest ones. Cannibalism has never been found in the Mongolian river populations of *Oreoleuciscus humilis*. The environments of lakes forming after the drought are rather homogeneous

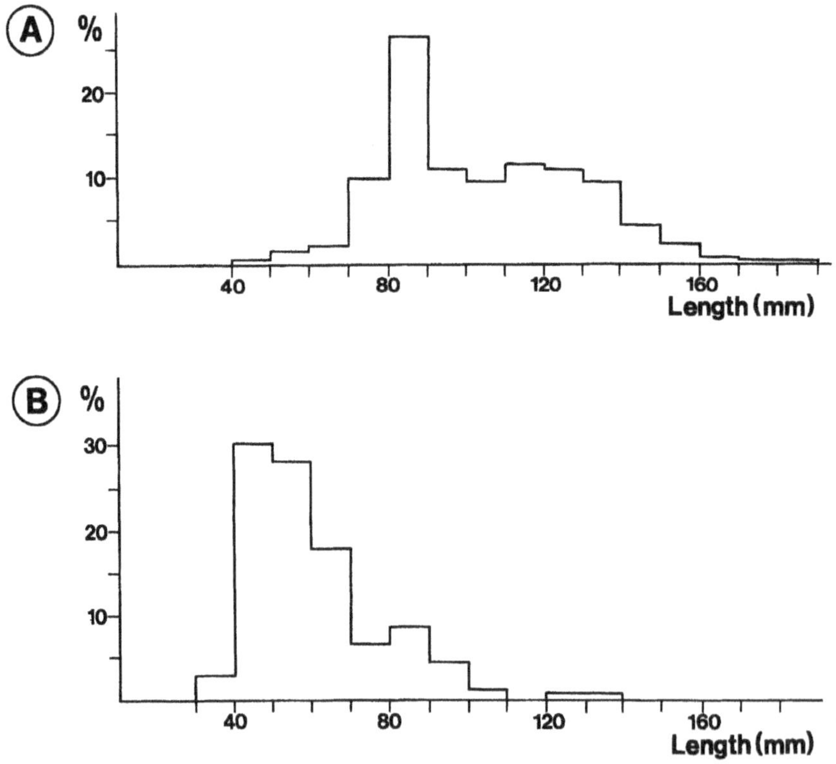

Fig. 6. Length-frequency distribution of dwarf Altai osman, *Oreoleuciscus humilis* in different habitats of the downstream part of the Taytzyn-Gol River in June 1988.

A – Pool (*n* = 595)

B – Raceway (*n* = 143)

Fig. 7. Taytzyn-Gol River. Riffle and pool.

and food supply for fish is very low. Moreover when the depth is small, and there is no shelter, the small individuals cannot avoid the big ones. It should be emphasized that the dwarf osmans

Fig. 8. Downstream of Tuyn-Gol River at the dry period.

Fig. 9. The Orog-Nur Lake at the beginning of the wet period (1990).

from the rivers to the lakes find conditions there favourable for differentiation into forms, but not for the permanent existence of sympatric forms. It is probable that the conditions of the ecotone zone are practically indispensable for coexistence of the two forms of *O. humilis*. In our case, existence of the littoral zone in a lake and a connection between the lake and a river (lake/river ecotone) is essential for preserving forms of Altai osman.

Conclusions

Mongolian Lake Valley, a closed watershed in Central Asia, is subjected to periodic droughts accompanying the disappearance of lakes and some parts of rivers.

Two forms of Altai osman, *Oreoleuciscus humilis*, the dwarf form and the lake form have been distinguished in this region according to differences in their morphology, life history strategies, rate of growth, diet, and fecundity.

During the drought both forms disappear from the lakes; populations of the dwarf form remain in the rivers.

After the lakes are filled again, the lake form originates anew from the river dwarf form.

During the restoration of the lakes the biggest individuals of the dwarf form become cannibals and their rate of growth increases.

The homogeneous environment promoting size differentiation of fish and scarcity of food for fish in the newly formed lakes are suggested to be the main causes of the described phenomena.

Acknowledgements

I would like to thank Dr A. Dulma (Inst. of General and Experimental Biology Mongolian Ac. Sci.) whose general assistance and cooperation made this study possible. I am grateful to B. Tuvshintogs (Mongolian State Univ.), Dr V. A. Kesminas (Inst. of Ecology, Lithuanian Ac. Sci.) and V. Ya. Ermokhin (Severtzov's Inst. Evolutionary Animal Morphology and Ecology Russian Ac. Sci.) for help in sampling fish in rivers. I am indebted to Dr M. V. Mina (Inst. of Developmental Biology, Russian Ac. Sci.) for his helpful comments, including language improvements.

References

Basanzhav, G., Yu. Yu. Dgebuadze, A. N. Dyemin, A. Dulma, V. Ya. Ermokhin, V. I. Lapin, I. N. Ryabov, P. Ya. Tugarina, 1983. Obzor vidov ikhtiophauny MNR. [Review of ichthyofauna of Mongolian People's Republic]. In V. E. Sokolov & M. I. Shatunovski (eds), Ryby MNR [The fishes of the Mongolian People's Republic]. Nauka; Moscow: 102–224 (in Russian).

Basanzhav, G., Yu. Yu. Dgebuadze, A. N. Dyemin, A. Dulma, V. Ya. Ermokhin, V. I. Lapin, B. Nansalma, O. N. Pugachyev, Zh. Perenleyzhamtz, I. N. Ryabov, P. Ya. Tugarina, 1985. Ekologia i hozyajstvennoe znachenie ryb MNR [Ecology and economical potential of the fishes of the Mongolian People's Republic]. V. E. Sokolov & M. I. Shatunovski (eds), Nauka; Moscow, 200 pp. (in Russian).

Bogutskaya, N. G., 1988. Topographia kanalov sejsmosencornoj sistemy karpovykh ryb podsemejstv *Leuciscinae*, *Xenocyprininae* i *Cultrinae*. [Topography of the seismosensory system canals in cyprinids of the subfamilies *Leuciscinae*, *Xenocyprininae* and *Cultrinae*]. Journal of Ichthyology, USSR. 28, 3: 367–382 (in Russian).

Borisovetz, E. E., Yu. Yu. Dgebuadze, V. Ya. Ermokhin, 1985-a. Morphometricheskij analis altajskikh osmanov (rod *Oreoleusciscus*, *Cyprinidae*, *Pisces*) vodoeyomov MNR: mnogomernyj podkhod. [Morphometric analysis of Altai osmans (*Oreoleuciscus*, *Pisces*, *Cyprinidae*) in water bodies of Mongolia: a multivariate approach]. Zoological Jounal (USSR), 64, 8: 1199–1212 (in Russian).

Borisovetz, E. E., Yu. Yu. Dgebuadze, V. Ya. Ermùokhin, 1985-b.

Klaster-analis predstavitelej roda *Oreoleuciscus* bassejna ozera Taytzyn-Tzagan-Nur. [Cluster analysis of fish genus *Oreoleuciscus* Taytzyn-Tzagan-Nur lake basin]. In Andreev V. L. & V. V. Sukhanov (eds), Automatical investigations and geographical data analysis, DVNC USSR Ac. Sci. Publ., Vladivostok: 172–187 (in Russian).

Dashdorzh, A., A. Dulma, K. Pivnicka, 1969. Contribution to the systematics of the genus *Oreoleuciscus* Warpachowski, 1989. *Cyprinidae*). Vestnik Ceskoslovenske spolecnosti zoologicke, 33, 4: 289–299.

Dulma, A., 1974. Biologiya ozer MNR. (Biology of Mongolian lakes). Essay of Doctor Thesis. Irkutsk: 28 pp. (in Russian).

Kozlov, P. K., 1905. Po Mongolii do granitz Tibeta. (Along Mongolia up to the Tibet borders). Sankt-Petersburg, 256 pp. (in Russian).

Kuznetzov, N. T., 1959. Gidrographiya rek MNR. (Hydrography of rivers of Mongolian People's Republic.). USSR Ac. Sci. Publ., Moscow: 238 pp. (in Russian).

Mayr, E., E. G. Linsley & R. L. Usinger, 1953. Methods and principles of systematic zoology. N.-Y. (McGraw Hill): X + 328.

Murzaev, E. M., 1952. Mongol'skaya Narodnaya Respublika. [Mongolian People's Republic.]. Geographgiz; Moscow: 470 pp. (in Russian).

Naiman, R. J., 1976. Productivity of a herbivorous pupfish population (*Cyprinodon nevadensis*) in warm desert stream. J. Fish Biol. 9: 125–137.

Nichols, J. T., 1930. Fish of the genus *Oreoleuciscus* in Gobi Desert Lakes. Copeia. 1: 16.

Penczak, T. & M. Molinski, 1984. Fish production in Oued Sebaou, a seasonal river in North Algeria. J. Fish Biol. 25: 723–732.

Potanin, G. N., 1893. Tungutsko-Tibetskaya okraina Kitaya i Tzentral' naya Mongoliya (Tungus-Tibet outlying districts of China and Central Mongolia). v.1. Sankt-Petersburg: 568 (in Russian).

Skreslet, S., 1973. The ecosystem of the arctic lake Nordlaguna, Jan Mayen Island. III. Ecology of Arctic char, *Salvelinus alpinus* (L.). Astarte 6: 43–54.

Svetovidova, A. A., 1965. Revizia roda altajskikh osmanov *Oreoleuciscus* Warp. (*Pisces*, *Cyprinidae*). (Revision of genus Altai Osmans *Oreoleuciscus* Warp. (*Pisces*, *Cyprinidae*)). Journal of Ichthyology, USSR. 5: 245–261. (in Russian).

Tzerensodnom, Zh., 1970. Mongol orny nuur. (Lakes of Mongolia). Ulan-Bator, 200 pp. (in Mongolian).

Vasil'eva, K. D., 1985. O taxonomicheskom statuse nekotoryskh form roda *Oreoleuciscus* (*Cyprinidae*). (On the taxonomic status of some forms of the genus *Oreoleuciscus* (*Cyprinidae*)). Journal of Ichthyology, USSR. 25: 196–211 (in Russian).

Vostokova, E. A., 1985. Lanshafty Doliny ozer MNR i ikh indikatzionnoe znachenie. (Landscapes of Mongolian Lake Valley and its significance as indicators). In Biogeographicheskie aspekty opustynivaniya. USSR Ac. Sci. Publ.; Moscow: 54–80 (in Russian).

Hydrobiologia **303**: 247–255, 1995.
F. Schiemer, M. Zalewski & J. E. Thorpe (eds), The Importance of Aquatic-Terrestrial Ecotones for Freshwater Fish.

Zonation of a lentic ecotone and its correspondence to life history strategies in fish

Josef Wanzenböck[1] & Katalin Keresztessy[2]
[1] *Institute for Zoology, University of Vienna, Althanstraße 14, A-1090 Wien, Austria;* [2] *Institute of Animal Husbandry, Biotechnological Laboratory, University of Agricultural Sciences, H-2103 Gödöllö, Hungary*

Key words: fish community, drainage ditales, endangered species

Abstract

A drainage channel network southeast of lake Neusiedl was investigated for fish species distribution. The network consists of relatively large and permanent channels providing a stable habitat for fish and smaller unstable channels drying up approximately once every five years. A consistent pattern of fish species distribution was found which could be interpreted with the help of a new triangular model of life history adaptations. In permanent channels so-called periodic and equilibrium species prevail whereas temporary channels are inhabited by opportunistic species exclusively. In the drainage channel network a number of locally endangered fish species which have disappeared from the adjacent lake during the last decades were found suggesting that the channel network functions as a refuge for these fishes.

Introduction

Ecotonial habitats have great temporal and spatial heterogeneity and often display gradients in ecological factors within a relatively narrow area. Therefore they offer good opportunities to test ecological concepts (Wiens *et al.*, 1985; Pieczynska, 1990). In aquatic-terrestrial boundary zones, ecotone size and configuration depend on shore slope and water level fluctuation as well as depositional and erosional processes. Water level fluctuation creates shifting boundaries and causes changes in the proportion of aquatic and terrestrial species of animals and plants (Pieczynska, 1990). These fluctuations create changing water and moisture gradients which exhibit regular patterns in aquatic-terrestrial ecotones and are ecologically of great significance (Cowardin *et al.*, 1979). With increasing distance from a permanent water body, aquatic habitats become less stable. Near permanent bodies of water, seasonally flooded areas are often found which have a more or less predictable wet-and-dry cycle. Further away, aquatic habitats are created only by periodic flooding and have a higher probability of drying up (Cowardin *et al.*, 1979). Thus, water bodies distant from the main water body are less stable and less predictable in providing habitat for aquatic life.

The idea that stability of ecological factors correlates with life history strategies of organisms has been formulated by various authors (McArthur & Wilson, 1967; Murphy, 1968; Pianka, 1970; Schaffer, 1974; Stearns, 1976; Winemiller, 1992). Older models of life history strategies were unidimensional: r-K continuum (McArthur & Wilson, 1967; Pianka, 1970) as well as the 'bet-hedging' model (Murphy, 1968; Schaffer, 1974). Winemiller recently (1992) developed a triangular life history model. He distin-

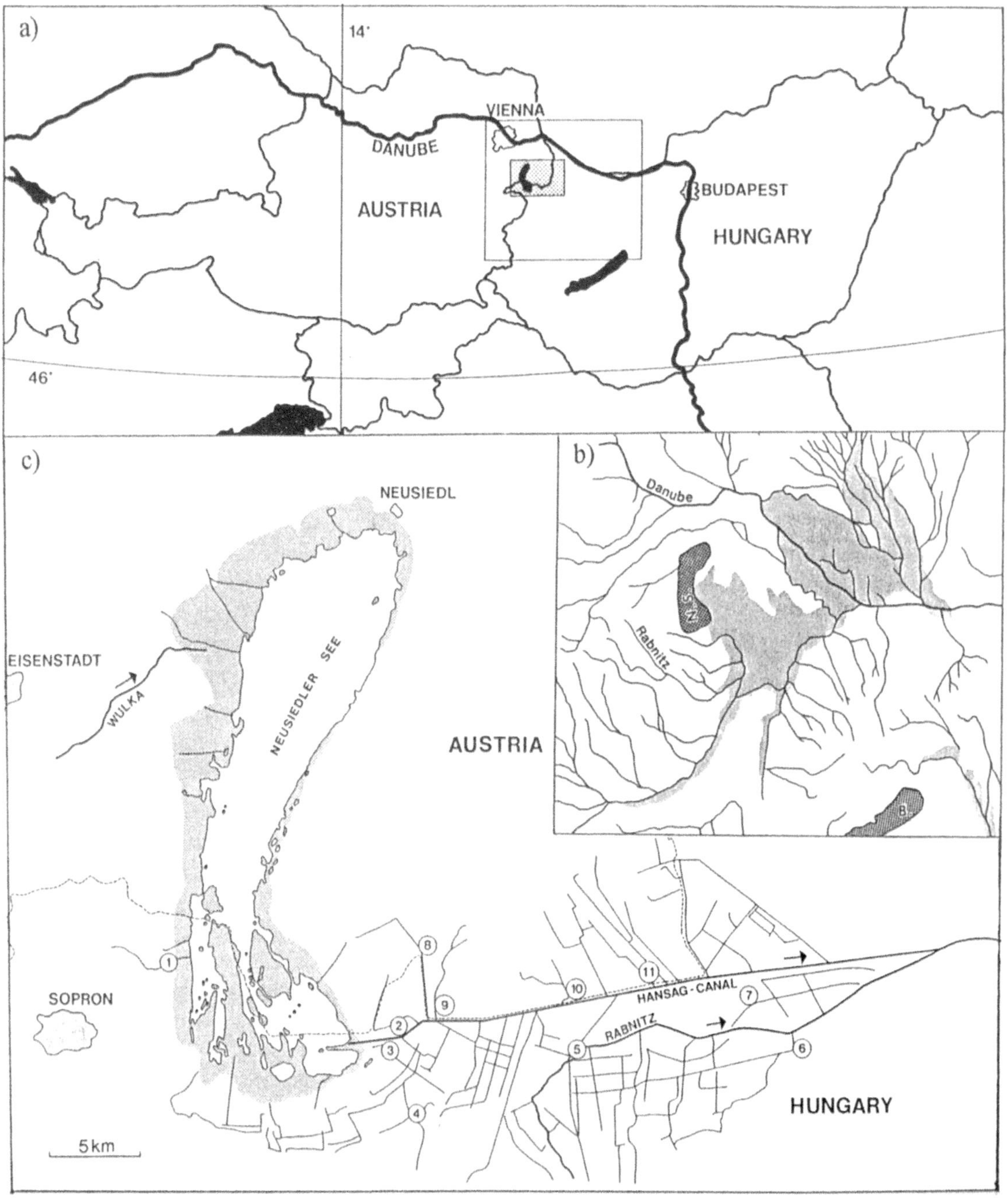

Fig. 1. Location of lake Neusiedl at the Austrian-Hungarian border (a), the regularly inundated areas in the vicinity of the lake before water control measures (drawn after a map in the Hungarian National Atlas) (b), and the location of eleven sampling stations within the current drainage channel network of the Hansag region (c).

guished three endpoints on the adaptive surface of life-history strategies based on trade-offs between three fundamental demographic axes (age at maturity, fecundity, and juvenile survivorship), despite reservations concerning the use of dimensions in defining life history strategies (Southwood, 1988; Stearns, 1992). The endpoints are represented by an 'opportunistic strategy', maximizing colonization capabilities in environments that change frequently or stochastically, a 'periodic strategy', which is favored in environments with large-scale cyclic or spatial (seasonal) variation, and an 'equilibrium strategy', which is favored in environments with low variation in habitat quality and strong biotic interactions. The model has been developed for fish but explains also life history adaptations of many other animal groups such as amphibians, reptiles, birds, mammals, insects, and many other invertebrates (Winemiller, 1992). The interaction of habitat type and stability, especially of terrestrial-aquatic transition zones, and fish species distribution has been the focus of a number of investigations in stream ecology (see Schlosser, (1991) for a review, and other articles in this volume). Lentic systems have received less attention up to now.

This study examines the extent to which Winemillers life history model can explain fish species composition in habitats of different stability within a lentic ecotone. A second aspect of the study was to evaluate the role of the ecotone as a refuge for locally endangered fish species that disappeared from the adjacent lake during the last decades.

Material and methods

Lake Neusiedl is situated on the Austrian-Hungarian border 50 km south-east of Vienna (Fig. 1a). The area south-east of lake Neusiedl, the so-called Hansag, was a huge swampland (about 300 km^2) connecting the shallow lake (mean $z = 1$ m) with the floodplains of the Danube and the Rabnitz and Raab rivers (Fig. 1b). In 1911 an artificial outlet of the lake, the Hansag-canal, was finished and since then the area was successively drained and converted to farmland. Figure 1c shows the locations of 11 sampling stations within this drainage channel network where fish populations were studied during a one-year period. Sampling stations were located at 9 drainage ditches of different sizes and distance from the lake and at two little brooks. At each sampling point a stretch of 100 m was electrofished with a 7.5 kW direct current electrofishing device four times during the study period. Fish larger than 10 cm were identified, weighed and measured in the field and then released. Smaller fish were subsampled, fixed with formalin, and brought to the laboratory for further analysis. Indigenous fish species were categorized according to Winemiller's life-history model based on literature data on age at maturity, fecundity and parental investment (a correlate of juvenile survivorship, see also Winemiller (1989)) (Deelder & Willemsen, 1964; Backiel & Zawisza, 1968; Thorpe, 1977; Kizina, 1987; Lelek, 1987; Terofal, 1984; Muus & Dahlström, 1981; Toner & Lawler, 1969; Ladiges & Vogt, 1965). Fish species in which females usually reach sexual maturity after two years, fecundity is more than 100000 eggs per average female spawning once per breeding season, and can be classified as nonguarders and open substrate egg-scatterers (*sensu* Balon, 1984) were considered relative periodic species (Table 1) (with one exception: *Stizostedion lucioperca* which is a guarder and nest spawner but was also considered a periodic species because of its late maturation and high fecundity). Species with essentially the same features but displaying fractional spawning and in some cases lower fecundities were placed between the periodic and the opportunistic endpoints. These were further divided between those tending more toward the periodic strategy because of lower individual growth rates and maturation after three years versus those tending more to the opportunistic strategy because of higher growth and sexual maturation after two years (species indicated with an asterix in Table 1). *Misgurnus fossilis* was classified as an opportunistic species because females can reach sexual maturity after one year, fecundity is low (Käfel, 1991) and it is a non-guarder

Table 1. Fish species found in the larger, permanent channels (2,6,7) and in the smaller, temporary channels (4,9,10) respectively, and their life-history category after Winemiller, 1991.

Species	Channel no.	Life history category
Permanent channels		
Esox lucius	2,6	periodic
Stizostedion lucioperca	2	periodic
Perca fluviatilis	2,6,7	periodic
Gymnocephalus cernua	2	periodic
Abramis brama	2,6,7	periodic
Ambramis ballerus	2	periodic
Rutilus rutilus	2,6	periodic
Scardinius erythrophthalmus	2,6,7	periodic
Blicca bjoerkna	6	periodic
Pelecus cultratus	2	periodic
Alburnus alburnus	2,6,7	periodic-opportunistic
*Tinca tinca**	2,6,7	periodic-opportunistic
*Carassius carassius**	6,7	periodic-opportunistic
*Carassius auratus gibelio**	2,6,7	periodic-opportunistic
*Misgurnus fossilis**	2,6,7	opportunistic
*Leucaspius delineatus**	2,6,7	opportunistic-equilibrium
*Proterorhinus marmoratus**	6	opportunistic-equilibrium
Rhodeus sericeus amarus	6,7	opportunistic-equilibrium
Temporary channels		
*Carassius carassius**	9	periodic-opportunistic
*Carassius auratus gibelio**	4,9,10	periodic-opportunistic
*Leucaspius delineatus**	9	opportunistic-equilibrium

(*sensu* Balon, 1984). The other species in Table 1 have similar features but display territoriality and egg guarding. They were therefore placed between the opportunistic and the equilibrium endpoints of the adaptive surface of life history strategies. This group was further divided into a group tending more to the equilibrium strategy (one species, the bitterling *Rhodeus sericeus amarus* which has very high juvenile survival due to hiding eggs in freshwater mussels) and the rest with comparably higher juvenile mortality tending more to the opportunistic strategy (species indicated with an asterix in Table 1). To substantiate this grouping we performed a cluster analysis using the literature data mentioned above. To elucidate the relationship between life history groups and habitat characteristics a redundancy analysis was made using the computer program CANOCO (Ter Braak, 1989). Habitat characteristics used in this analysis were channel size (area of cross-section) and the nominal variable 'permanent' *vs* 'temporary'.

Results

Table 2 gives an overview of our catches at the 11 sampling points. Twenty six fish species were found during the study. At sampling points close to the lake (sites 1, 2, 3) a fish community very similar to the one in the lake was found. In the channels further away the fish community changes and a number of small sized, locally endangered fishes were found. Figure 2 shows the development of the fish community in lake Neusiedl during the last decades (after Herzig-Straschil, 1989) in comparison to the species found during our study. Several of the locally endangered fish species that disappeared from the lake were found in the drainage channel network of the Hansag region. The sampling point at the Rabnitz brook (site 5) shows a quite distinct fish community with many rheophilic fish species found only in this place.

To address the question of channels with different stability, we compared three larger, deeper, and relatively permanent channels (sites 2, 6, 7) with three smaller, temporary channels (sites 4, 9, 10) which had completely dried up in the summer of 1990 and were filled again during our study. According to long term water level measurements of the lake, the smaller channels dry up approximately once every five years. Table 1 shows the fish species caught in the permanent and the temporary channels respectively and their classification according to Winemiller's life-history model. The result of the cluster analysis, which produced species groups identical to our main groups, is shown in Fig. 3. The majority of fish species caught in the permanent channels belong to a group characterized by a relative periodic life-history strategy or can be placed between the periodic and the opportunistic endpoints (15 species out of 20). The four species placed between the periodic and the opportunistic endpoints were further divided as outlined above: one species tending more to the periodic strategy and three species tending more to the opportunistic strategy (marked by an asterix in Table 1). Only one species was classified as opportunistic and four species were placed between the opportunistic and

Table 2. Numbers of individual fish caught at 11 different sampling stations located in the drainage channel network of the Hansag region (for sampling station numbers refer to Fig. 1c).

Species	Sampling stations										
	1	2	3	4	5	6	7	8	9	10	11
Anguilla anguilla	5	8	1	0	0	0	0	0	0	0	0
Esox lucius	0	2	0	0	0	18	0	0	0	0	2
Stizostedion lucioperca	0	9	0	0	0	0	0	0	0	0	0
Perca fluviatilis	0	1	0	0	0	17	3	0	0	0	0
Pelecus cultratus	0	4	0	0	0	0	0	0	0	0	0
Alburnus alburnus	0	156	12	0	44	1	1	0	0	0	0
Abramis brama	0	74	11	0	5	16	1	0	0	0	0
Blicca bjoerkna	0	0	0	0	0	1	0	0	0	0	0
Rutilus rutilus	15	60	90	0	86	21	0	0	0	0	0
Tinca tinca	0	2	3	0	2	23	26	0	0	0	0
Scardinius erythrophthalmus	2	13	14	0	10	66	7	3	0	0	0
Gymnocephalus cernua	0	8	0	0	1	0	0	0	0	0	0
Abramis ballerus	0	1	0	0	0	0	0	0	0	0	0
Carassius auratus gibelio	8	39	105	55	2	88	101	104	3	3	0
Carassius carassius	0	0	0	0	0	26	76	0	0	0	0
Misgurnus fossilis	0	1	0	0	5	26	7	0	0	0	0
Leucaspius delineatus	0	4	0	0	1	3	1	0	1	0	0
Rhodeus sericeus amarus	0	0	6	0	91	145	117	0	0	0	0
Proterorhinus marmoratus	0	0	0	0	8	1	0	0	0	0	0
Pseudorasbora parva	0	2	5	0	15	14	6	0	0	0	0
Leuciscus cephalus	0	0	0	0	104	0	0	0	0	0	0
Leuciscus leuciscus	0	0	0	0	10	0	0	0	0	0	0
Gobio gobio	0	0	0	0	22	0	0	0	0	0	0
Gobio albipinnatus	0	0	0	0	94	0	0	0	0	0	0
Chondrostoma nasus	0	0	0	0	2	0	0	0	0	0	0
Barbus barbus	0	0	0	0	1	0	0	0	0	0	0
Total	30	384	247	55	503	466	346	107	4	3	2
Grand Total	2147										

the equilibrium endpoints. The latter were further divided into species nearer to an opportunistic strategy (three species) and those nearer to an equilibrium strategy (one species, the bitterling *Rhodeus sericeus amarus*). In this way, two main groups were distinguished: a relative opportunistic group of species (marked with an asterix in Table 1) and a periodic-equilibrium group. In the permanent channels, species of the periodic-equilibrium group outnumber the opportunistic group (13:7). However in the temporary channels only three species were found, all of which belong to the opportunistic group. This pattern is repeated on the level of individuals. Figure 4 shows percentages of individuals belonging to the periodic-equilibrium or the opportunistic group of species respectively. In the permanent channels 66% of the individuals belong to species categorized as either periodic or equilibrium whereas 33% of the individuals were assigned to species of the opportunistic group. In the temporary channels all the fishes caught belong to opportunistic species.

The redundancy analysis of life history groups and habitat characteristics revealed a highly significant relationship. The eigenvalues of the ordination axes are 0.46, 0.35, and 0.03 respectively, the 'species-environment' correlation coefficients being 0.85 and 0.26. A Monte Carlo permutation test showed significance of the first ordination axis ($p<0.05$).

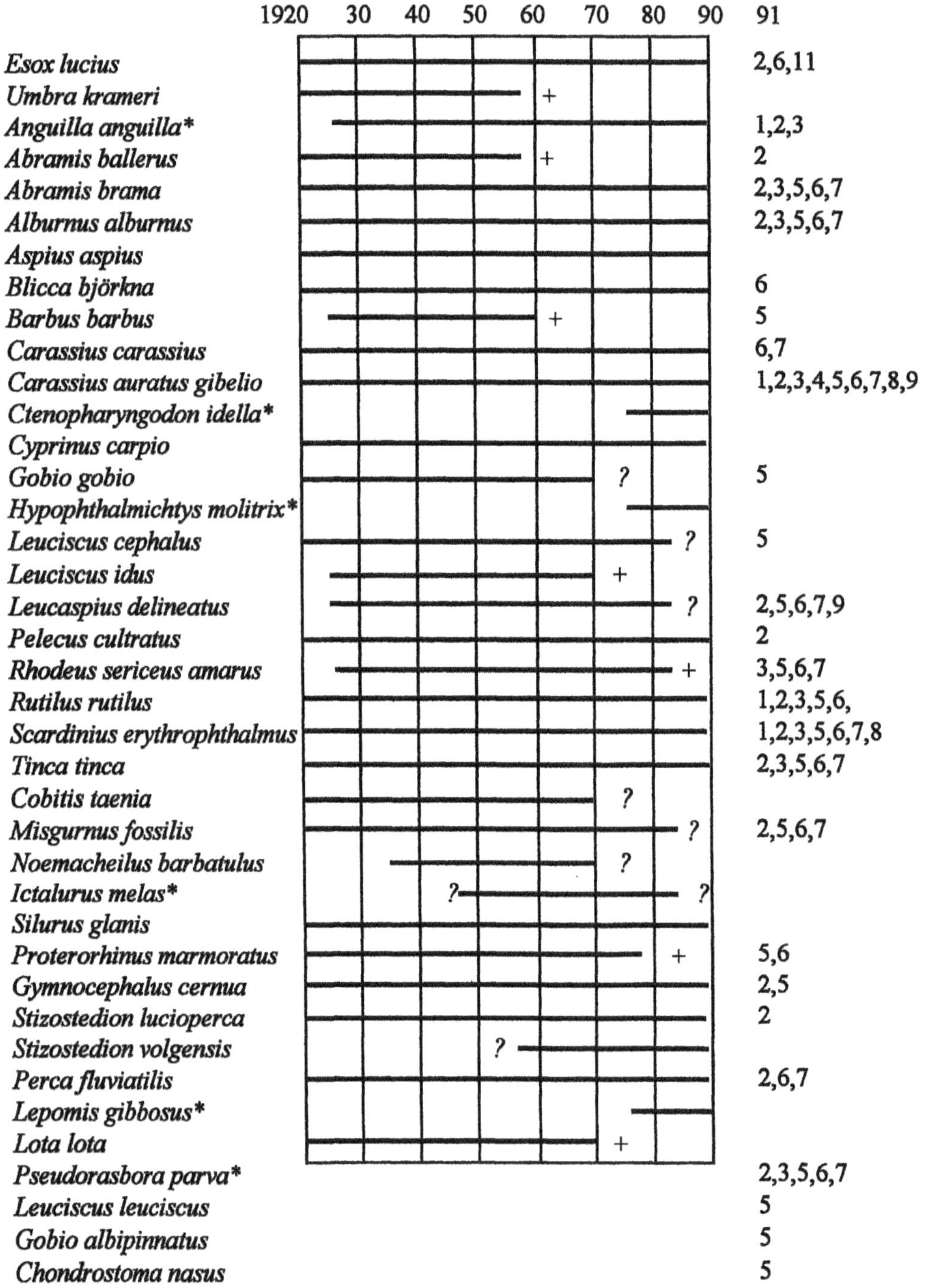

Fig. 2. The development of the fish community of the lake since 1920 (after Herzig-Straschil 1989) in comparison to the species caught in the drainage channel network in 1991 (numbers on the right side refer to station numbers in Fig. 1c). —— occurrence confirmed,? occurrence uncertain, + species extinct, * introduced species.

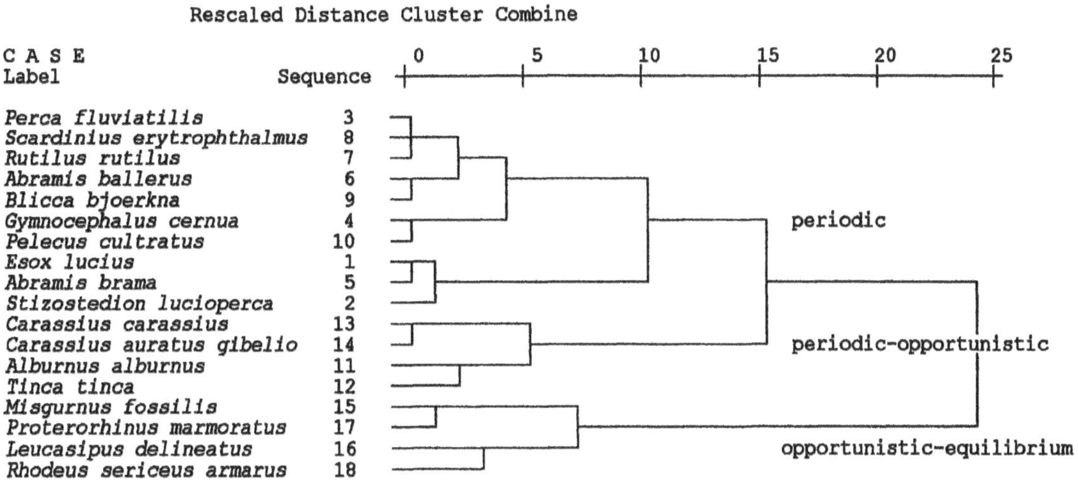

Fig. 3. Dendrogramm showing the result of a cluster analysis of the native fish species based on life history characteristics.

Discussion

The finding of most of the locally endangered fish species in the drainage channel network of the Hansag region confirms the idea that this ecotone functions as a refuge for these fish species. The disappearance of these fishes from the lake has been attributed to the heavy stocking of non-native eel (*Anguilla anguilla*) (Waidbacher, 1985; Herzig-Straschil, 1989), a predator of small fishes (Lelek & Pelz, 1986; Tesch, 1991). In the drainage channels, eels are either absent or found in low numbers (maximum was four eels per 100 m) compared to the lake (where on average 100 eels per 100 m shoreline are found, Mikschi pers. comm.). However, other factors like eutrophication of the lake as well as the loss of natural water level fluctuations might influence the observed distribution of locally endangered fish species.

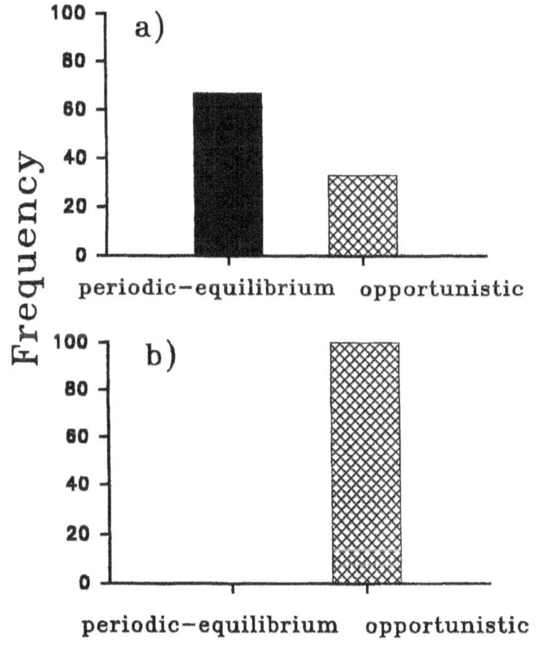

Fig. 4. Relative frequency (% of all individuals sampled) of individual fish belonging to the periodic or equilibrium groups of species or to the opportunistic group in (a) permanent channels and in (b) temporary channels.

A clear trend was found with regard to the correlation of life history features of fish species and their distribution in habitats of different size and stability. Species classified as periodic or equilibrium dominate in relatively large and permanent channels, whereas relatively small and temporary channels were inhabited by opportunistic species exclusively. Channel size might be one factor influencing fish species distribution. However, two brooks (# 1 and 5) were comparable in size with the smaller channels but provide a more stable habitat for fishes. In these brooks, fish species of both groups, opportunistic as well as periodic-equilibrium species, were found.

The small channels were dried up in the summer of 1990, an incidence happening approximately once every five years. This situation leads to a complete destruction of the fish community in these habitats, which are recolonized from larger, deeper channels with rising water levels. Opportunistic species are supposed to have relatively high colonization abilities (Winemiller, 1992) due to their low age at maturity, small clutch size, and low parental investment per progeny (allowing several spawnings per year) leading to their prevalence in the smaller channels. Especially wild goldfish (*Carassius auratus gibelio*) dominated in newly flooded channels. This species has many of the opportunistic characters already mentioned and is known to reproduce gynogenetically leading to extremely low parental investment and consequently superior colonization abilities. In fact this species colonized large parts of the Danube system within the last three decades (Holcik, 1980).

In general, the observed pattern in fish species distribution corresponds well with the life history model proposed by Winemiller (1992). Nevertheless, the results appear unidimensional along a 'bet-hedging' axis, which is indicated in this model running from the opportunistic endpoint to a line connecting the periodic and equilibrium endpoints. However, in this study only two types of habitats were distinguished: relatively permanent and relatively temporary drainage channels. Since also the permanent channels show strong seasonal variation in many important ecological factors (e.g. temperature, water level, vegetation) relatively periodic species are expected in these habitats according to the triangular life history model. Habitats lacking seasonal variation in which 'true' equilibrium species in Winemiller's sense are expected, were not present and they might be rare in temperate regions. There, the scarcity of suitable habitats corresponds with the scarcity of species which can be classified as equilibrium species according to life-history features. This is expected to be more pronounced in Europe than in North America when differences in recolonization processes after the last ice age, *i.e.* reduced recolonization by southern species due to the barrier of the Alps, are taken into account. Most of the families of teleost fishes exemplified in Winemiller (1992, Table 1) as having many equilibrium species are predominantly tropical or subtropical. *Cyprinidae*, either 'large minnows' or 'small minnows', which are categorized as relatively periodic and opportunistic strategists respectively represent the majority of species dealt with in the present study (Table 1). In fact we placed only one species, the bitterling (*Rhodeus sericeus amarus*), near an equilibrium life history strategy and it was only found in the permanent channels. Altogether a distinction between fish communities, dominated by periodic species on the one hand, and equilibrium species on the other hand, was neither possible nor was it expected according to habitat features. Insofar the studied ecotone provides a special case in Winemiller's triangular life-history model because of the absence of very stable habitats in which equilibrium species are expected. However, the usefulness of new models might be tested just as much if applied to special cases as if applied to general cases. The triangular life history model was found most useful for assigning fish species to different life history strategies because it provides a relatively objective system with measurable characters. The resulting scheme of life histories of the fish species caught was helpful in interpreting distribution patterns in the studied aquatic-terrestrial, lentic ecotone. In general life-history adaptations to habitats of different stability are important factors influencing species distribution in addition to other factors already investigated (Weinstein *et al.*, 1980; De Nie, 1987; Townsend & Peirson, 1988; Leslie & Timmins, 1990; Tonn, 1990; Winemiller & Leslie, 1992).

Acknowledgements

The study was financed by the Austrian Bundesministerium für Wissenschaft und Forschung within an East–West research program. We sincerely appreciate the help of the numerous people involved in field work or sample analysis and of those who provided hydrological data. For discussion and criticism we are indebted to F. Schiemer and two referees.

References

Backiel, T. & J. Zawisza, 1968. Synopsis of biological data on the bream *Abramis brama* (Linnaeus, 1758). FAO Fisheries Synopsis No. 36. FAO Rome.

Balon, E. K., 1984. Patterns in the evolution of reproductive styles in fishes. In G. W. Potts & R. J. Wootton (eds), Fish Reproduction. Strategies and Tactics. Academic Press, London: 35–53.

Cowardin, L. M., V. Carter, F. C. Golet & E. T. LaRoe, 1979. Classification of wetlands and deepwater habitats of the United States. United States Fish and Wildlife Service FWS/OBS 79/31.

Deelder, C. L. & J. Willemsen, 1964. Synopsis of biological data on the pike-perch *Lucioperca lucioperca* (Linnaeus) 1758. FAO Fisheries Synopsis No. 28. FAO Rome.

DeNie, H. W., 1987. The decrease in aquatic vegetation in Europe and its consequences for fish populations. EIFAC/CECPI Occasional paper No 19, 52 pp.

Herzig-Straschil, B., 1989. Die Entwicklung der Fischfauna des Neusiedler Sees. Vogelschutz in Österreich 3: 19–22.

Holcik, J., 1980. Carassius auratus (Pisces) in the Danube river. Acta Sc. Nat. Brno. 14: 1–43.

Käfel, G., 1991. Autökologische Untersuchungen an *Misgurnus fossilis* im March-Thaya Mündungsgebiet. Dissertation University of Vienna, 109 pp.

Kizina, L. P., 1987. Some data on the biology of the genus *Carassius* from the lower reaches of the Volga delta. Journal of Ichthyology 1987: 31–40.

Ladiges, W. & D. Vogt, 1965. Die Süßwasserfische Europas. Verlag Paul Parey, Hamburg, Berlin, 250 pp.

Lelek, A., 1987. The Freshwater Fishes of Europe, Vol. 9, Threatened Fishes of Europe. Aula-Verlag, Wiesbaden, 343 pp.

Lelek, A. & G. R. Pelz, 1986. Untersuchungen zur ökologischen Bedeutung von Aalen (*Anguilla anguilla*) und Aalbesatzmaßnahmen. Cour. Forsch.-Inst. Senckenberg 85: 57–64.

Leslie, J. K. & C. A. Timmins, 1990. The community of young fish in drainage ditches in southwestern Ontario. Arch. Hydrobiol. 118: 227–240.

McArthur, R. H. & E. O. Wilson, 1967. Theory of island biogeography. Princeton University Press, Princeton.

Murphy, G. I., 1968. Pattern in life history and environment. Am. Nat. 102: 390–404.

Muus, B. J. & P. Dahlström, 1981. Süßwasserfische Europas. BLV Verlagsgesellschaft mbH., München, 224 pp.

Pianka, E. R., 1970. On r- and K-selection. Am. Nat. 104: 592–597.

Pieczynska, E., 1990. Lentic aquatic-terrestrial ecotones: their structure, functions, and importance. In R. J. Naiman & H. Decamps (eds), The Ecology and Management of Aquatic-Terrestrial Ecotones. Man and the Biosphere Series 4. Unesco Paris & The Parthenon Publishing Group, 103–140.

Schaffer, W. M., 1974. Optimal reproductive effort in fluctuating environments. Am. Nat. 108: 783–790.

Schlosser, I. J., 1991. Stream fish ecology: a landscape perspective. BioScience 41: 704–712.

Steams, S. C., 1976. Life-history tactics: a review of the ideas. Quat. Rev. Biology 51: 3–47.

Stearns, S. C., 1992. The Evolution of Life Histories. Oxford University Press, 384 pp.

Southwood, T. R. E., 1988. Tactics, strategies and templets. Oikos 52: 3–18.

Ter Braak, C. J. F., 1989. CANOCO – an extension of DECORANA to analyze species-environment relationships. Hydrobiologia 184: 169–170.

Terofal, F., 1984. Süßwasserfische in europäischen Gewässern. Mosaik Verlag, München, 287 pp.

Tesch, F. W., 1991. *Anguilla anguilla*. In H. Hoestlandt (ed.), The Freshwater Fishes of Europe. Vol. 2. Clupeidae, Anguillidae. Aula Verlag, Wiesbaden: 389–437.

Thorpe, J., 1977. Synopsis of biological data on the perch *Perca fluviatilis* Linnaeus, 1758 and *Perca flavescens* Mitchill, 1814. FAO Fisheries Synopsis No. 113 FAO Rome.

Toner, E. D. & G. H. Lawler, 1969. Synopsis of biological data on the pike *Esox lucius* (Linneus, 1758). FAO Fisheries Synopsis No. 30 (Rev. 1) FAO Rome.

Tonn, W. M., 1990. Climate change and fish communities: a conceptual framework. Trans. am. Fish. Soc. 119: 337–352.

Townsend, C. R. & G. Peirson, 1988. Fish community structure in lowland drainage channels. J. Fish Biol. 32: 283–295.

Waidbacher, H., 1984. Fischereibiologische Untersuchungen am Neusiedler See unter besonderer Berücksichtigung des Aals. In Naturraumpotential Neusiedler See. Wissenschaftliche Arbeiten Burgenland: 469–525.

Weinstein, M. P., S. L. Weiss & M. F. Walters, 1980. Multiple determinants of community structure in shallow marsh habitats, Cape Fear River Estuary, North Carolina, USA. Mar. Biol. 58: 227–243.

Wiens, J. A., C. S. Crawford & J. R. Gosz, 1985. Boundary dynamics: a conceptual framework for studying landscape ecosystems. Oikos 45: 421–427.

Winemiller, K. O., 1989. Patterns of variation in life history among South American fishes in seasonal environments. Oecologia 81: 225–241.

Winemiller, K. O., 1992. Life-history strategies and the effectiveness of sexual selection. Oikos 63: 318–327.

Winemiller, K. O. & M. A. Leslie, 1992. Fish assemblages across a complex, tropical freshwater/marine ecotone. Envir. Biol. Fishes 34: 29–50.

Hydrobiologia **303**: 259–264, 1995.
F. Schiemer, M. Zalewski & J. E. Thorpe (eds). The Importance of Aquatic-Terrestrial Ecotones for Freshwater Fish

Land/Inland water ecotones: intermediate habitats critical for conservation and management

F. Schiemer[1], M. Zalewski[2] & J. E. Thorpe[3]
[1]*Institut für Zoologie der Universität Wien, Vienna. Austria*
[2]*Institute of Applied Ecology, University of Łódź, Łódź, Poland*
[3]*SOAFD Freshwater Fisheries Laboratory, Pitlochry. Scotland. UK*

Key words: aquatic resources, management, top down effects, life history strategies, population genetics

Introduction

In the 20th century the utilisation of natural resources, the emission of pollutants and the degradation of landscape has increased beyond a critical level. Only during the last two decades has the predictive ability of environmental sciences gained strength and become the basis for modern conservation and environmental management procedures. This is especially true for aquatic ecology, where a series of new concepts and theories arose (e.g. the biomanipulation concept Shapiro *et. al.* (1975); role of land/water ecotones — Naiman *et al.* (1988); river continuum concept — Vannote *et al.* (1980); flood pulse concept — Junk *et al.* (1989)).

In the 21st century, freshwater will be one of the most precious and most threatened resources, and the restoration and management of inland waters will have high priority; this will require the application of environmentally responsive engineering and technology.

Control of environmental quality will require several levels of resolution:

1. at the catchment area level, e.g. maintenance of landscape patchiness, pollution control,
2. at the ecotone level (e.g. control of non-source pollution and siltation, maintenance and restoration of biodiversity),
3. at the level of the aquatic ecosystems (e.g. biomanipulation of lakes and reservoirs, Shapiro & Wright, 1984; self-purification processes of rivers, Zalewski *et al.*, 1992).

The last two aspects especially are strongly dependent on the structure and function of aquatic ecotones.

Ecotones are interfaces, whose properties, scale and significance depend upon the particular ecological problem under discussion. As transition zones, their extent depends on the properties which are in transition. Consequently, the concept is multifaceted, dynamic, and difficult to constrain structurally within Holland's definition (1988). Naiman *et al.* (1988) noted that ecotones represent regions of high variance along environmental gradients. This important statistical property is a key factor in the applicability of the concept. It emphasises the localised relative diversity of structure and functions, and thereby implies that they have particular importance for the conservation of the biological resources associated with them. The primary goal of biological conservation is the maintenace of genetic continuity of the component species within systems; this ensures the maximum responsiveness of those species to environmental change. Maintenance of such genetic variety is facilitated by maintening environmental variety. Just as the nature of individual variation gives meaning to average measures in population biology, it is the nature of the variation of their component properties which gives significance to ecotones.

Hypotheses about the importance and influence of ecotonal habitats on freshwater fish biology were articulated during a meeting in Krakow, Poland, in 1990 (Zalewski *et al.*, 1991). The ecotone concept has been treated with considerable latitude in this volume. The term is used in a rather broad sense in order to focus on various significant aspects of terrestrial/aquatic interfaces on fish and fish populations.

Besides its relevance for practical applications, the ecotone approach has been profitable in creating opportunities to test general ecological hypotheses, e.g. on the relationships between habitat stability and life-history strategies; on the correlation of environmental complexity with genetic diversity; on the relevance of complementary landscape structures for population dynamics of species.

Ecotones — their forms and functions for fish

In freshwaters, from the fish's point of view, ecotones take several forms: as littoral zones in ponds, lakes and reservoirs, or as bank structures and flood plains in rivers. In addition to the lateral land/water margins, vertical transitions, between the bed sediments and stepwise longitudinal transitions, e.g. between riffles, cascades and pools, are significant.

Ecotones are important for fish as spawning grounds and as habitats — particularly during the early life stages. The structural qualities of the shoreline determine the degree of cover available to avoid predators and wash out effects during floods, as well as to provide conditions for overwintering.

Marginal zones usually provide the highest food availability, both by autochthonous production (e.g. Wetzel, 1990) and allochthonous material from the adjacent terrestrial zones. Beyond these direct functions as habitat and resource, the ecotones mediate the conditions for fish development by functioning as a filter for light, nutrients, sediments etc.

For management purposes a detailed and quantitative understanding of the various fish/ecotone interactions is required. Significant attributes of ecotone quality and their descriptors have been discussed in this volume, especially in the papers of Kolasa & Zalewski, Schlosser, and Bretschko.

One such attribute is ecotone morphology, i.e. shape, extent and composition. Shape refers the to sinuosity, contortion and complexity of wetlands as well as lake and river margins; this is for example expressed by indices of shoreline and littoral development, or by the areal extent of littoral zones and flood plains. Composition considers the vegetational structure (e.g. guild composition, height, density and canopy closure).

Besides the quality of local ecotone structure, the network (form and connectivity) of habitat patches and ecotones within a landscape context is significant:

Most fish species will not only exploit different parts of their habitat differentially to cover various requirements, e.g. feeding, shelter, spawning, but also require a sequence of habitat types over the course of their ontogenetic development. Compound effects of landscape structure must therefore be taken into consideration. Schlosser (t.v.) distinguished the following factors as particularly decisive for population dynamics and community composition of fish:

(a) habitat complementation, which refers to the spatial proximity of non-substitutable habitat types and their connectivity (Dunning *et al.*, 1992),
(b) habitat supplementation (i.e. facultative habitat use by a species),
(c) source and sink areas (zones differently endowed as spawning and nursery zones; sources are zones of high recruitment from which less well endowed sink areas are stocked),
(d) neighbourhood effects, i.e. the permeability between differently endowed patches.

Further descriptors of ecotones consider their functions as sources or sinks of material and as filters or barriers between the patches they connect. This not only holds true for energy, light, inorganic and organic nutrients, siltation (see Rabeni & Smale, t.v.) or toxic substances, but also with regard to ecotones functioning as a habitat for interacting biota: prey, predators or carriers of fish parasites and diseases.

Critical attributes of ecotones and their variance depend on spatial and temporal scales. In order to quantify their role, these attributes must be measured in a manner appropriate to the specific biological problem at hand; they must then be related to the appropriate performance measures such as species' occurrence, developmental rates and production, population structure and biomass, genetic diversity, community structure, life history strategies a.s.o..

Just as the physical qualities of ecotones change temporally, so do the physiological and behavioural responses of fishes towards them:

Fish vary dramatically in size and requirements during ontogeny. Associated with this variation is a complex array of life-cycles and habitat-use patterns (see Schlosser, t.v.).

Drastic ontogenetic niche shifts occur during the early life history and during the reproductive period; these shifts are often bound to the lateral fringes of water bodies. A good match between the narrow requirements during the critical period and the structural and functional properties of the ecotones is decisive for population dynamics (Schiemer *et al.*, 1991; Schiemer & Zalewski, 1992).

The hierarchical structure of spatial and temporal scalings addressed above is particularly obvious in the case of flood plains. They basically represent ecotones between rivers and terra firma. Their boundaries contain a hierarchy of transition zones such as the river margins, the interlink between river and backwaters, the interception of backwaters with wetland areas.

During early life history the structural properties of the shoreline, in the dimension of a few metres, and their short-term changes (e.g. in hours) due to continuous water level fluctuations determines habitat quality. Resolution of the river-backwater continuity is significant on a larger scale (km), while habitat complementation (see above) on the order of 10–100 km may be relevant for certain anadromous species (see Schiemer & Zalewski, 1992, Agostinho & Zalewski, t.v.).

The resolution of the appropriate temporal and spatial scalings at which ecotonal properties are relevant for specific requirements of characteristic species is not only decisive for successful research, it is also a major research objective in itself.

Ecotone effects on fish in rivers

The energy of flowing water and the obstructions caused by geomorphology, boulders and large organic debris derived from riparian areas are significant in creating lateral and longitudinal heterogeneity; this in turn influences fish diversity, biomass and production in rivers.

The heterogeneity and structural qualities of the shoreline and the river bed are responsible for the diversity of the fish fauna and the habitat value for different ontogenetic stages. Structure provides not only refuges for fish, but also microhabitats for food organisms, velocity patterns and organic drift. This is of direct significance for energy expenditures in station maintenance, for foraging and for escape from predators. Behaviourally, the visual isolation conferred by increased structural complexity of a stream bed allows it to be occupied by higher numbers of otherwise territorially aggressive individuals (Kalleberg, 1958); this contributes to higher population and production levels. Small-scale habitat characteristics are apparently more important for small than for larger fish: this holds true during ontogenetic development (see, e.g. Schiemer & Zalewski, 1992) but is also evident in the habitat use by small riverine fish, where small-scale heterogeneity and short-term changes in flow are significant (Zweimüller, t.v.).

The positive relationship between channel morphology and heterogeneity as well as fish diversity, both inter- and intraspecifically, is well documented; it has been tested by revitalisation programmes of degraded river systems (e.g. Jungwirth *et al.*, t.v.).

The significant effect of riparian vegetation for creating habitat diversity and determining instream processes has been demonstrated by Penczak, t.v. Concomitant losses in fish diversity and production are observed when streamside vegetation is removed, but can be restored if the vegetation zone is allowed to regenerate.

Power & Power's contribution (t.v.) pinpoints the gradient of low to high ecotone significance in various climates. In the high arctic, seasonal ice cover controls the transfer of material from land to water; it also excludes fish for long periods from marginal habitats and forces them into pools. Thus, marginal features are significant at different times.

The significance of source and sink areas and neighbourhood effects in rivers have been clearly demonstrated by various authors. Schlosser, for example, was able to demonstrate that beaver ponds function as a population source for interconnecting streams in North America. Schiemer *et al.* (1991) have shown that in the large regulated rivers of Europe only limited areas of the inshore zones provide nurseries for rheophilic species; these areas function as source areas. The population decline of former mass fishes in large European river systems, as for example *Chondrostoma nasus L.*, is a result of insufficient source areas.

Ecotone effects in lakes and reservoirs

As in rivers, the littoral zone of stagnant water bodies is functionally very important (see Wetzel, 1992), and ecotone effects are multiple and complex. They influence fish populations decisively via spawning success, as habitats, as shelter against predation or wave action and as feeding zones. Fish assemblages, population density and structure, i.e. the strength of year classes, are not only dependent on the extent and structure of the marginal zones, but also on the magnitude, timing and predictability of water level fluctuations.

The dependence upon ecotones as habitats is particularly high during early life history and decreases often with age. Inshore zones are often protective, fry-supportive habitats with a high and diverse food availability, allowing rapid growth in order to survive the critical phase in development (Matena, Simonian *et al.*, t.v.). Littoral development generally correlates with fish yield.

The strongly differing form and extent of ecotones in reservoirs – from geometrically regular artificial embankments, steep former river banks, to extensively flooded shallow areas – in combination with different patterns of water level fluctuations, provide good 'experimental' conditions to study ecotone effects on fish populations (Duncan & Kubecka, t.v.). The extent of vegetated margins or seasonally flooded vegetation is decisive for the success of phytophils and phytolithophils, which usually represent the main spawning type. In European reservoirs with poor ecotone development, perciforms (esp. *Perca, Stizostedion, Gymnocephalus*) prevail. They are less dependent on marginal structures and have a pelagic larval phase. A correlation could be established between the extent of marginal zones and the relationship of cyprinids to percids (see Fig. in Duncan & Kubecka, t.v.); this reflects spawning success primarily.

In the artificial and geometrically regular drinking water reservoirs of London, the littoral zone is very weakly developed and the reservoirs are characterized by low fish abundance. Water level fluctuations disturb the spawning success, especially of species spawning at the fringes.

Water levels in reservoirs usually exhibit short-term and seasonal fluctuations. Oscillations in year classes may be a consequence. High water levels in spring and early summer guarantee success, while low water levels depress recruitment. Short-term fluctuations of approximately 50 cm per day can eliminate whole year classes of a species.

Top down effects of fish

Top down effects by fish can a) either influence ecotones themselves or b) affect other compartments of the aquatic ecosystem. Via recruitment these effects can be dependent on ecotones:

Fish feeding and mechanical action are known to influence the vegetational structure of the littoral zones. Such influences can initiate feedback mechanisms amplifying or dampening system processes. Species feeding on macrophytes or benthos in the littoral zones can enhance or depress ecotonal structure, depending on population density of fish. One can speculate that intermediate densities result in the highest degree of structure.

An interesting example was presented by Kolasa & Weber (t.v.), who found that foraging activity of carp creates local passages and opens water patches among emergent plants. The heterogeneity so created may in turn enhance survival and growth of carp fry and juveniles (Fig. 1/1).

On the other hand, complex top down effects on the dynamics of lake and reservoir systems and in turn on water quality may occur. The spawning success and population dynamics of species having different trophic ecologies, — zooplanktivorous, herbivorous or piscivorous — can have far-reaching effects via feeding activity, mechanical action, and sediment reworking. Feedback loops can be initiated for example via internal loading, food chain control or phytoplankton/macrophyte antagonism. By size-specific feeding on zooplankton, roach or perch for example exert the well known top-down effects on phytoplankton and trophic conditions. This can change the delicate balance between macrophytic and phytoplankton growth (Fig. 1/2). Spawning success of benthivorous species such as bream can enhance the internal loading by high population levels of adults (Fig. 1/3).

Finally, the spawning success of piscivorous fish such as pike or pike-perch will have controlling effects on other species (Fig. 1/4).

Relevance of the ecotone concept in practice

The increasing intensity of land use is reflected in degradation of land/water ecotones and ecological functions. In certain fields the ecotone approach is apparently particularly important, e.g. in river engineering, in the biomanipulation of reservoirs and lakes, and in nature conservation with regard to maintaining biodiversity. The book presents examples of ways in which the ecotone approach can be applied:

Many river systems have been simplified by man for limited usages such as drainage channels, ship transport channels, or for power generation. A consequence of this loss of physical diversity has been widespread losses of biological diversity. However, as our understanding of the dependence of biota on the ecotonal habitats of rivers has increased, rational restoration programmes have been devised and tested;

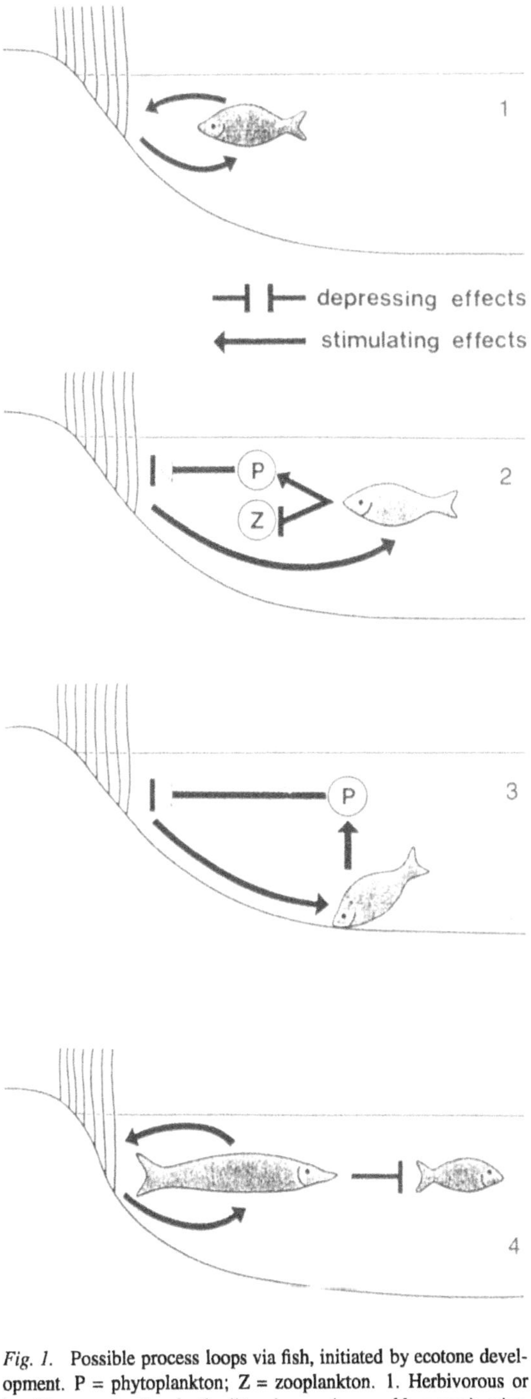

Fig. 1. Possible process loops via fish, initiated by ecotone development. P = phytoplankton; Z = zooplankton. 1. Herbivorous or benthivorous feeding in the littoral zone (e.g. rudd or carp), stimulating ecotone structure. 2. Effects of zooplanktivores (e.g. roach or perch), enhancing phytoplankton. 3. Effects of benthivores (e.g. bream), enhancing internal nutrient loading. 4. Piscivores (e.g. pike), controlling populations of other fish species.

one result is the reinstatement of some of the relevant physical variation to particular stream and river systems, with dramatic increases in biological variety, fish diversity, biomass, and production (Jungwirth *et al.*, Muhar *et al.*, t.v.).

Biomanipulation, i.e. the control of food chain relationships to influence the trophic state of water bodies has become a major management tool for water quality control. In Europe, for example, good recruitment of cyprinids and percids — dependent on spawning success and larval survival in the littoral zone — tends to reduce the ability of large zooplankton such as *Daphnia* to prevent phytoplankton blooms. Control of water level fluctuations provides opportunities for manipulating stocks by drying eggs or disturbing spawning disturbance. The manipulation of land/water ecotones may become a major tool in controlling of water quality by regulating the predators of zooplankton (Zalewski *et al.*, t.v.).

Adaptive characteristics of fish

The adaptive qualities of species with regard to the characteristics of land/water ecotones (e.g. form, heterogeneity, stochasticity of processes) have rarely been addressed to date. Such adaptive characteristics may span from metabolic or behavioural mechanisms to complex life history strategies. Winemillers (1992) life history model, for example, distinguishes three strategies (opportunistic, periodic, equilibrium) on three demographic characteristics of a species: age at maturity, fecundity and juvenile survivorship, could be successfully related to a cline of aquatic habitat size and stability in a large wetland system. The range was from large, permanent channels where equilibrium species dominated to relatively small and temporary channels where opportunistic species prevailed (Wanzenböck & Kerteszy, t.v.).

Population genetics with regard to ecotonal form and function is a key research field (see Coelho & Zalewski, t.v.). 'Environmental heterogeneity' is thought to be a major factor in maintaining and structuring genetic diversity (Nevo, 1988), and fitness in natural populations is a positive function of heterozygosity (Frankel & Soule, 1981). In most types of freshwater ecosystems, fish diversity depends greatly on boundary environments; they are the most diversified, maintain population genetic diversity, and are thus sites for active microevolution (Salo *et al.*, 1986; Naiman *et al.*, 1988).

In conclusion: This issue is presenting the state of investigations on often neglected aspects of linkages between fish and the riparian zones. The dependence of fish on structural and functional attributes of the land/water ecotones proves to be complex but has far-reaching effects. Insight into these relationships opens new perspectives for maintaining biodiversity, applying fish as environmental indicators and — via top-down — in the regulation of aquatic system functions.

References

Agostinho, A. & M. Zalewski, 1995. The dependence of fish community structure and dynamics on floodplain and riparian ecotone zone in Parana River, Brazil, Hydrobiologia 303 (Dev. Hydrobiol. 105): 141–148.

Bretschko, G., 1995. River/land ecotones: scales and patterns. Hydrobiologia 303 (Dev. Hydrobiol. 105): 83–91.

Coelho, M. M. & M. Zalewski, 1995. Evolutionary adaptations by fish to ecotonal complexity in spatially variable landscapes – a perspective. Hydrobiologia 303 (Dev. Hydrobiol. 105): 223–228.

Dgebuadze, Y., 1995. The land/inland water ecotones and fish population of Lake Valley (West Mongolia). Hydrobiologia 303 (Dev. Hydrobiol. 105): 237–247.

Duncan, A. & J. Kubecka, 1995. Land/water ecotone effects in reservoirs on the fish fauna. Hydrobiologia 303 (Dev. Hydrobiol. 105): 11–30.

Dunning, J. B., B. J. Danielson & H. R. Pulliam, 1992. Ecological processes that affect populations in complex landscapes. Oikos 65: 169–175.

Frankel, O. H. & M. E. Soule, 1981. Conservation and evolution. Cambridge University Press, Cambridge, 327 pp.

Holland, M. M., 1988. SCOPE/MAB technical consultations on landscape boundaries. In: Di Castri (ed.). A new look at ecotones: Merging international projects on landscape boundaries. Biology International. Special issue 17: 47–106.

Junk, W. J., P. B. Bayley & R. E. Sparks, 1989. The flood pluse concept in river-floodplain systems. In: D. P. Dodge (ed.), Proceedings of The International Large River Symposium. Can. Spec. Publ. Fish. Aquat. Sci. 106: 110–127.

Jungwirth, M., S. Muhar & S. Schmutz, 1995. The effects of recreated instream and ecotone structures on the fish-fauna of an epipotamal river. Hydrobiologia 303 (Dev. Hydrobiol. 105): 195–206.

Kolasa, J. & L. Weber, 1995. Relationship between the spatial scale and biotic variability in a wetland ecotone. Hydrobiologia 303 (Dev. Hydrobiol. 105): 61–67.

Kolasa, J. & M. Zalewski, 1995. Notes on ecotone attributes and functions. Hydrobiologia 303 (Dev. Hydrobiol. 105): 1–7.

Kirchhofer, A., 1995. Morphological variability in the ecotone – an important factor for the conservation of fish species richness in Swiss rivers. Hydrobiologia 303 (Dev. Hydrobiol. 105): 103–110.

Matena, J., 1995. The role of ecotones as feeding grounds for fish fry in a Bohemian water supply reservoir. Hydrobiologia 303 (Dev. Hydrobiol. 105): 31–38.

McQueen, D. J., M. R. S. Johannes, J. R. Post, T. J. Stewart & D. R. S. Lean, 1989. Bottom-up and top-down impacts on freshwater pelagic community structure. Ecol. Monogr. 59: 289– 309.

Muhar, S., S. Schmutz & M. Jungwirth, 1995. River restoration concepts-goals and perspectives. Hydrobiologia 303 (Dev. Hydrobiol. 105): 183–194.

Naiman, R. J., H. Decamps, J. Pastor & C. A. Johnston, 1988. The potential importance of boundaries to fluvial ecosystems. J. N. Am. Benthol. Soc. 7: 289–306.

Nevo, E., 1988. Genetic diversity in nature. Patterns and theory. In: M. K. Hecht & B. Wallace (eds). Evol. Biol. 23: 217– 246.

Penczak, T., 1995. Effects of removal and regeneration of bankside vegetation on fish population dynamics in the Warta River, Poland. Hydrobiologia 303 (Dev. Hydrobiol. 105): 207–210.

Power, G. & M. Power, 1995. Ecotones and fluvial regimes in arctic lotic environments. Hydrobiologia 303 (Dev. Hydrobiol. 105): 111–124.

Rabeni, C. F. & M. A. Smale, 1995. Effects of siltation on stream fishes and the potential mitigating role of the buffering riparian zone. Hydrobiologia 303 (Dev. Hydrobiol. 105): 211–219.

Salo, J., R. Kalliola, I. Hakkinen, Y. Makinen, P. Niemela, M. Puhakka & P. D. Coley, 1986. River dynamics and the diversity of Amazon lowland forests. Nature 322: 254–258.

Schiemer, F., T. Spindler, H. Wintersberger, A. Schneider & A. Chovanec, 1991. Fish fry associations: Important indicators for the ecological status of large rivers, Verh. int. Ver. Limnol. 24: 2497–2500.

Schiemer, F. & M. Zalewski, 1992: The importance of riparian ecotones for diversity and productivity of riverine fish communities. Neth. J. Zool. 42: 323–335.

Schlosser, I. J., 1995. Critical landscape attributes that influence fish population dynamics in headwater streams. Hydrobiologia 303 (Dev. Hydrobiol. 105): 71–81.

Shapiro, J., V. Lamarra & M. Lynch, 1975. Biomanipulation: an ecosystem approach to lake restoration. Limnology Research Centre, University of Minnesota 143: 1–32.

Shapiro, J. & D. I. Wright, 1984. Lake restoration by biomanipulation: Round Lake, Minnesota — the first two years. Freshwat. Biol. 14: 371–383.

Simonian, A., I. Tátrai, P. Bíró, G. Paulovits, L. G.-Tóth & G. Lakatos, 1995. Biomass of planktonic crustaceans and the food of young cyprinids in the littoral zone of Lake Balaton. Hydrobiologia 303 (Dev. Hydrobiol. 105): 39–48.

Wanzenböck, J. & K. Keresztessy, 1995. Zonation of a lentic ecotone and its correspondence to life history strategies in fish. Hydrobiologia 303 (Dev. Hydrobiol. 105): 249–257.

Wetzel, R. G., 1990. Land-water interfaces: metabolic and limnological regulators. Verh. int. Ver. Limnol. 24: 6–24.

Vannote, R. L., G. W. Minshall, K. W. Cummins, J. R. Sedell & C. E. Cushing, 1980. The River Continuum Concept. Can. J. Fish. aquat. Sci. 37: 130–137.

Winemiller, K. O., 1992. Life-history strategies and the effectiveness of sexual selection. Oikos 63: 318–327.

Zalewski, M., P. Frankiewicz & M. Nowak, 1995. Biomanipulation by ecotone management in a lowland reservoir. Hydrobiologia 303 (Dev. Hydrobiol. 105): 49–60.

Zalewski, M., W. Puchalski, P. Frankiewicz & M. Nowak, 1991. The relation between primary production and fish biomass distribution in an upland river system. Verh. int. Ver. Limnol. 24: 2493–2496.

Zweimüller, I., 1995. Microhabitat use by two small benthic stream fish in a 2nd order stream. Hydrobiologia 303 (Dev. Hydrobiol. 105): 125–138.